PROGRESS IN COSMOLOGY

ASTROPHYSICS AND
SPACE SCIENCE LIBRARY

A SERIES OF BOOKS ON THE RECENT DEVELOPMENTS
OF SPACE SCIENCE AND OF GENERAL GEOPHYSICS AND ASTROPHYSICS
PUBLISHED IN CONNECTION WITH THE JOURNAL
SPACE SCIENCE REVIEWS

VOLUME 99
PROCEEDINGS

PROGRESS IN COSMOLOGY

PROCEEDINGS OF THE OXFORD INTERNATIONAL SYMPOSIUM
HELD IN CHRIST CHURCH, OXFORD, SEPTEMBER 14–18, 1981

Edited by

A. W. WOLFENDALE

Physics Department, University of Durham, England

D. REIDEL PUBLISHING COMPANY

DORDRECHT : HOLLAND / BOSTON : U.S.A.

LONDON : ENGLAND

Library of Congress Cataloging in Publication Data

Oxford International Symposium (1981 : Christ Church, Oxford)
    Progress in cosmology.

    (Astrophysics and space science library , v. 99. Proceedings)
    Includes index.
    1. Cosmology–Congresses. 2. Nuclear astrophysics–
Congresses. I. Wolfendale, A. W. II. Title. III. Series.
QB980.O93   1981   523.1   82–9095
                                       AACR2
ISBN-13: 978-94-009-7875-1   e-ISBN-13: 978-94-009-7873-7
DOI: 10.1007/978-94-009-7873-7

Published by D. Reidel Publishing Company,
P.O. Box 17, 3300 AA Dordrecht, Holland.

Sold and distributed in the U.S.A. and Canada
by Kluwer Boston Inc.,
190 Old Derby Street, Hingham, MA 02043, U.S.A.

In all other countries, sold and distributed
by Kluwer Academic Publishers Group,
P.O. Box 322, 3300 AH Dordrecht, Holland.

D. Reidel Publishing Company is a member of the Kluwer Group.

TABLE OF CONTENTS

PREFACE

When my colleague Dr. Paul Kent asked me which branch of Physics
was most lively and which would lend itself best to a small high
quality Symposium, I had no hesitation in answering 'Cosmology'.
It seemed very timely that a meeting should take place which would
bring together scientists interested in all branches of Astronomy,
including Cosmic Rays, and Elementary Particles too and endeavour to
put at least some of the pieces of the jigsaw together.

The vast majority of the papers presented were later produced
in appropriate camera-ready form and are published in this volume.
I am very grateful to the authors for their ready cooperation.

Grateful thanks are also extended to the Board of Management
of the Foster-Wills and Theodor Heuss Scholarships, Oxford
University and the Deutscher Akademischer Austauschdienst (German
Academic Exchange Service) who funded the Symposium. The Director
of the German Academic Exchange Service, Frau M.E. Schmitz and her
colleague Mrs. Susan Putt, organized the whole meeting in a most
exemplary fashion.

Finally, on behalf of all participants and guests, sincere
thanks are offered to Paul Kent as Convenor for initiating the
Symposium, arranging the social events and organizing accommodation
in such magnificent surroundings. Christ Church was the home of
Lewis Carrol and we were ever mindful - and appropriately so - of
Alice.

A.W. Wolfendale

Durham, February 10th, 1982

A. W. Wolfendale (ed.), Progress in Cosmology, vii.
Copyright © 1982 by D. Reidel Publishing Company.

A NEW VIEW OF BARYON SYMMETRIC COSMOLOGY BASED ON GRAND UNIFIED THEORIES

F W. Stecker, Laboratory for High Energy Astrophysics
NASA/Goddard Space Flight Center, Greenbelt, MD. 20771, U.S.A.

Cosmology is often wrong but never in doubt.

L. Landau

Every man takes the limits of his own field of vision for the limits of the world.

A. Schopenhauer

1. INTRODUCTION

     It has been proposed by Heisenberg (1967) and Fritzsch and Minkowski (1975) that the fundamental Lagrangian of nature should be completely symmetric and that all observed asymmetries are due to asymmetries in the vacuum state.   Thus, all asymmetries would arise from spontaneous symmetry breaking of the ultimate grand unified theory. The full symmetry of the theory is expected to hold in the interactions above some critical temperature T_c. Below that critical temperature, multiple vacuum states can be arrived at by the process of spontaneous symmetry breaking. Each "state" corresponds to a unique set of vacuum expectation-values of scalar fields (or their equivalent) which, having been randomly determined through a dynamical instability, themselves determine a new, selfcontained gauge-field theory. Thus, a theory of nature is arrived at randomly from a number of equally probable theories, the original Lagrangian not having uniquely determined the "low-temperature" physics which we observe at our accelerators.

     The symmetry breaking process has been compared to spontaneous magnetization of a piece of ferromagnetic material cooled below the Curie temperature.   In this case, it is well known that domains are formed, each of which having its own randomly determined direction of magnetization. We are comfortable with the ferromagnetism case because we accept the fact that the original symmetry in the physics is still reflected by the fact that there is no overall preferred direction of magnetization so that, in the absence of an external magnetic field, a "large enough" sample will possess no net magnetization even though the spontaneous magnetization

1

A. W. Wolfendale (ed.), Progress in Cosmology, 1–20.
Copyright © 1982 by D. Reidel Publishing Company.

will be quite evident on the scale of an individual domain, destroying the symmetry locally.

It thus becomes possible to envision that in remote regions of the presently observable universe, a different self-contained field theory may hold (as defined by the set of vacuum expection values of scalar fields) or, at least, may have held at some point in the evolution of the universe. This is what the process of spontaneous symmetry breaking implies, provided that the various "domains", over which the randomly determined parameters of the broken gauge theory hold, were not in causal contact at the time the dynamical instability occurred. Because of the finite light-travel time and age of the universe at the time of symmetry breaking, we are of necessity, dealing with field theories which held over finite regions (horizon sizes) of the universe. These differences in the "laws of nature" which involve particle physics at very high energy can, for the most part, be expected to show up only locally in very subtle and sophisticated accelerator experiments, thus giving us the comfortable intuitive feeling of the uniqueness of all physical laws which has successfully guided us in the past. But the past has also taught us that intuition derived from more familiar situations (classical macrophysics, low velocity physics) can be misleading when extrapolated to less familiar, more subtle, or newly considered phenomena.

There is at least one phenomenon of truly cosmic significance for which it is clear that the concept of "locally asymmetric" physics from spontaneous symmetry breaking should be considered. That phenomenon is the scenario for the creation of "baryon asymmetries" in the early big-bang from spontaneous symmetry breaking of grand unified gauge theories. When considered in the framework described above, we see the real possibility that the created baryon asymmetries are "local" in cosmic sense; the spontaneous symmetry breaking process (in this particular case of CP symmetry) may lead to the creation of separate domains of baryon and antibaryon excess with various real observational and theoretical consequences. Indeed, as we will see, various astrophysical data such as the cosmic γ-ray background spectrum, cosmic-ray $\bar{p}$ flux measurements, recent determinations of a low primordial He abundance, and galaxy clustering can be interpreted as favoring this point of view.

2. UNIFIED GAUGE FIELD THEORIES

The various fields describing the forces of nature can be represented by the symmetries they possess in terms of the transformations of the quantum systems they produce which leave the Lagrangian invariant. The generators can be related to generalized charges. For example, in the case of QED, conservation of charge can be derived from the symmetry with respect to a one parameter phase transformation called a gauge transformation, with the generator being electric charge. The symmetry group is the unitary group U(1). The electromagnetic field $A_\mu(x)$ is introduced by requiring invariance under local gauge transformations $\lambda(x)$ and requiring that the derivatives of the charged fields transform in the same way as the fields themselves. This leads to the introduction of a gauge covariant derivative with an additional term involving eA_μ so that A_μ enters the theory through the kinetic energy term in the Larangian.

More complex gauge fields can be constructed from generators which preserve the form of the Lagrangian under more complex symmetry groups involving larger numbers of parameters, i.e., group spaces of higher dimension. These generators obey Lie algebras. An example of importance to the unified field theory of electromagnetic and weak interactions, is the gauge group SU(2), the unitary group whose fundamental representation consists of two-dimensional (traceless) matrices of determinant + 1. For this group, the generators can be represented by the familiar Pauli spin matrices. The demand for local gauge invariance under SU(2) transformations, as in the case of QED, requires the introduction of a new gauge field B_μ and coupling constant g (instead of e) in the covariant derivative.

In the electroweak theory of Glashow, Weinberg and Salam (GWS) the gauge group is a product SU(2) x U(1). In the quantum gauge theory of strong interactions, QCD (quantum chromodynamics), the generalized charges are referred to as colors. In GWS, the four transformation parameters result in the four gauge bosons γ (photon), $W^\pm$, Z^0 the heavy bosons which carry the weak charged and neutral currents. For an SU(n) theory, there are n^2-1 free parameters. In QCD or color SU(3) there are $3^2-1 = 8$ gluons which carry the force. In the simplest grand unified theory, viz. SU(5), there are a total of 24 gauge bosons, γ, $W^\pm$, Z^0, the 8 gluons and 12 new superheavy bosons, $X^{4/3}$, $Y^{1/3}$ of all three colors together with their antiparticles (Georgi and Glashow 1974). It is these bosons which are responsible for the "leptoquark" force which can transform quarks into leptons and vice versa, violating baryon number and producing an excess of matter (or antimatter) out of the primordial thermal radiation. (For further discussion, see, e.g., Stecker 1980b, Langacker 1981).

3. SPONTANEOUS SYMMETRY BREAKING

Of course, in our world of "low temperature" physics much of the symmetry of the unified theories is badly broken, leaving only $SU(3)_c$ and $U(1)_{EM}$. This is reflected in the large masses of all of the gauge bosons except γ and the gluons (which are massless) and the corresponding weakness of the weak and leptoquark interactions. The broken symmetries are incorporated into the theory by keeping the full symmetry in the Lagrangian but allowing the gauge bosons to obtain their masses "spontaneously" as the result of introducing new scalar (or "Higgs") fields which have a non-zero vacuum expectation value. One big advantage of the Higgs mechanism is that it allows the construction of a theory which is renormalizable, i.e., for which the calculations of observables give finite results. The way the Higgs mechanism works is as follows. Consider for example, a real scalar field whose contribution to the Lagrangian takes the form

$$L_s = \frac{1}{2} (\partial_\mu \phi)(\partial^\mu \phi) - V(\phi) \tag{1}$$

where the potential term is an even function $V(\phi) = V(-\phi)$. Consider, e.g., a potential of the form

$$V(\phi) = \frac{1}{2} \mu^2 \phi^2 + \frac{1}{4} \lambda \phi^4 \tag{2}$$

where $\lambda > 0$ so that the energy is bounded from below. In the case $\mu^2 < 0$, $V(\phi)$ has minima at

$$<\phi> = \pm \left(\frac{-\mu^2}{\lambda}\right)^{1/2} \equiv v \tag{3}$$

which gives, by definition, the vacuum expectation value for ϕ. The Lagrangian gives the equation of motion for scalar particles of mass $\sqrt{2\lambda}\, v$ excited near a ground state v. Note that for $\phi = \phi - v$, $V(\phi) \neq V(-\phi)$ and the symmetry is broken.

If ϕ couples to fermions with a coupling of the Yukawa form

$$L_Y = f \phi \bar{\psi} \psi \tag{4}$$

the Higgs field ϕ gives fermions masses of order fv. Thus, without explicitly introducing masses into the Lagrangian, the Higgs mechanism produces masses in the theory which are proportional to v, i.e., $m_f \sim fv$, $m_\phi \sim \sqrt{2\lambda}\, v$, $m_B \sim gv$.

For a more detailed discussion of this mechanism of spontaneous symmetry breaking, see, e.g., Albers and Lee (1973) and Beg and Sirlin (1974). So far we have spoken of vacuum expectation values $<\phi>$ of the scalar fields in a zero-temperature theory with the symmetries of the Lagrangian broken by the Higgs mechanism. The cosmological implications come in when we consider what happens as T increases to temperatures $T \gtrsim <\phi>$. In this case some, or all, of the symmetry in the theory may be restored (e.g. Weinberg 1974, Linde 1979) i.e., $<\phi>_T \to 0$ for $T > T_c$ (some critical temperature) and the corresponding masses go to zero. A direct analogy can be made here with the theory of superconductivity, where the Cooper pairs play the role of Higgs particles and the photon acquires an effective mass for $T < T_c$ which disappears at $T > T_c$ (the Meissner effect). In the finite temperature case, the Higgs fields have a thermal distribution of excitations and the vacuum expectation value is replaced by the operator Gibbs average. In the simple case of equation (2) the resulting potential acquires an effective quadratic term

$$-\mu^2_{eff}(T) \simeq -\mu^2 + \sigma T^2 \tag{5}$$

and critical temperature $T_c = |\mu| \sigma^{-1/2}$ where $\mu_{eff} = 0$ in the case $\sigma > 0$. In general, σ is a function of the coupling constants of the model.

4. BARYON PRODUCTION IN THE EARLY UNIVERSE

In the early big-bang, if the dynamics of the universe is dominated by the energy density of the thermal radiation, the temperature of the universe $T \propto 1/\sqrt{t}$, where t is the age of the universe. (The exception is when the expansion is dominated by the energy density of the Higgs field. That case will be discussed later.)

The critical temperature for symmetry breaking at the electroweak level, i.e., $SU(2)_L \times U(1)_Y \to U(1)_{EM}$ is usually considered to be of order $1/\sqrt{G_F} \simeq 300$ GeV, but as one can see from equation (5), T_c depends on the specific parameters of the theory. In fact, it is possible that $T_c \gg 1/\sqrt{G_F}$ as we will discuss later. The characteristic temperature scale for grand unification is given by the energy scale at which the coupling constants for the electroweak gauge groups and strong gauge group become comparable. This is given from renormalization group theory to be of order $\sim 10^{15}$ GeV, above which for the SU(5) theory only one coupling constant, associated with this simple gauge group, exists. Thus, it is at this temperature level, $T \sim m_x$, that baryon generation processes will be of importance.

A scenario for baryon production through the decay of these superheavy gauge and Higgs bosons has been given by Weinberg (1979). He considered the decay of these "X-bosons" into two channels $X \to ql$ and $X \to \bar{q}\bar{q}$ with branching ratios r and 1-r respectively, together with the antiparticle decays $X \to \bar{q}\bar{q}$ and $X \to qq$ with branching ratios $\bar{r}$ and $1-\bar{r}$.

The three conditions for production of a baryon excess in the early universe are (1) baryon (quark) nonconservation, (2) nonconservation of C (charge conjugation) and CP (C x parity) and (3) thermal disequilibrium (Sakharov, 1967). We have seen that grand unification supplies condition (1). The expansion of the universe supplies condition (3). The need for condition (2) can clearly be seen in the Weinberg scenario. The baryon number generated in the X and X decays is

$$\Delta B = \frac{1}{2} \left[\frac{1}{3}r - \frac{2}{3}(1-r) - \frac{1}{3}\bar{r} + \frac{2}{3}(1-\bar{r}) \right] = \frac{1}{2}(r-\bar{r}) . \qquad (6)$$

If CP is conserved, $r = \bar{r}$ and no baryon excess is generated. It should also be noted that the sign of the CP violation determines the sign of $r - \bar{r}$ and therefore the sign of ΔB. Thus, <u>whether a baryon excess or an antibaryon excess is created by this process depends on the sign of the CP violation parameter</u>. The result is a baryon-to-photon ratio

$$\eta \equiv \frac{n_B}{n_\gamma} \sim (10^{-3}-10^{-2}) \Delta B \qquad (7)$$

where ΔB is given by equation (6). From astrophysical observations, one obtains $10^{-10} \stackrel{<}{\sim} \eta \stackrel{<}{\sim} 10^{-8}$. Nanopoulos and Weinberg (1979) conclude that the decays of the superheavy scalar bosons are most relevant for cosmological baryon production. They estimate that $10^{-8} \varepsilon \stackrel{<}{\sim} \Delta B \stackrel{<}{\sim} 10^{-6} \varepsilon$. The parameter ε, is a parameter characterizing the strength of CP violation. Nanopoulos and Weinberg estimate $10^{-9} \stackrel{<}{\sim} |\eta| \stackrel{<}{\sim} 10^{-3}$ immediately after the

era of baryon production, the sign being undetermined. (Numerous other
authors have also worked on the problem of estimating η . See, e.g., Kolb
and Wolfram, 1980a; Langacker 1981 and references therein.)

5. CP VIOLATION AND COSMOLOGICAL IMPLICATIONS

It follows from the discussions of the previous section that the sign
of the baryon number excess, which determines whether matter or antimatter
is created, depends on the sign of the CP violation parameter. In the
scenarios usually considered, CP violation of one sign only is put into
the model explicitly in the Lagrangian via complex Yukawa couplings
between the fermions and scalar fields, i.e., L_Y of the form in equation
(4) with f complex, or in complex self couplings of the scalar fields,
i.e., λ complex in the potential term $\frac{1}{4}\lambda\phi^4$. However, it is also
possible for the CP violation to arise from the mechanism of spontaneous
symmetry breaking. Such a mechanism has been proposed to explain the
smallness of the CP violation implied by the small electric dipole moment
of the neutron (Mohapatra and Senjanovic 1978). Furthermore, if CP is
broken spontaneously, the amount of CP violation is finite and calculable,
whereas the presently popular baryon production scenarios invoke a "hard"
CP violation, leading to infinite renormalizations of the CP parameter
which thus become incalculable undetermined free parameters. With spontaneous CP violation the Lagrangian is CP invariant (f and λ real), but
the scalar fields themselves take on complex vacuum expectation values
which produce the CP violation. In this second case, the CP violation is
not put in by hand ad hoc. We start out with a completely CP symmetric
theory with the symmetry of the Lagrangian reflected in the state of the
universe at the highest temperatures. This being the case, owing to the
finite age of the universe t_u, regions separated by distances greater than
~ ct_u are not, and never were during the course of the expansion, in
causal contact. Thus, if spontaneous symmetry breaking of CP occurred at
a time t_{CP}, it would have occurred independently and with random signs in
regions separated by distances larger than ~ ct_{CP}. The symmetry of the
Lagrangian becomes hidden on a small scale. However, there will be no
preferred direction on a global (universal) scale. One may expect that
spontaneous symmetry breaking processes in the early big-bang will most
likely break baryon symmetry in localized regions of the universe but will
preserve the overall global matter-antimatter symmetry of the initial
state. Thus, present ideas of unified gauge theories with spontaneous CP
symmetry breaking can lead naturally to an overall baryon-symmetric
cosmology as suggested by Brown and Stecker (1979). Kolb (1981) has
pointed out an interesting fact relevant to the question of domain size
and structure from percolation theory. He notes that the effective domain
size will be much larger than the causal horizon when the symmetry is
broken, owing to the statistics of the problem. Thus, a spectrum of size
scales will result, including large scale domain structure.

Senjanovic and Stecker (1980) have considered mechanisms of
spontaneous soft CP violation within the context of the specific grand
unified theories involving the SU(5) and SO(10) gauge groups. They

discuss two distinct classes of models, viz., those with only one source
of CP violation independent of temperature for SU(5) and those in which
the CP violation at the super-heavy mass scale for SO(10) has nothing to
do with the observed CP violation at "low temperatures" in the K^0 $-K^0$
system. They conclude that independently of the particular model, the
domain picture of the universe emerges naturally in theories of soft CP
violation.

In the minimal SU(5) model with only one Higgs multiplet, CP
violation has to be put in "by hand" in the Lagrangian in the form of
complex Yukawa couplings, since the vacuum expectation value of the Higgs
field can always be redefined to be real by means of a gauge
transformation. Choosing such a hard CP violation, yields a baryon-photon
ratio which is unacceptably small compared to that determined by
astrophysical observation (Barr, Segre and Weldon 1979; Yildiz and Cox
1980). It is therefore necessary for consistency to increase the number
of 5-dimensional Higgs multiplets. Increasing this number to three
results in a realistic grand unified theory based on SU(5) which allows
for soft CP violation at high temperatures. Two of the Higgs fields
acquire vacuum expectation values with a relative phase which cannot be
transformed away, since they carry the same U(1) quantum number.
Senjanovic and Stecker consider a Higgs sector with three 5-dimensional
multiplets with the following pattern of symmetry breaking at the
electroweak level (T $\lesssim$ 300 GeV):

$$\langle\chi\rangle = \begin{pmatrix} 0 \\ 0 \\ 0 \\ 0 \\ \rho \end{pmatrix}, \quad \langle\phi_1\rangle = \begin{pmatrix} 0 \\ 0 \\ 0 \\ 0 \\ v_1 \end{pmatrix}, \quad \langle\phi_2\rangle = \begin{pmatrix} 0 \\ 0 \\ 0 \\ 0 \\ v_2 e^{i\theta} \end{pmatrix} \qquad (8)$$

It can be shown that at T >> 300 GeV the symmetry will still be
broken, with $\langle\chi\rangle$ = 0 but with $\langle\phi_1\rangle$ and $\langle\phi_2\rangle$ nonvanishing. This follows
from having the coefficient μ_{eff} of the quadratic terms in the Lagrangian
for $V(\phi_1,)$ and $V(\phi_2)$ of the form given by equation (5) with $\sigma < 0$ at T ~
300 GeV. Then, noting that σ is a slowly varying function of T, owing to
the logarithmic temperature dependence of the coupling "constants"
(obtained from renormalization group theory), in some cases $\sigma(T)$ becomes
positive for $T_c \gtrsim m_x$. Thus, spontaneous soft CP breaking at the
electroweak level can be effective even at baryon production temperatures.

The Higgs potential as a function of θ can, in general, be written as

$$V(\theta) = A + B \cos \theta + C \cos 2\theta \qquad (9)$$

where A, B, and C are independent of θ. Obviously, for an appropriate
range of parameters, the minimum of the Higgs potential lies at $\theta_o \neq 0$
with $\cos \theta_o = -B/4C$, so that we always have two solutions, θ_o and $-\theta_o$.

The value of $r-\bar{r}$ is proportional to $\sin \theta$. Now since $\theta = \pm \theta_o$ (the
solution of the minimization of the potential), one obtains from equation
(7)

$$\eta \propto \pm \sin \theta o \qquad\qquad (10)$$

The renormalization group analysis suggests the possibility that at even higher temperatures $T > m_x \simeq 10^{15}$ GeV, the symmetry was unbroken. Then as the temperature decreased below the mass scale of the superheavy gauge bosons, we expect that separate domains were generated with θ_o and $-\theta_o$ phases. Therefore from equation (10) it is obvious that one is bound to expect domains with matter and antimatter excesses in the universe. Senjanović and Stecker also considered a recently suggested model (Harvey, Ramond and Reiss 1980), based on the SO(10) grand unified theory. The idea is that a 126-dimensional representation of Higgs fields can be shown to be able to acquire a complex vacuum expectation value for a range of parameters of the Higgs potential. Therefore, one can have CP violation at the unification temperature scale completely independent of the nature of the light (electroweak) Higgs sector. (This situation is to be contrasted with the SU(5) theory, where the heavy Higgs ($\sim m_x$) sector is chosen to be a 24-dimensional, or adjoint representation, whose vacuum expectation value is always real.) Again, as in the previous example, one can show that $\pm \theta_o$ are solutions which minimize the potential.

6. DOMAIN GROWTH AND HORIZON GROWTH

The above discussion suggests that the initial domains were formed at a time when the temperature of the universe was comparable to the masses of the superheavy gauge or Higgs bosons involved in the symmetry breaking. A particularly promising mechanism for producing domains on an astronomically relevant scale has been suggested by Sato (1981). This mechanism depends on the fact that the expansion of the universe can be drastically altered from the standard radiation-dominated relationship if the energy density of the Higgs field is larger than that of the thermal radiation.

In the early, high temperature universe, using the Robertson-Walker metric, the Einstein equations reduce to

$$\left(\frac{\dot{R}}{R}\right)^2 = \frac{\kappa}{R} + \frac{\Lambda}{3} + \frac{8\pi G \varepsilon}{3} \simeq \frac{8\pi G}{3} (\varepsilon_r + \varepsilon_v) \ . \qquad\qquad (11)$$

For $\varepsilon_r \gg \varepsilon_v$ with $\varepsilon_r \propto T^4$, equation (16) yields the standard result $T \propto 1/\sqrt{t}$. However, when $\varepsilon_v \gg \varepsilon_r$ and for temperatures not near the critical temperatures for symmetry breaking, $\varepsilon_v(T) \simeq$ const., and it follows from equation (16) that the universe expands exponentially. This rapid expansion is a result of the large negative pressure of the vacuum (Bludman and Ruderman 1977; Kolb and Wolfram 1980b; Guth 1981). The result is an exponential stretching of the domains of CP coherence from their initial size, provided a first order (discontinuous) phase transition is involved. In the Sato scenario, the universe then supercools below T_c to a T_{c1} whereupon the transition becomes second order (continuous) or possibly driven,(cf. Witten 1981) whereupon a rapid universal phase transition releases an energy density ε_v. The universe then reheats to temperatures where X-particles are produced, which

subsequently decay to give baryon and antibaryon asymmetries on a
macroscopic scale. These exponentially stretched domains of baryon and
antibaryon excess may evolve further (Omnes 1972) leading to the formation
of matter and antimatter galaxies in separate regions of the universe
(Stecker and Puget 1972). This picture is outlined in Figure 1.

The symmetry breaking mechanisms which we have been discussing can
lead to the formation of various topological structures such as monopoles,
strings and domain walls, which could affect the dynamics and isotropy of
the universe. The problem of monopole formation has received the most
attention since, for simple grand unification scenarios, the production of
these particles would result in the universe having a mass density many
orders of magnitude higher than astronomical observations allow
(Zel'dovich and Khlopov 1978; Preskill 1979). Some suggestions for
solving the monopole problem involve the exponential stretching process
discussed in the last section and multiple phase transition (symmetry
breaking) scenarios (Langacker and Pi 1980). The breaking of discrete
symmetries can lead to domain wall formation, and it has been argued that
such walls, if formed, must disappear at an early stage in order to be
consistent with the observed homogeneity of the universe (Zel'dovich,
Kobzarev and Okun 1974). Clearly, the exponential stretching mechanism
which has been invoked to solve the monopole problem could also alleviate
the wall problem while providing a mechanism for domain growth. Vilenkin
(1981) has considered the dynamics of walls and strings and discussed
several mechanisms for wall disappearance, one of which again involves
multiple symmetry breaking. He has also found that domain walls do not
reflect light but do repel nonrelativistic particles. Such a repulsion
might play a role in keeping matter and antimatter apart at some stage in
the early universe. Using an idea reminiscent of the suggestion of
Vilenkin (1981), Kuz'min, Tkachev and Shaposhnikov (1981) have
demonstrated a method by which domain walls may vanish. Choosing a model
based on three Higgs multiplets, similar to that discussed previously,
these authors show how the CP asymmetries operative at the baryon
production stage may be restored as the universe cools, resulting in the
dissipation of the domain walls.

7. GALAXY FORMATION

Models of galaxy formation from "primordial turbulence" have always
been attractive as a way of accounting for galaxy formation as well as for
observed parameters such as the angular momenta and spatial distribution
of galaxies. However, in that work, turbulence was introduced in ad hoc
manner and, furthermore, such turbulence would be strongly damped out in
the cosmic plasma because of the very high viscosity of the blackbody
radiation field which remains coupled to the plasma until the
neutralization ("recombination") epoch.

In the baryon symmetric cosmology scenario, this viscous dissipation
is constantly fought by continuing radiation pressure from annihilation on
the boundaries of matter and antimatter regions, which regenerates the

SIMPLEST BARYON SYMMETRIC BIG-BANG SCENARIO

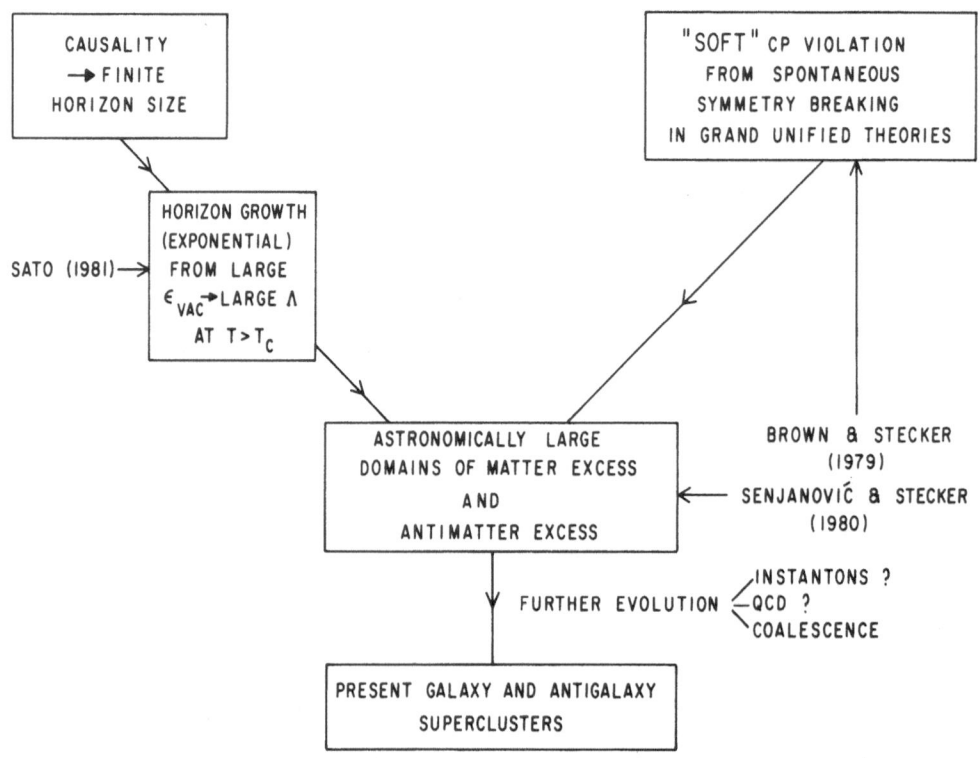

Figure 1. New framework for baryon symmetric big-bang cosmology.

turbulence. Radiation pressure from the annihilation, being directed
generally away from the boundaries can drive mass fluid motions as well as
causing further coalescence until the separate regions reach the size of
galaxy clusters.

At the recombination epoch, the viscosity dropped drastically and the
turbulent fluid motions became supersonic. Thus, both "small-scale"
turbulence and density fluctuations could start to build up in the
decoupled atomic fluid and later contract to form galaxies. In this
scenario annihilation pressure can provide a continuous source of
generating turbulence (Stecker and Puget 1972).

Barrow and Turner (1981) have proposed another scenerio for galaxy
formation based on having exponential domain growth in combination with
hard and soft CP violation. This picture results in a desired spectrum of
isothermal fluctuations leading to galaxy formation and preserves an all
matter universe which the authors desire. However, note that if we
eliminate the element of hard CP violation (undesirable if only because it
leads to problems with the neutron electric dipole moment) we still obtain
isothermal fluctuations - but in this case we again arrive at a baryon

symmetric domain cosmology.

8.   THE COSMIC γ-RAY BACKGROUND RADIATION

One of the most significant consequences of baryon symmetric big-bang cosmology lies in the prediction of an observable cosmic background of γ-radiation from the decay of π^0-mesons produced in nucleon-antinucleon annihilations. This is also perhaps at present the most encouraging aspect of this cosmology, since it satisfactorily explains the observed energy spectrum of the cosmic background γ-radiation as no other proposed mechanism does (with the possible exception of hypothetical point sources).

For high redshifts z, when pair production and Compton scattering become important, it becomes necessary to solve a cosmological photon transport equation in order to determine the γ-ray background spectrum. This integro-differential equation takes account of γ-ray production, absorption, scattering, and redshifting (Stecker, Morgan and Bredekamp 1971).

Figure 2 shows the observational data on the γ-ray background spectrum. The dashed line marked X is an extrapolation of the X-ray background component. The theoretical curve marked "annihilation" is the calculated annihilation spectrum (Stecker 1978). The excellent agreement between theory and data is apparent. This striking evidence has been a prime motivation for studying BSDC. Other recent attempts to account for the γ-ray background radiation spectra by diffuse processes give spectra which are, in one way or another, inconsistent with the observations, generally by being too flat at the higher energies.

It is possible that the γ-ray background is made up of a superposition of point sources. However, since only one extragalactic source has been seen at energies above ~1 MeV, this remains a conjecture. Such a hypothesis must be tested by determining the spectral characteristics of extragalactic sources and comparing them in detail with the characteristics of the background spectrum. It presently appears, e.g., that Seyfert galaxies may have a characteristic spectrum which cuts off above a few MeV, so that they could not account for the flux observed at higher energies.

9.   ANTIMATTER IN THE COSMIC RADIATION

Measurements of cosmic-ray antiprotons can give us important information about cosmic-ray propagation and also provide a test for primary cosmological antimatter. Gaisser and Levy (1974) pointed out that observation of a cosmic $\bar{p}$ flux without the low energy cutoff characteristic of secondary antiprotons whould be a signal of a primary component of antiprotons in the cosmic rays. Buffington, Schindler and

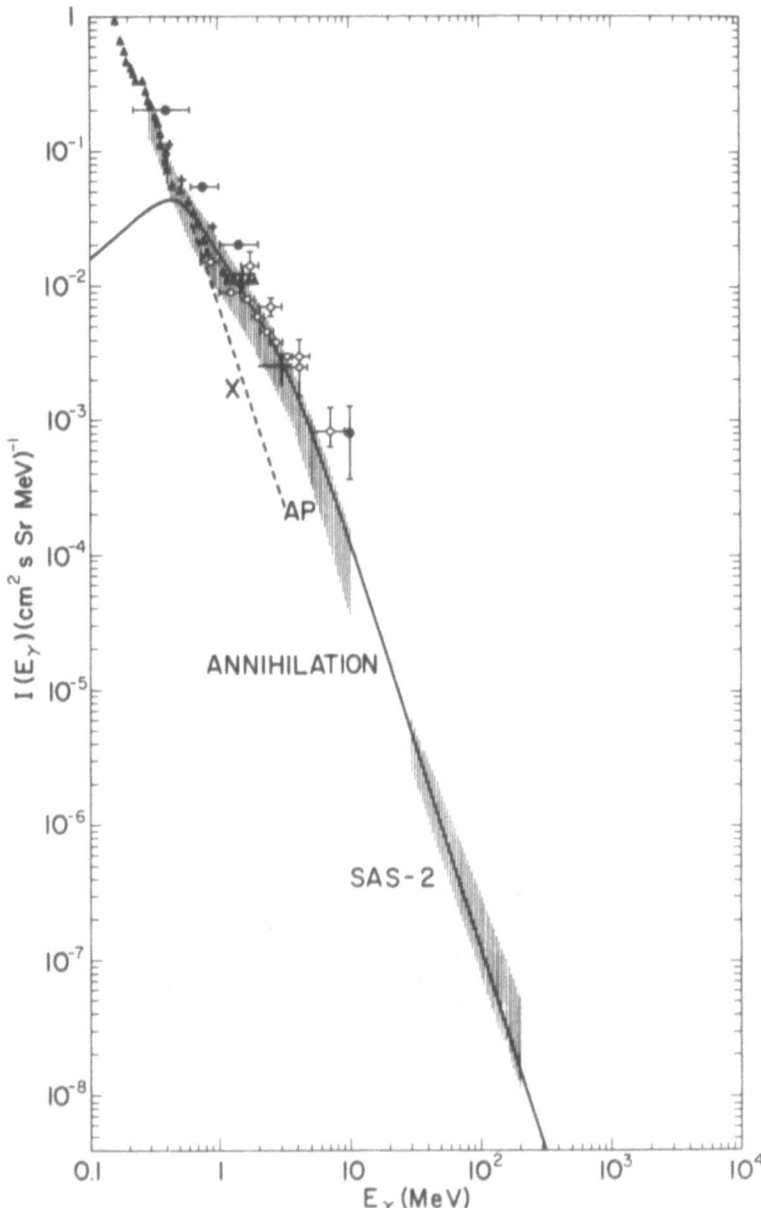

Figure 2. Data on the cosmic γ-ray background radiation from Apollo 15
and the SAS-2 satellite. Also shown are data from balloon experiments and
theoretical curves.

Pennypacker (1981), observing at energies well below the secondary cutoff,
appear to see just such a signal of primary antiprotons. Data on p̄ fluxes

at higher energies (Bogomolov, et al. 1979, Golden, et al. 1979) give
measured values a factor of 4-10 above the fluxes expected for a standard
"leaky box" type propagation model with the primaries passing through ~ 5
g/cm^2 of material (Stecker, Protheroe and Kazanas 1981 and references
therein).

The magnitude of the secondary $\bar{p}$ component depends critically upon
how cosmic rays are stored in and propagate through the Galaxy. The
simplest model describing the propagation of cosmic rays in the Galaxy is
the leaky box model. The closed galaxy model gives a higher $\bar{p}/p$ ratio
than the leaky box model. In the version of the closed galaxy model
proposed by Peters and Westergaard (1977) the sources of cosmic rays are
located in the spiral arms of the Galaxy, from which they slowly leak out
into an outer containment volume which comprises part of the disk and the
surrounding halo, a region which we will refer to here collectively as
"the halo". The outer boundary of the halo constitutes a closed box from
which cosmic rays cannot escape. Depletion of cosmic rays in the halo is
then solely due to nuclear interactions and energy losses. The halo thus
contains an "old component" while the spiral arms also contain a "young
component" of cosmic rays.

An important parameter of the closed galaxy model is K, the ratio of
the mass of gas in the galaxy as a whole to that in the spiral arms.
Peters and Westraard attempted a fit to the observed secondary to primary
ratios for values of K in the range 50 to 500. The rate of production of
antiprotons in the halo has been calculated for values of K ranging from
50 to 500. We show the resulting $\bar{p}/p$ ratios in Figure 3. As can be seen
from the figure, the closed galaxy model predictions are compatible with
the high energy data but predict a $\bar{p}$ flux which is still more than a
decade below that observed by Buffington, et al.

There are various problems associated with the closed galaxy model in
any case. It cannot account for the shape of the cosmic ray proton
spectrum at high energies (Ormes and Balasubrahmanyan, private
communication.) The model also requires confinement of a young component
to a spiral arm region containing the Sun. Such a picture does not appear
to be consistent with analysis of the non-thermal radio data (Price 1974;
Brindle et al. 1978) or a detailed analysis of the galactic γ-ray data
(Stecker 1977). Finally, it should be stressed that there are no physical
reasons for arguing that the Galaxy should be substantially closed to
cosmic-ray leakage.

It is difficult to see how the high flux of antiprotons below the low
energy cutoff characteristic of secondary antiprotons can be explained by
a secondary galactic component. Figure 3 also shows the prediction for
the leaky box model. If this model provides the correct description of
galactic cosmic ray confinement and propagation, then the spectrum of an
additional primary antiproton component making up the deficit $\bar{p}$ flux would
have roughly the same shape as the galactic proton spectrum. The ratio of

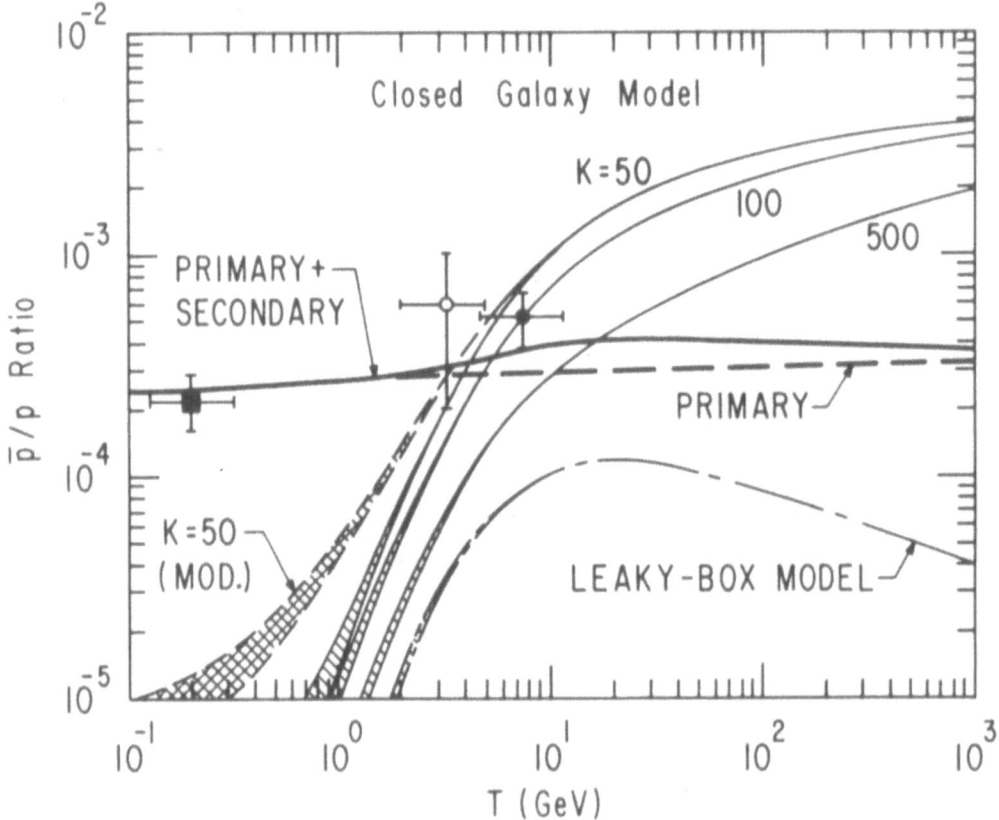

Figure 3. The predicted p̄/p ratio for the closed galaxy model and leaky
box model compared with the observed ratio. The curve labled K=50 Mod
indicates the effect of solar modulation with a mean energy loss of 600
MeV on the closed galaxy model prediction for K=50. Key to data: (■)
Buffington, et al. (1981); (o) Bogomolov et al. (1979); (●) Golden et al.
(1979). The heavy line shows the effect of adding an extragalactic p̄
component to the leaky box model prediction as discussed in the text.
(From Stecker, et al. 1981).

the extragalactic p̄ flux to the galactic proton flux would then be
$(3.2 \pm 0.7) \times 10^{-4}$. This is plotted as the heavy dashed line in Figure 3.
The reduction in the p̄/p ratio below this value at low energies is due to
the combined effects of "galactic modulation" (ionization energy losses,
nuclear interactions and p̄ annihilation) and solar modulation. The p̄/p
ratio for the sum of this extragalactic component plus the secondary
(leaky box model) component is shown by the heavy line of Figure 3.

     The inconsistency of the observed cosmic ray antiproton spectrum and
intensity with the calculated secondary flux, as well as the fact that
p̄/p≃ const. independent of energy, may both be indications of a possible
primary extragalactic origin. Using rough energetics arguments (Ginzburg
and Syrovatskii 1964) one can estimate that leakage from normal galaxies
would produce an extragalactic cosmic ray component with a flux
$(I_{ex}/I_{gal})_{NG} = \xi_{NG} \simeq 10^{-5} - 10^{-4}$. For active galaxies, these estimates

yield $\xi_{AG} \simeq 10^{-3}$. If we assume that half of the extragalactic flux is from antimatter sources, the resulting estimate for $\bar{p}/p \simeq 1/2\xi_{AG} \simeq 5\times10^{-4}$ is interestingly quite close to the measured values.

In discussing the $\bar{p}$ data, we should note the upper limits on the fluxes of antinuclei. The best 95% confidence upper limits at present are $\bar{\alpha}/\alpha \lesssim 1.5\times10^{-4}$ at 4.33 GeV/c (Badhwar, et al. 1978) barely consistent with $\bar{\alpha}/\alpha \simeq \bar{p}/p$, and $\bar{\alpha}/\alpha < 2.2\times10^{-5}$ in the low energy range of 130-370 MeV/ nucleon (Buffington, et al. 1981) indicating that $\bar{\alpha}/\alpha < \bar{p}/p$ in this energy range. This latter upper limit is consistent with $\bar{\alpha}/\alpha = \xi_{NG}/2 = 5\times10^{-6}$ -5\times10^{-5}$ (see above). Note that we can only argue that $\bar{\alpha}/\alpha = \bar{p}/p$ for cosmic ray poduction in <u>normal galaxies,</u> since we are comparing extragalactic fluxes with fluxes produced by processes in our own galaxy. It is conceivable that cosmic-ray α's produced in the cores or jets of <u>active galaxies</u> are broken up by collisions with matter or photons. Thus, the observed $\bar{p}$'s could come from active antimatter galaxies without accompanying $\bar{\alpha}$'s, but with the expected $\bar{\alpha}/\alpha \sim 10^{-5}$ from normal antimatter galaxies. In this case, future cosmic-ray experiments may soon detect $\bar{\alpha}$'s! Antimatter active galaxies containing regions of high photon or matter density may not be detectable as γ-ray sources, however, they may be directly determined to be antimatter sources through their production of cosmic ray $\bar{\nu}_e$'s (Learned and Stecker 1979).

In a matter-antimatter symmetric domain cosmology it is posible for the helium formed in the first three minutes of the big-bang to have been partially or totally destroyed by photodisintegration by annihilation γ-rays. This process has been suggested to account for the recent observations of low He abundances in less evolved galaxies, implying a low value for the primordial helium abundance as compared to theory (Stecker 1980a, 1981). If this is indeed the case, active galaxies and quasars during the "bright phase" (Berezinsky and Smirnov 1975) may have had very little He to accelerate.

Let us now consider the propagation of extragalactic cosmic rays. Not much is known regarding the physical parameters involved and one has to resort to rough estimates. A diffusion model can be considered as a first approximation to the problem (Ginzburg and Syrovatskii 1964). The mean distance cosmic rays diffuse in time t_u is $\langle R \rangle \simeq (2Dt_u)^{1/2}$ where $D = (1/3)\,\ell v$ is the diffusion coefficient and $t_u \sim 10^{10}$ years. Since $v \sim 10^{10}$ cm s^{-1} the largest uncertainty lies with the determination of the length scale. The length ℓ is of the order of the scale of inhomogeneity of the intergalactic magnetic field, which is not less than the intergalactic particle mean free path, i.e. $\ell \gtrsim (n\sigma)^{-1}$. In an ionized gas with $T \sim 10^6$-10^8K and $n_e \sim 10^{-7}$-10^{-5} cm^{-3}, the corresponding lower limit for the mean diffusion distance is then in the range 0.5 to 500 Mpc. Thus, extra- galactic cosmic rays can reach our galaxy in a Hubble time from other clusters or superclusters which may consist of antimatter galaxies and contain cosmic ray sources. The estimates are admittedly quite uncertain, especially since they depend on the topology of intergalactic field lines which the cosmic rays follow, since their gyroradii are expected to be small compared to ℓ. (See also Kiraly, these proceedings.)

10. "CELL" STRUCTURE OF THE UNIVERSE

     Not only do galaxies form clusters, but also these clusters of
galaxies are not uniformly distributed; they cluster into superclusters.
Between the superclusters are large voids—regions with a very low
(possibly zero) space density of galaxies (Joeveer and Einasto 1978;
Gregory and Thompson 1978; Chincarini and Rood 1979; Shanks, these
proceedings). The existence of these holes is the kind of structure which
can arise from a BSDC. The cosmic background γ-radiation originating from
supercluster boundary annihilations should exhibit angular fluctuations
which can best be studied with a high-resolution detector such as the 100
MeV spark chamber detector proposed for a future satellite "Gamma Ray
Observatory".

11. FUTURE TESTS USING HIGH ENERGY COSMIC RAY NEUTRINOS

     Several suggestions have been made recently for using high-energy
neutrino astronomy to look for antimatter elsewhere in the universe
(Learned and Stecker 1979; Berezinsky and Ginzburg 1981, Stecker and Brown
1981). These suggestions are all based on the fact that cosmic ray pp and
pγ interactions favor the secondary production of π^+'s over π^-'s, whereas
for $\bar{p}p$ and $\bar{p}\gamma$ interactions the situation is reversed. The subsequent
decay of the pions results in equal amounts of ν_μ's and $\bar{\nu}_\mu$'s of almost
equal energies. However, π^+ decay leads to ν_e production, whereas π^-
decay leads to $\bar{\nu}_e$ production. A production mechanism of particular
importance in this context because of its large inherent charge asymmetry
involves the photoproduction of charged pions by ultrahigh energy cosmic
rays interacting with the universal 3K blackbody background radiation.
The most significant reactions occur in the astrophysical context
principally through the Δ resonance channels because of the steepness of
the ultrahigh energy cosmic ray spectrum.

     There is a significant and potentially useful way of distinguishing
ν_e's from $\bar{\nu}_e$'s, namely through their interactions with electrons. The
$\bar{\nu}_e$'s have an enhanced cross section (resonance) through formation of weak
intermediate vector bosons such as the W^-. For electrons at rest in the
observer's system, the resonance occurs for cosmic $\bar{\nu}_e$'s of energy
$M_W^2/2m_e = 6.3 \times 10^3$ TeV for $M_W \simeq 80$ GeV corresponding to $\sin^2\theta_W \simeq 0.23$ in
the GWS model.

     If one entertains the possibility of higher mass intermediate vector
bosons, $\bar{\nu}_e + e^- \rightarrow B^-$, and correspondingly higher resonance energies, a
feasible test for cosmic antimatter may be at hand.

     The cosmic and atmospheric fluxes for $\bar{\nu}_e$'s, based on cosmic ray
production calculations have been given by Stecker (1979). Assuming that

there is no significant enhancement in the flux from production at high
redshifts, the integral $\bar{\nu}_e$ spectrum from $\gamma\bar{p}$ interactions is expected to be
roughly constant at 10^{-18} to 10^{-17} $\bar{\nu}$'s cm^{-2} sr^{-1} up to an energy of ~ 2 X
10^7 TeV, above which it is expected to drop steeply. It is expected that
the largest competing background flux of $\bar{\nu}_e$'s will be prompt $\bar{\nu}_e$'s from the
decay of atmospherically produced charmed mesons. A cosmic $\bar{\nu}_e$ signal may
be heavily contaminated by prompt atmospheric $\bar{\nu}_e$'s at the W resonance
energy. The cosmic flux is expected to dominate the higher energies so
that the existence of higher mass bosons B$^-$ may be critical to any
proposed test for cosmic antimatter using diffuse fluxes (Brown and
Stecker 1981, Stecker and Brown 1981). (There is now experimental
evidence for M$_W$ > 100 GeV as suggested by composite models (Fitzsch, these
proceedings.))

An acoustic deep underwater neutrino detector may provide the best
hope for testing for cosmic antimatter by studying the diffuse background
neutrinos. The practical threshold for such devices appears to be in the
neighborhood of 10^3 - 10^4 TeV (Bowen and Learned 1979). For higher mass
resonances B$^-$, the relevant neutrino resonance energy $E^B \propto M^2$ and the
effective detection volume $V_{eff} \propto M_B^6$. Considering that the incident flux
is expected to be roughly constant up to energies ~ 2 X 10^7 TeV, one gains
much in looking for higher mass resonances at higher energies. Acoustic
detectors of effective volume >> 10 km^3 (10^{10} tons) may be economically
feasible and event rates of ~ 10^2 - 10^4 yr^{-1} may be attained in time.

The asymmetry in the production of charged pions in matter versus
antimatter sources is reflected in cosmic-ray pp and $\bar{p}p$ interactions as
well as pγ and $\bar{p}\gamma$ interactions. Through the principal decay mode, this
asymmetry is again reflected in a ν_e - $\bar{\nu}_e$ asymmetry and thus in the
characteristics of events produced in deep underwater neutrino detectors. For ν-sources, these effects may be measurable at energies ~ 1-10
TeV with optical detectors (Learned and Stecker 1979). The possibility
that pγ and $\bar{p}\gamma$ interactions in quasars and active galaxies would produce
significant fluxes of $\bar{\nu}_e$'s , detectable through the W$^-$ resonance, has been
suggested by Berezinsky and Ginzburg (1981) as a way of looking for cosmic
antimatter. Hopefully, this interesting suggestion will be explored in
more detail as our understanding of the nature of cosmic ray production in
compact objects increases.

12. CONCLUSION

In grand unified theories, a scenario has been developed for the
evolution of the early universe wherein the matter which eventually forms
the galaxies arises as a "baryon excess" owing to baryon number non-
conserving interactions at ultrahigh energies. This scenario requires
that CP symmetry be broken. Although the nature of CP breaking, even at
low energies, has not yet been established, there are several reasons to
prefer spontaneous CP symmetry breaking. Aside from the philosophical
consistency with the whole concept of spontaneous symmetry breaking which
is the key to unified gauge theories, there are several important

technical reasons (beyond the scope of this paper) for suspecting that CP is broken spontaneously. This mechanism explains naturally why CP violation is small (a problem with the alternative Kobayashi-Maskawa model of explicit hard CP violation). It leads in a natural way to natural flavor conservation (a generalization of the GIM mechanism). It provides a solution to the strong CP problem as well (e.g. Mohapatra and Senjanović 1978).

Spontaneous breaking of CP leads to a domain structure in the universe with the domains evolving into separate regions of matter excess and antimatter excess. The creation of these excesses subsequent to a period of exponential horizon growth (a dynamical effect of the Higgs fields) can result in a universe in which matter galaxies are formed in some regions and antimatter galaxies are formed in others. There is no need for a separation mechanism since the regions containing the excesses come into being as separate regions. There are advantages in this model in explaining various astrophysical data such as the cosmic γ-ray background spectrum, cosmic ray $\bar{p}$ flux measurements, a low primordial He abundance and strong galaxy clustering. (It should, of course, be kept in mind that all present scenarios for the early big-bang based on present specific grand unification models have problems with topological singularities and that these ideas may be drastically revised if it becomes necessary to attribute a composite nature to quarks and leptons.)

It should be kept in mind that any positive observational data (e.g., cosmic-ray $\bar{p}$'s, $\bar{\alpha}$'s, $\bar{\nu}_e$'s) supporting the existence of large amounts of antimatter in the universe will be evidence of the spontaneous nature of CP violation at high energies, in accord with our earlier discussion. Thus, astrophysical tests which can distinguish between an all matter cosmology and a baryon symmetric domain cosmology can tell us something important and fundamental about the nature of particle physics at extremely high energies. (See also Stecker 1978, 1980b.)

REFERENCES

Abers,E. S. and Lee, B. W., 1973, Phys. Lett. 9C, 1.

Badhwar, G. D., et al., 1978, Nature 274, 137.

Barr, S., Segre, G. and Weldon, H. A., Phys. Rev. D20, 2949.

Barrow, J. D. and Turner, M. S. 1981, EFI Preprint No. 81-02.

Bég, M. A. B. and Sirlin, A. 1974, Ann Rev. Nucl. Sci. 24, 79.

Berezinsky, V. S. and Ginzburg, V. L., 1981, Mon. Not. Roy. Astr. Soc. 194, 3.

Berezinsky, V. S. and Smirnov, A. Yu., 1974, Astrophys. Space Sci. 32, 461.

Bludman, S. and Ruderman, M. A., 1977, Phys. Rev. Lett. 38: 255.

Bogomolov, E.A., et al., 1979, Proc. 16th Intl. Cosmic Ray Conf. Kyoto 1, 330.

Bowen, T. and Learned, J. G., 1979, Proc. 16th Intl. Cosmic Ray Conf., Kyoto 10, 386.

Brown, R. W. and Stecker, F. W., 1979, Phys. Rev. lett. 43: 315.

Brown, R. W. and Stecker, F. W., 1981, Proc. 1980 DUMAND Summer Workshop, ed. V. Stenger, U. Hawaii, 2, 240.
Buffington, A., Schindler, S. M. and Pennypacker, C. R. 1981, Astrophys. J. 248, 1179.
Chincarini, G. and Rood, H. J., 1979, Astrophys. J. 230, 648.
Fritzsch, H. and Minkowski, P., 1975, Ann. Phys. 93, 193.
Gaisser, T. K. and Levy, E. H., 1974, Phys. Rev. D10, 1731.
Georgi, H. and Glashow, S. L., Phys. Rev. Lett. 32, 438.
Ginzburg, V. L. and Syrovatskii, S. I., 1964, Origin of Cosmic Rays (MacMillan: New York).
Glashow, S. L., 1960, Phys. Rev. 118, 316.
Golden, R. L., et al., 1979, Phys. Rev. Lett. 43, 1196.
Gregory, S. A. and Thompson, L. A., 1978, Astrophys. J. 222, 784.
Guth, A. H., 1981, Phys. Rev. D23, 347.
Harvey, J. A., Ramond, P. and Reiss, D. B., 1980, Phys. Lett. 92B, 309.
Heisenberg, W., 1967, Einfuerung in die einheitliche Feldtheorie der Elementarteilchen, Hirzel, Stuttgart.
Joeveer, M. and Einasto, J. 1978, in The Large Scale Structure of the Universe (ed. Longair, M. S. and Einasto, J.) I.A.U. Symp. No. 79 (Dordrecht: Reidel Pub. Co.) p. 241.
Kolb, E., 1981, Proc. Johns Hopkins Workshop on Current Problems in Particle Theory (May 1981), in press.
Kolb, E. W. and Wolfram, S., 1980a, Nucl. Phys. B172, 224.
Kolb, E. W. and Wolfram, S., 1980b, Astrophys. J. 239, 428.
Kuzmin, V. A., Tkachev, I. I. and Shaposhnikov, M. E., 1981, Piśma ZhETF 33, 557.
Langacker, P. and Pi, S. Y., 1980, Phys. Rev. Lett. 45, 1.
Langacker, P., 1981, Phys. Rpts., in press.
Learned, J. G. and Stecker, F. W., 1980, Proc. Neutrino 79 Intl. Conf., Bergen, Norway (ed. Haatuft, A. and Jarlskog, C.) 2: 461.
Linde, A. D., 1979, Rep. Prog. Phys. 42: 389.
Mohapatra, R. N. and Senjanović, G., 1978, Phys. Lett. 79B: 283.
Mohapatra, R. N. and Senjanović, G., 1979, Phys. Lett. 89B: 57.
Nanopoulos, D. V. and Weinberg, S., 1979, Phys. Rev. D20, 2484.
Omnes, R., 1972, Phys. Rep. 3C, 1.
Peters, B. and Westegaard, N. J., 1977, Astrophys. and Space Sci. 48, 21.
Sakharov, A. D., 1967, JETP Lett. 5, 24.
Sato, K., 1981, Phys. Lett. 99B, 66.
Senjanović, G. and Stecker, F. W., 1980, Phys. Lett. 96B, 285.
Stecker, F. W., 1975, Phys. Rev. Lett. 35, 188.
Stecker, F. W., 1978, Nature 273, 493.
Stecker, F. W., 1979, Astrophys. J. 228, 919.
Stecker, F. W., 1980a, Phys. Rev. Lett. 44, 1237.
Stecker, F. W., 1980b, Proc. Tenth Texas Symposium on Relativistic Astrophysics, Ann. N. Y. Acad. Sci., in press.
Stecker, F. W., 1981, Phys. Rev. Lett. 46, 517.
Stecker, F. W. and Brown, R. W., 1981, Proc. 1980 DUMAND Summer Workshop, ed. Stenger, V., U. Hawaii, 2, 248.
Stecker, F. W., Morgan, D. L. and Bredekamp, J. H. 1971, Phys. Rev. Lett. 27, 1469.
Stecker, F. W. and Puget, J. L., 1972, Astrophys. J. 178, 57.

Stecker, F. W., Protheroe, R. J. and Kazanas, D., 1981, <u>Proc. 17th Intl.</u>
    <u>Cosmic Ray Conf.</u> (Paris), in press.
Vilenkin, A., 1981, Phys. Rev. <u>D23</u>, 852.
Weinberg, S., 1974, Phys. Rev. <u>D9</u>, 3357.
Weinberg, S. 1979, Phys. Rev. Lett. <u>42</u>, 850.
Witten, E., 1981, Nucl. Phys. <u>B177</u>, 477.
Yildiz, A. and Cox, P., 1980, Phys. Rev. <u>D21</u>, 906.
Zeldovich, Ya. B., Kobzarev, I. Yu. and Okun, L. B., 1974, Zh. Eksp. Teor.
    Fiz. <u>67</u>, 3.

COSMOLOGY AND PARTICLE PHYSICS

John D. Barrow. Department of Physics, University of
California, Berkeley, California 94720, USA   and
Astronomy Centre, University of Sussex Brighton BN1 9QH, UK*

* Present address

A brief overview is given of recent work that integrates cosmology and
particle physics. The observational data regarding the abundance of
matter and radiation in the Universe is described. The manner in which
the cosmological survival density of stable massive particles can be
calculated is discussed along with the process of cosmological nucleosynthesis. Several applications of these general arguments are given
with reference to the survival density of nucleons, neutrinos and unconfined fractionally charge particles. The use of nucleosynthesis to
limit the number of lepton generations is described together with the
implications of a small neutrino mass for the origin of galaxies and
clusters.

1.   INTRODUCTION

At present cosmology and elementary particle physics are interacting very strongly. This interaction has been stimulated in part by
the attempts of particle physicists to constrain the many new theoretical
developments arising from unified gauge theories and in part by the
desire of cosmologists to construct a coherent model of the early Universe
right back to the Planck epoch $t_p = G^{-1/2} \sim 10^{-43} s$.

In this brief overview I would like to give a resumé of some of the
basic observational facts concerning the large scale structure of the
Universe;   provide some standard recipes for using cosmology to constrain
aspects of high energy physics and then give several specific examples
which are of current and mutual interest to high energy physicists and
astronomers.

2.   HOT BIG-BANG COSMOLOGY

The following is a summary of some basic observations concerning the
expansion and matter content of the Universe:

A. W. Wolfendale (ed.), Progress in Cosmology, 21–38.

2.1   Expansion:

The red-shifting of the spectra of distant galaxies was first interpreted as a systematic Doppler effect by Hubble. The red-shift has a
straightforward explanation if the Universe is expanding with velocity
v over a length scale r where, to first order,

$$v = H_o r \qquad\qquad (1)$$

Hubble's constant H_o is presently measured in the range 75 ± 25 Kms^{-1}
Mpc^{-1}. We shall parameterize it by $h_o \equiv H_o/100$ Kms^{-1}Mpc^{-1}.
Measurements of H_O imply the age of the Universe to be less than $\sim10^{10}$
h_o^{-1}yr.

2.2   Baryon Content:

The composition of the visible Universe is predominantly baryons
(rather than antibaryons or radiation). It is conventional to record
this density in units of a 'critical' density, ρ_{cr}, where

$$\rho_{cr} \equiv \frac{3H_o^2}{8\pi G} = 2 \cdot 10^{-29} h_o^2 \text{ gm cm}^{-3} \qquad\qquad (2)$$

The critical density is the largest average density that allows universal expansion to infinity. In general we can write

$$\rho \equiv \Omega\rho_{cr} \qquad\qquad (3)$$

so Ω is the ratio of the potential to kinetic energy in the Universe.
The number density of baryons (B) is just

$$n_B = 1.1 \cdot 10^{-5} \Omega_B h_o^2 \text{ cm}^{-3} \qquad\qquad (4)$$

Direct observational determinations of Ω_B are beset by a host of
selection effects and subtleties of interpretation (Faber and Gallagher
1979):   Observations of binary galaxies and the motion of galaxies in
small groups imply that the density of all gravitating material lies in
the range

$$0.04 \lesssim \Omega \lesssim 0.13 \qquad\qquad (5)$$

Direct observations of galaxies account for a luminous component with

$$0.006 \lesssim \Omega \lesssim 0.7 \qquad\qquad (7)$$

It is safe to conclude only that Ω_B lies somewhere in the range
$\sim10^{-2}$ - 1 if the cosmological constant is zero.

2.2.1 Baryon distribution: The distribution of luminous matter is observed to be extremely clumpy and there is evident clustering of material into objects of preferred sizes; stars, galaxies and so forth. The two-point correlation for galaxies (the excess probability, over random, of finding a galaxy at a distance r from another chosen at random), $\xi(r)$, is measured as (see Peebles 1980)

$$\xi(r) = \left(\frac{r_o}{r}\right)^{1.8}; \quad r_o \sim 4h_o^{-1} \text{ Mpc} \qquad (8)$$

This implies the existence of a characteristic non-linear mass scale dividing well-developed from weak clustering[2]. This scale is

$$M_* \sim 5 \cdot 10^{14} \, \Omega_B h_o^{-1} \, M_\odot \qquad (9)$$

2.2.2 Composition: The portion of the Universe we have seen is predominantly composed of hydrogen ($\sim$ 75% by mass). The cosmologist is principally interested in the remaining mass, almost all of which is in the form of helium-4. The abundance of helium-4 is a crucial datum because it appears to have been produced by nucleosynthesis in the early stages of the Universe's history, $t_{ns} \sim 1\text{-}10^3$s. Measurements of the helium-4 (and deuterium) abundances allow deductions to be made about events which influence the efficiency of this primordial nucleosynthesis.

Typical observational determinations for Y, the helium-4 mass fraction, are (Talent 1980): (i) galactic HII regions with normal metal abundance (and so a high level of stellar processing), Y = 30 ± 2%; (ii) HII regions with low metals (Z < 0.02), Y = 23 ± 2%. (iii) Globular clusters, Y = 22 ± 4% (Renzini 1979). These and other data imply a primordial (unprocessed) value for the helium mass fraction in the range 20-24%.

The deuterium mass fraction in interstellar material is measured to be $\sim 2.10^{-3}$% (Laurent et al 1979) and the primordial mass fraction probably lies in the range $\sim 2\text{-}8 \cdot 10^{-3}$%.

2.3 Radiation Content: The bulk of the radiative energy lies in the microwave band and has an almost thermal spectrum corresponding to a temperature 2.9 ± 0.1 K. This contributes a present energy density which is about four orders of magnitude lower than that of the baryonic material and the number density of photons is, if the expansion were exactly isotropic, given by

$$n_{\gamma o} = 548 \left(\frac{T_{\gamma o}}{3K}\right)^3 \text{ cm}^{-3} \qquad (10)$$

In contrast to the distribution of matter, the radiation is extremely uniform and isotropic. Temperature anisotropies are less than one part in 10^4 over a wide variety of angular scales (Silk 1980). The approximation of large scale homogeneity and isotropy used when modelling the present structure of the observable Universe is justified by this evidence.

The existence of the microwave background radiation introduces a new dimensionless number into physics: the baryon to photon ratio of the Universe, given by (4) and (10) as

$$\left(\frac{n_B}{n_\gamma}\right)_o = 2.0 \cdot 10^{-8} \; \Omega_B h_o^2 \left(\frac{3K}{T_{\gamma o}}\right)^3 \tag{11}$$

Using $T_{\gamma o} < 3K$, $\Omega_B > 0.04$ and $h_o > 0.5$ we have

$$\left(\frac{n_B}{n_\gamma}\right)_o \geqslant 2.0 \cdot 10^{-10} \tag{12}$$

A closely related quantity is the entropy per baryon s/n_B (Boltmann's constant is $k_B \equiv 1$) and

$$\frac{n_B}{s} = 0.14 \; \frac{n_B}{n_\gamma} \tag{13}$$

If we regard the early universe ($t < 10^{10}$ s) as a plasma then the number of particles in a Debeye sphere, N_D, is determined by the entropy per baryon and is $N_D \sim 8.1 \cdot 10^3 \, (s/n_B)^{1/2} \gg 1$ and so there should exist good collective behaviour.

3.   METHODOLOGY

The mode of interaction between particle physics and cosmology is very simple: (1) pick physical processes predicted to occur in the 'standard' hot big bang model which are sensitive to adaptions at high energy (for example the addition of extra species of particle or types of interaction) and whose consequences are amenable to observational check; (2) introduce these new parameters into the model of the very early Universe and determine the observational consequences; (3) delineate the allowed adaptions to the particle physics theory. A general word of caution is best introduced at this stage: when doing high energy physics in the hot big bang model we are really confronted with two sets of unknowns:
     (A) Various particle parameters: unknown masses, lifetimes, inter-action strengths and so forth.
     (B) Various cosmological parameters: the closeness of the early Universe to exact homogeneity and isotropy, the abundance of mini black holes, gravitational waves and so on.

For obvious reasons, particle physicists would like to assume that category (B) is empty and the standard Friedmann model holds right back to $t_p \sim 10^{-43}$s. We have direct observational support for this assumption back to $\sim 10^{12}$s (the last scattering time of the 3K photons) and indirect support back [8] to $\sim$ 1s (the determination of the relative number of neutrons and photons available for the nucleosynthesis of helium-4). For the purposes of this talk we shall assume Friedmann behaviour, but we shall indicate how deviations from this dogma affect some of the predictions concerning particle parameters.

The assumption of isotropy and homogeneity (the 'Cosmological Principle') reduces the Einstein equations from ten coupled partial differential equations in four variables to a single ordinary differential equation for the expansion scale factor R(t). This is Friedmann's equation (Weinberg 1972)

$$\frac{\dot{R}^2}{R^2} = \frac{8\pi G\rho}{3} - \frac{k}{R^2} + \Lambda \qquad (14)$$

In the early Universe the spatial curvature term kR^{-2} is unimportant, and so we shall neglect it and also, for the time being, the cosmological constant, Λ.

In its early stages ($t \lesssim 10^{11}$s) the Universe is in a quasi-neutral plasma state of thermal equilibrium with density

$$\rho = \frac{g(T)}{30} \pi^2 T^4 \qquad (15)$$

where $g(T) = \Sigma g(\text{bosons}) + (7/8)\Sigma g(\text{fermions})$ is the total effective number of degrees of freedom at temperature $T \propto R^{-1}$. The temperature-time adiabat for the early evolution is given by solving (14) and (15) to give

$$\frac{t}{1s} = 2.4 \cdot 10^{-6} g^{-1/2} \left(\frac{1\text{GeV}}{T}\right)^2 \qquad (16)$$

or, in Planck units, $m_p \equiv G^{-1/2}$,

$$t = \frac{0.3 m_p}{g^{1/2} T^2} \qquad (17)$$

In this isotropic Friedmann Universe the temperature-time relation depends only on fundamental constants. This is no longer the case in anisotropic and inhomogeneous cosmological models: there is a contribution from the initial conditions in these cases.

For bosons we have $n \sim \zeta(3)T^3/\pi^2$ and typical interactions of strength α have cross-sections $\sigma \sim \alpha^2 T^{-2}$ so the mean free path of particles is λ where

$$\lambda = (\sigma n)^{-1} \sim (g\alpha^2 T)^{-1} \qquad (18)$$

Since the average inter-particle spacing is $n^{-1/3} \sim T^{-1}$ we have that

$$\frac{\text{Mean Free Path}}{\text{Inter-particle spacing}} \sim (g\alpha^2)^{-1} \gg 1 \qquad (19)$$

and so the ideal gas approximation holds as a description of events in the early Universe.

As we go backwards in time to higher temperatures the particle separations and interaction lengths become smaller than hadronic dimensions (for example the size of causally connected regions – the horizon size – at $\sim 10^{-35}$s.is just 10^{-12}fm!), and so the fundamental constituents of the Universe must be modelled by point-like objects – fermions (q,l) and gauge or scalar bosons.

The simple t(T) relation (16), or (17), allows us to exploit the Universe as an expanding theoretical laboratory with infinite energetic capability. Let us now look at two typical types of calculation which have wide applicability. They involve determining the survival probability of hypothetical massive particles and probing the stability of primordial nucleosynthesis.

3.1   Massive Particle Survival:

No hypothetical particle, whatever its nature, can contribute more than $\sim 2\rho_{cr}$ to the present total density of the Universe if $\Lambda = 0$ otherwise contradiction with observations of the decceleration parameter and the radioactive age of the earth would arise (Peebles 1971). To calculate the present survival density of massive, stable particles with some interaction rate $\langle \sigma v \rangle$ we must write down the Boltzmann equation for the evolution of the total number of these particles with mass m, N(m), in the expanding Universe, Chiu (1966), Zel'dovich (1965), Steigman (1979),

$$\dot{N} = -N^2 \langle \sigma v \rangle - \frac{3R\dot{N}}{R} + \Pi(t) \tag{20}$$

$$= -\left\{ \begin{array}{c} \text{destruction} \\ \text{of} \\ \underline{mm} \end{array} \right\} - \left\{ \begin{array}{c} \text{dilution by} \\ \text{adiabatic} \\ \text{expansion} \end{array} \right\} + \left\{ \begin{array}{c} \text{production} \\ \text{of} \\ \underline{mm} \end{array} \right\}$$

If we "turn-off" the expansion then detailed-balancing determines the production rate $\Pi(t)$ in terms of the equilibrium distribution that must arise when we "turn-off" the expansion and set $\dot{R} = 0$.

$$\Pi(t) = N_{eq}^{2} \langle \sigma v \rangle \tag{21}$$

Solving (20) we see that when the particles are relativistic they are as abundant as photons and behave as though massless

$$\frac{N(m)}{N(\gamma)} \sim 1; \quad T > m \tag{22}$$

but if the $\overline{mm}$ remain in equilibrium when they become non-relativistic (T < m) then

$$\frac{N(m)}{N(\gamma)} \sim \left(\frac{m}{T}\right)^{3/2} \quad \exp(-m/T); \quad T < m \tag{23}$$

The deficiency of massive particles relative to photons just reflects the cost in energy of making $m\bar{m}$ pairs from the photon sea when $T < m$. Clearly, if massive particles were always in equilibrium none would remain today. However, in practice, the equilibrium always ceases at some temperature T_* when the expansion rate of the Universe becomes faster than the interaction rate $\langle\sigma v\rangle$. The massive particles gradually cease to interact and the $N(m)/N(\gamma)$ ratio freezes in at the value of (23) when $T = T_*$.

In general, the present survival density of massive particles is

$$\rho_o(m) = 3.4 \ 10^{-56} \ \frac{10^{-15} \text{cm}^3 \text{s}^{-1}}{\langle\sigma v\rangle} \left(\frac{m}{T_*}\right) \ g^{-1/2}(T_*) \ \text{gm cm}^{-3} \tag{24}$$

where the temperature T_* is determined implicitly by

$$\frac{T_*}{m} + \frac{1}{2} \ \ln\left(\frac{T_*}{m}\right) = 45.4 + \ln(g^{1/2}X) \tag{25}$$

$$X \equiv \left(\frac{m}{1\text{GeV}}\right) \ 10^{15} \left(\frac{\langle\sigma v\rangle}{\text{cm}^3 \text{s}^{-1}}\right) \tag{26}$$

Four specific examples are instructive:

3.1.1. Nucleon Survival: Suppose the Universe were created with zero net baryon number and this baryon number was conserved. The strong interaction would keep nucleons and antinucleons in equilibrium until $t \sim 10^{-4}$ s. For these interactions $\langle\sigma v\rangle \sim 10^{-15}$ cm^{3}s^{-1}, and so they remain in equilibrium for a significant period after the nucleons become non-relativistic, $T_* \sim 0.02m_N$. The final abundance of nucleons relative to photons is consequently very low; according to (23) we find

$$\frac{n(N)}{n_\gamma} = \frac{n(\bar{N})}{n_\gamma} \sim 10^{-18} \tag{27}$$

and the current energy density of nucleon survivors would be minute

$$\rho_{B_o} \sim 10^{-40} \text{ gm. cm}^{-3} \tag{28}$$

This argument, leading to a prediction (27) in gross disagreement with observation (12), was first framed by Chiu (1966) and Zel'dovich (1965) and clearly shows the simple Friedman model Universe cannot be baryon symmetric. It indicates that there is an anomalously large

density of nucleon survivors in the Universe because there was some form
of a baryon asymmetry built into the Universe prior to $\sim 10^{-5}$s. For a
discussion of this see Barrow (1980) and references therein.

3.1.2.   Massive Neutrinos: The same analysis can be employed to
calculate the survival density of massive neutrinos if $\langle \sigma v \rangle$ is determined by the weak interaction. The weak interaction rate is $\sim 10^{-50}$
$(T/1K)^5$ and is faster than the Universal expansion rate whilst T > 1MeV.
Those neutrinos with $m_\nu \lesssim 1$MeV will become collisionless whilst they are
still relativistic and will (up to statistical factors) remain as
abundant as photons. Today we will observe, (Gershtein and Zel'dovich
(1966), Szalay and Marx (1976), Cowsik and McClelland (1972); g_ν is
the total number of helicity states for all light neutrinos):

$$(\rho_\nu)_o = 6 \cdot 10^{-31} m_\nu \quad (\text{eV})^{-1} \text{ gm cm}^{-3} \qquad (29)$$

If there are three neutrino generations of equal mass then, the average
neutrino mass must be less than $\sim 60 h_o^2$eV when $\Lambda = 0$.

Heavier neutrinos ($m_\nu \gtrsim 1$MeV) will remain in equilibrium whilst they
are non-relativistic until $T_* \sim 0.05 m_\nu$ and their survival density is
calculated as (14)

$$(\rho_\nu)_o = 5 \cdot 10^{-29} m^{-2} \quad (\text{GeV})^2 \text{ gm cm}^{-3} \qquad (30)$$

In this case the constraints on Ω imply $m_\nu \gtrsim 2$GeV. Note that the heavier
the neutrino the less it contributes to the total density of the Universe
if it is heavier than ~ 1MeV. This is because the $\nu\bar{\nu}$ destruction cross-
section is $\propto m_\nu^2$ in non-relativistic equilibrium.

3.1.3.   Unconfined Quarks: If confinement is not absolute then (24)
allows the quark abundance inherited from the big bang to be calculated.
This was originally done by Zel'dovich, Okun and Pikelner in 1965 under
the assumption that all quarks were free. They concluded that a severe
conflict existed between particle physics and cosmology. For $m_q \sim 100 m_N$
and $\langle \sigma v \rangle_q \sim \langle \sigma v \rangle_N$ they calculated a present quark:nucleon ratio

$$\frac{n_q}{n_B} \sim 10^{-10} \qquad (31)$$

This was 10-20 orders of magnitude in excess of current observational
limits. Currently published limits vary from $< 10^{-19} - 10^{-27}$, but
given the lack of understanding of quark chemistry and likely places for
free quarks to reside, they probably should not be taken too seriously.

Wagoner and Steigman (1979) have recently updated the earlier analyses to accommodate partial confinement. They employ a q-$\bar{q}$ potential

$$\frac{U(r)}{1\text{GeV}} = -0.026 \left(\frac{1\text{fm}}{r}\right) + 1.18 \left(\frac{r}{1\text{fm}}\right) \tag{32}$$

but assume the potential reaches a limiting value at $U = 2m_q$ when $r = r_c$; that is, gluon strings are imagined to snap when $r > r_c$. This could be the case if gluons possessed a mass $\sim r_c^{-1}$.

The quark-hadron phase transition will occur at some characteristic temperature T_{qh}. This temperature cannot be calculated exactly but must exceed the temperature at which hadrons begin to overlap $(\frac{4\pi r^3}{3} n_\pi \sim 1)$. If $r_\pi \sim 0.56$ then this lower bound lies in the range 170 $\stackrel{-}{-}$ 310 MeV. Above these energies composite hadrons are no longer well-defined entities. The cosmic medium is envisaged to change into a 'soup' of quarks and gluons during the first 10^{-5}s of expansion history. An upper bound on T_{qh} is provided by the energy at which the quark-gluon fluid becomes non-ideal because of colour interactions $(rdU/dr \sim T)$. This ensures $T_{qh} \stackrel{<}{\sim} 360\text{--}440$ MeV. As T falls below ~ 440 MeV quarks can begin to bind into hadrons, but as the number so bound increases, the screening of the colour force is reduced so the average quark feels a stronger potential. The potential was modelled so that those with energy exceeding m_q remain free. While they remain in equilibrium via $q + \overline{q} \leftrightarrow \gamma$, the abundance of free quarks will be $(g_q \sim 30)$

$$n_q = n_{\overline{q}} = \frac{g_q}{2\pi^2} \left(\frac{m_q}{T}\right)^2 T^3 \; e^{-m_q/T} \tag{33}$$

when T falls below ~ 170 MeV interactions between quarks $(q + q \leftrightarrow N + \overline{q},$ $\overline{q} + \overline{q} \leftrightarrow \overline{N} + q, \; q + \overline{q} \leftrightarrow \gamma)$ will continue at a rate given by $<\sigma v>_q \sim \pi r_c^2 \sim 4.10^{24} m_q^2 (\text{GeV}) \text{ fm}^3\text{s}^{-1}$. The Boltzmann equation then gives the final abundance of free quarks as

$$\left|\frac{n(q) + n(\overline{q})}{n_B}\right|_o \sim 10^{-10} \left(\frac{1\text{GeV}}{m_q}\right)^3 \quad \text{for } m_q \stackrel{<}{\sim} 8 \text{ GeV} \tag{34}$$

$$\sim 10^9 \left(\frac{m_q}{T_{qh}}\right)^2 \exp\left(\frac{-m_q}{T_{qh}}\right) \text{for } m_q \stackrel{>}{\sim} 8\text{GeV} \tag{35}$$

Taken at face value the measurements of the Stanford group (La Rue et al 1979) indicate the existence of fractional charge in niobium and if this is identified with free quarks it gives $[n(\overline{q}) + n(q)]/n_B \sim 10^{-20}$ and so implies $13 < m_q < 27\text{GeV}$. Direct accelerator searches (Jones, 1977) give a lower limit of $m_q > 4\text{GeV}$.

It is interesting to point out the manner in which the quantitative conclusions regarding particle survival depend upon the unverified assumption of exactly isotropic expansion in the early stages of the Universe $(t < 0.1s)$: In general, anisotropic expansion increases the volumetric expansion rate of the Universe by providing it with more degrees of freedom and so makes it easier for particle distribution to pass out of equilibrium earlier. <u>More</u>   massive particles therefore survive than in the isotropic model. This is an interesting feature because if any of these hypothetical particles are ever discovered, their cosmological abundance will provide an indirect means of ascertaining the cosmological expansion dynamics prior to ~0.1s. As a specific example, it can be shown that if the Universe expands anisotropically ($R \propto T^{-1} \propto t^{1/3}$), until a time t_A exceeding ~10^{-4}s then a baryon symmetric Universe would contain a baryon (or antibaryon) to photon ratio in excess of (27):

$$\frac{n(N)}{n_\gamma} = \frac{n(\bar{N})}{n_\gamma} \sim 10^{-18} \left(\frac{t_A}{10^{-4}s}\right)^{\frac{1}{2}} \tag{36}$$

This estimate is based upon the simplifying assumption that there are no complicated non-equilibrium effects associated with the particles once they become collisionless. These would probably decrease n_N/n_γ below the value in (36).

3.1.4. <u>Three-body Annihilations</u>:   The Boltzmann equation (20) contains an implicit assumption - that massive particles annihilate primarily via <u>two</u>-body collisions. This is accurate in the cases we have considered but it is possible to imagine species of massive particles that only annihilate via three-body encounters. The existence of such particles - called.theta (θ) particles - was recently suggested by Okun (1980). The annihilations of θ's will be proportional to N^3 rather than N^2 as in (20) and, if stable, they are found to contribute a total density of (Dolgov 1980),

$$\rho_o (\theta) \sim 8.4 \cdot 10^{-27} \left(\frac{m_\theta}{1 \text{ GeV}}\right)^{1/2} \quad \text{gm cm}^{-3} \tag{37}$$

3.2 Primordial Nucleosynthesis:

According to the hot big bang model there exists a narrow niche of cosmic history during which the fusion of light elements is possible (Peebles 1971, Weinberg 1972). Above ~5.10^9 K the radiation sea in the Universe will be so energetic that any light nuclei will be immediately photo-disintegrated. Below ~8.10^8K nucleon kinetic energies are too weak to surmount Coulomb barriers and fall within range of the strong nuclear binding forces.

When $T \underset{\sim}{>} 10^{10}$K neutrons and protons will be present with a relative equilibrium   abundance ($T_{10} \equiv 10^{-10}$ T/K)

$$\frac{n}{p} = \exp\left(\frac{-\Delta m}{T}\right) \sim e^{-1.5T_{10}^{-1}} \tag{38}$$

(Δm is the neutron-proton mass difference) which is maintained by the weak interactions.

$$n + e^{+} \leftrightarrow p + \bar{\nu}_{e}$$

$$n + \bar{\nu}_{e} \leftrightarrow p + e^{-}$$

$$n \leftrightarrow p + e^{-} + \bar{\nu}_{e}$$

(39)

These interactions proceed at a rate $\sim 10^{-50}(T/1K)^{5} s^{-1}$ only until the cosmological expansion rate proceeds more rapidly. This occurs when the temperature is T_f, defined by

$$t_{wk}(T_f) \equiv t_{ex}(T_f) \qquad (40)$$

and the final ratio of neutrons and protons available for nucleosynthesis is (aside from some small correction for subsequent beta decays)

$$\left(\frac{n}{p} \right)_{f} = \exp \left(\frac{-\Delta m}{T_f} \right) \qquad (41)$$

Now T_f is calculated by equating the cosmological expansion and weak interaction rates, and is

$$T_f = 0.7g^{1/6} \quad (MeV) \qquad (42)$$

Thus, the larger g, the number of spin states in equilibrium at this temperature, so the higher T_f, the greater the n/p ratio available for nucleosynthesis and the higher the resulting helium-4 abundance because essentially all the available neutrons are fused into helium-4 via

$$n + p \rightarrow D + \gamma$$
$$D + D \rightarrow {}^{3}He + n \rightarrow {}^{3}H + p$$
$$\searrow {}^{3}H + p$$
$${}^{3}H + D \rightarrow {}^{4}He + n$$

(43)

and so the final helium mass fraction is of order

$$Y = \frac{2n}{n + p} \qquad (44)$$

At the time when the n/p ratio is frozen-in, $T_f \sim 1 MeV$, the total effective number of degrees of freedom contributed by three lepton generations (e, ν_e), (μ, ν_μ) and (τ, ν_τ) is

$$g = g_\gamma + \frac{7}{8}\{g_e + 3g_\nu\} = \frac{43}{4} \qquad (45)$$

(the muon and tau plus their antiparticles have long since annihilated at $T \sim m_\mu$ and m_τ respectively.) This yields $T_f \sim 1.04$ MeV $\sim 0.90 \cdot 10^{10}$ K and n/p ~ 0.19 giving Y ~ 30%.

Clearly the addition of extra degrees of freedom (massless species and relativistic massive particles) to the cosmis medium when T ~1MeV at t ~ 1s will increase the final helium abundance __exponentially__. This was first realized by Shvartsman in 1969.

Recently Olive et al. (1981b) have updated Wagoner's nucleosynthesis code and carefully re-evaluated the rough argument sketched above in the light of the latest observational data on Ω (assuming $\Lambda = 0$), Y and the neutron half-life, τ_n. The following is a selection of their results:

(i)   If $h_0 > 0.5$, $\Omega_B > 0.04$ (derived from observations of binaries and small groups, Faber and Gallagher 1979), and $\tau_n > 10.13$ min. then Y<25% (27%) implies the total number of neutrino types is < 4 (< 6).

(ii)   If the gravitating material perceived dynamically in binary galaxies and small groups is not primarily in nucleonic form then the next best lower limit on Ω_B comes from observations in the solar neighbourhood, (Faber and Gallagher 1979), $\Omega_B > 0.00414h_0^{-1}$. This does __not__ allow any limits to be placed on the number of neutrino types if the cosmic helium abundance exceeds 21%. This occurs because such a low nucleon density makes nuclear captures of $p + n \rightarrow D + \gamma$ so inefficient that, although the extra neutrino species create a higher n/p ratio at T_f, the additional neutrons do not all get synthesized into helium-4. This is the weakest possible conclusion regarding the number of neutrino types.

(ii)   If we adopt less conservative, and so less certain, constraints on Ω_B inferred from hot gas in galaxy clusters ($\Omega_B > 0.007h_0^{-3/2}$) and the cosmic deuterium abundance then Y < 25% (27%) implies <6(8) two-component neutrinos, respectively. These limits apply to massless left-handed neutrinos with only one possible helicity state. If any neutrinos possess a small (~ few eV) rest mass they affect nucleosynthesis in the same way as massless neutrinos because they are relativistic at the time of synthesis.

The importance of these results is that in cases (i) and (iii) they are considerably more restrictive than those obtained by demanding that unified gauge theories possess asymptotic freedom (Vaughn 1979). The standard renormalization group equation for the running gauge coupling constant $\bar{g}$ has the 1-loop approximation

$$16\pi^2 \ \frac{d\bar{g}}{dt} = -b\bar{g}^3 \qquad\qquad (46)$$

where $t = \ln\lambda^2$ and λ is a parameter which sets the scale of momentum. The coefficient b is given in terms of the gauge group representation of the theory which can be asymptotically free only if b > 0. For example, in SU(n) the number of lepton generations (copies of the representation) must be less than $11n/2(n-3)$; in SU(5) this limit reduces to 13 and is considerably weaker than the comological limits (i) and (iii) above.

This basic technique can be extended to limit demographic explosions amongst other particle populations of hypothetical superweak particles, degenerate neutrinos and gravitons, (see Shvartsman 1969 and Olive et al 1981a).

It is also worth noting that a relaxation of the assumption that the the cosmological dynamics are homogeneous and isotopic at T_f complicates the conclusions that can be drawn from analyses of nucleosynthesis. Small anisotropies will increase the cosmological expansion rate and precipitate a neutron-proton freeze out at a higher T_f and so of greater magnitude. The effect is similar to adding extra neutrino species. Thus in the presence of small anisotropies even stronger limits on neutrino species would be possible. However, when the anisotropies become large considerable complications arise. If we assume, as is always done in the literature, (Thorne 1968, Barrow 1975), that the neutrino momentum distribution is isotropic even though the cosmological expansion is anisotropic then the expansion rate starts to affect the nuclear interactions and the helium abundance steadily falls. It reaches ~ 25% again for some very large anisotropy but the accompanying expansion rate is too rapid to photo-distingate all the deuterium made as a by-product of helium and it has too high an abundance to be in accord with observation. However, in this high anisotropy case the existing calculations are probably in error. When the anisotropy is very large the collisionless neutrinos will possess an extremely anisotropic momentum distribution and those moving parallel to the axis of slowest expansion (or even contraction for Kasner type anistropy) will heat-up relative to the equilibrium electron-photon gas. Eventually, they will become so energetic that they will again interact with the neutrons and protons and a new equilibrium could be established between the neutrons and protons. The influence of these non-equilibrium effects and the accompanying non-equilibrium effects produced by the neutrino stresses in the anisotropic dynamics are complicated but lead to a reduction in the helium production relative to that in a naive anisotropic model with the same anisotropy energy (Barrow 1981b).

4.   APPLICATIONS

We have just seen how cosmology can help particle physicists to constrain the masses and multiplicities of new particles. By way of compensation high energy physicists may well be able to help cosmologists solve their "missing mass" problem and shed some light on the origin of galaxies and clusters.

This possibility has emerged following two independent experimental claims for non-zero neutrino rest mass. Lyubimov et al. (1980) report that the Kurie plot for $^3H \rightarrow {}^3He + e^- + \bar{\nu}_e$ decay exhibits a feature consistent with a $\bar{\nu}_e$ mass in the range

$$14eV < m(\bar{\nu}_e) < 46eV \qquad (47)$$

Almost simultaneously Reines et al. (1980) claimed to observe neutrino oscillations consistent with a neutrino mass-difference squared of

$$10^{-3} (eV)^2 \underset{\sim}{<} (\delta m)^2 \underset{\sim}{<} 1 (eV)^2 \qquad (48)$$

By the reasoning of §3.1.2, if we suppose there exist three neutrino generations (ν_e, ν_μ, ν_τ) then the present neutrino number density will be $n_\nu \sim 450(T_\gamma/3K)^3$. If we measure the neutrino mass in units of 30eV and suppose $m(\nu_e) = m(\nu_\mu) = m(\nu_\tau)$ then the present contribution to the total density of the Universe will be ($m_\nu \equiv 30m_{30}eV$)

$$\Omega_\nu = 1.01 m_{30} h_o^{-2} \qquad (49)$$

If we employ the observational upper bound $\Omega(Total) < 2$ and assume the cosmological constant, Λ, to be zero this implies

$$m_{30} < 2h^2 \qquad (50)$$

and the 'age' of the Universe becomes (Doroshkevich et al (1980), Zel' dovich and Sunyaev (1980)),

$$t_o = \frac{1.1 \cdot 10^9}{h_o + 0.84 m_{30} 1/2} \quad yr \qquad (51)$$

So far we have ignored the possibility that the cosmological constant in (14) could be non-zero. If we define the deceleration parameter of the Universe as

$$q_o = - \left(\frac{\ddot{R}}{RH_o^2} \right)_o \qquad (52)$$

then (14) and its differential give two algebraic relations between Λ, q_o and Ω_o.

$$\frac{3\Omega_o}{2} = \frac{k}{R_o^2 H_o^2} + 1 + q_o \qquad (53)$$

$$\Omega_o = \frac{2\Lambda}{3H_o^2} + 2q_o \qquad (54)$$

These relations show why $\Lambda \neq 0$ can considerably weaken cosmological limits on elementary particle masses (Barrow, 1981a). The total mass density in the Universe could be bounded above by the observed value of the deceleration parameter using (7). However, if $\Lambda > 0$ it is possible for the matter density to be considerably higher than when $\Lambda = 0$ without adversely increasing q_o.

Various sets of observational data can be used to find a consistent range of values for q_o and Ω_o when $\Lambda \neq 0$. Lower limits we have on the age of the Universe from the solar age together with redshift-magnitude and quasar data (Gunn and Tinsley 1975, Zimmerman and Hellings 1980, Sandage and Yahil 1980, and Petrosian 1973) imply that even if $\Lambda \neq 0$

$$\rho_o < 1.8 \times 10^{-28} h_o^2 \quad gm \cdot cm^{-3} \qquad (55)$$

This relaxes the limit on light neutrino masses to

$$m_\nu \lesssim 279 \, h_o^2 \text{ eV} \tag{56}$$

and all other elementary particle mass limits are similarly weakened.

It is possible that this loop-hole in the mass constraints can be closed by using quasar Lyman-α absorption line data but a number of extra assumptions may be necessary about the source evolution, see Tytler (1981).

These light (< 1MeV) neutrinos become non-relativistic when their temperature falls below $T_\nu \sim m_\nu$. After the time when this occurs, (t~ $0.1(m_p/m_\nu)^2 \, t_p \sim 10^{10} m_{30}^{-2} s.$), the kinetic energy of the collisionless neutrinos falls because of the adiabatic expansion of the Universe ($v_\nu \propto R^{-1}$). Today the RMS velocity of neutrinos uninvolved in clustering should be ~$6m_{30}^{-1}$ Kms^{-1}. This is considerably lower than the velocity dispersions within galaxies and clusters for $m_\nu > 0.1$eV and so massive neutrinos will be susceptible to gravitational instability and will part-icipate in clustering, Schramm and Steigman (1981), Bond et al (1980).

When the neutrinos are relativistic (t << 10^{10}s.) the scale over which inhomogeneities in their spatial distribution can be damped by collisionless damping is the distance neutrinos can travel in the age of the Universe — the 'horizon' size; this encompasses a mass, Bond et al (1980), Bisnovaty-Kogan et al (1980),

$$M_J \sim \rho_\nu (ct)^3 \propto \frac{1}{Gt^2} \cdot t^3 \propto t \text{ for } T_\nu > m_\nu \tag{57}$$

When the neutrinos become non-relativistic the extent of damping decreases rapidly and is defined by the scale over which thermal and gravitational energies balance (that is, the Jeans mass); this gives a mass scale

$$M_J \sim \frac{T_\nu^2}{2m_\nu} \propto t^{-4/3} \text{ for } T_\nu < m_\nu \tag{58}$$

Inhomogeneities in ρ_ν are erased on mass scales below M_J. From (57) and (58) it is clear that the maximum damping scale is that attained at the moment of 'de-relativisation' $M_J(T_\nu = m_\nu)$. If we denote by t_ν the time at which neutrinos become non-relativistic then the maximum damping scale is just

$$M_* \sim \rho_\nu(t_\nu) t_\nu^3 \tag{59}$$

Now since $\rho_\nu \sim T_\nu^4 \sim (Gt_\nu^2)^{-1}$ we have a simple estimate for M_* in terms of the Planck mass, m_p, and the mass of the (heaviest) neutrino.

$$M_* \sim G^{-3/2} m_\nu^{-2} \sim \left(\frac{m_p}{m_\nu}\right)^2 m_p \tag{60}$$

Keeping account of all the dimensionless factors it is found that, (Bond et al 1980, Bisnovaty-Kogan et al 1980),

$$M_* = 4.10^{15} \, m_{30}^{-2} \, M_\theta \tag{61}$$

   This is the characteristic scale of neutrino clustering in the
early Universe. Structure in the neutrino distribution on scales
smaller than M_* would have been damped-out during the first million
years of the Universe's history. This characteristic mass is much
larger than the extent of galaxies ($\sim 10^{11} - 10^{12} M_\theta$) but is interestingly
close to that of galaxy clusters and the characteristic clustering scale
determined by Peebles (1980), see eqn (9).

   The implications of these calculations for theories of galaxy
formation are considerable and the detailed evolution of a baryon-
neutrino mixture is currently being investigaged in detail by a variety
of theorists.

   Unfortunately, the details of the collapse and fragmentation into
objects of galactic size involves many complicated considerations -
shocks, violent relaxation, cooling, fragmentation, coalescence, segregation of neutrinos and baryons and so forth. Yet oneinteresting and
simple consequence for the problem of galaxy formation has so far emerged: (25,28)

   Inhomogeneities of size less than $\sim 10^{17} M_\theta$ must wait until the
Universe cools to $\sim 4.10^3$K before their thermal coupling with the
photons is broken. Only after this moment are the baryon inhomogeneities
free to enhance by gravitational instability (their density contrast
grows as $\delta\rho/\rho \propto t^{2/3}$). However, neutrino imhomogeneities of similar
scale are gravitationally unstable as soon as they become non-relativistic.
This occurs considerably earlier than baryon-photon decoupling ($t_{dec} \sim$
10^{13}s) and so much smaller initial inhomogeneities can be employed
to grow galaxies and clusters if they are present in the neutrino, rather
than the baryon, distribution. In such a situation the initial baryon
fluctuations can be very small but will later enhance very rapidly as
they 'feel' the gravitational effect of the collisionless neutrino inhomogeneities. Such an idea is attractive because current limits on the
small- scale isotropy of the 2,9K microwave radiation place very strong
constraints on the amplitude of baryon inhomogeneities at the moment when
the 3K radiation decoupled from them. So strong are these limits that
they may even rule out the existence of primordial baryon inhomogeneities
large enough to evolve into the observed galactic structures! However,
if the gravitational inhomogeneities are carried by the neutrinos, which
are not thermally coupled to the photon radiation, then large neutrino
inhomogeneities can co-exist with a very low level of microwave background anisotropy. (25) For example: a universe having $\Omega_\nu = 1$ and
$\Omega_B = 0.03$ can grow galaxies by the present with a small-scale microwave
anisotropy of amplitude $\Delta T/T \sim 3.10^{-5}$ whereas if $\Omega_\nu = 0$ and $\Omega_B = 0.03$
the present microwave anisotropy is unacceptably large, $\Delta T/T \sim 6.10^{-2}$.

ACKNOWLEDGEMENTS

   This work was supported by a Miller Fellowship and I would like to
thank J.R. Bond, A. Szalay and J. Silk for discussions.

REFERENCES

Barrow, J.D.: 1976, Mon. Not. Roy. Astron. Soc. 175, pp. 359-370.
Barrow, J.D.: 1980, Surv. High Energy Phys. 1, pp. 183-212.
Barrow, J.D.: 1981a, Physics Lett . B., pp. 00.
Barrow, J.D.: 1981b, in preparation.
Barrow, J.D. and Silk, J.: 1981, Astrophys. J., 250 pp.00.
Bisnovaty-Kogan, G.S. and Novikov, I.D.: 1980,Sov. Astron. 24, pp.516-518.
Bisnovaty-Kogan, G.S. Lukash, V.N. and Novikov, I.D. 1980. Proceedings
    of the IAU, Liege, Belgium 1980, Reidel, Dordrecht.
Bond, J.R., Efstathiou, G. and Silk, J.: 1980,Phys.Rev.Lett. 45 pp.1980-84.
Chiu, H.Y.: 1966, Phys. Rev. Lett. 17, pp 712-714.
Cowsik, R. and McClelland, J.: 1972, Phys.Rev.Lett. 29, pp. 669-672.
Doroshkevich, A.G.,Khlopov, M.Y., Sunyaev, R.A., Szalay, A.S. and Zel'
    dovich, Y.B.: 1981, Proc. N.Y. Acad. Sci. (in press).
Faber, S.M. and Gallagher, J.S. 1979, Ann. Rev. Astron. Astrophys.17,
    pp. 135-187.
Gershtein, S.S. and Zel'dovich, Y.B.: 1966, Sov. Phys. JETP Lett. 4,
    pp. 174-177.
Gunn, J.E., and Tinsley, B.M.: 1975, Nature 257, pp.454-458.
Gunn, J.E., Lee, B.W., Lerche, I., Schramm, D.N. and Steigman, G.: 1978,
    Astrophys. J., 223, pp. 1015-1031.
Hut, P.: 1977, Physics Lett. B. 69, pp.85-88.
Jones, L.W.: 1977, Rev. Mod. Phys. 49, pp. 717-752.
Kozik, V.A.: 1980, Physics Lett. B. 94, pp. 266-269.
LaRue, G.S., Fairbank, W.M. and Phillips, J.P.: 1979, Phys. Rev. Lett. 42,
    pp. 142-145; 42, pp. 1029-1033.
Laurent, C., Vidal-Madjar, A. and York, D.G.: 1979, Astrophys. J. 228,
    pp. 923-941.
Lee, B.W. and Weinberg, S.: 1979, Phys. Rev. Lett. 39, pp. 165-169.
Lyubimov, V.A., Novikov, E.G., Nozik, V.Z. Tretyakov, E.F. and Kozik, V.S.:
    1980, Physics Lett. B. 94, pp.266-270.
Okun, L.B.: 1980, Sov. Phys. JETP, Lett. 31, pp. 144-147.
Olive, K.A., Schramm, D.N. and Steigman, G.: 1981a, Nuclear Phys. B.180,
    pp. 497-515.
Olive, K.A., Schramm, D.N. Steigman, G., Turner, M.S. and Yang, J.: 1981b,
    Astrophys. J. 246, pp. 557-568.
Peebles, P.J.E.: 1971, Physical Cosmology, Princeton: Princeton U.P.
Peebles, P.J.E.: 1980, The Large Scale Structure of the Universe, Princeton:
    Princeton U.P.
Petrosian, V.: 1973, in Confrontation of Cosmological Theories with Obser-
    vation, ed. Longair, M.S., Dordrecht: Reidel.
Reines, F., Sobel, H.H. and Pasierb, E.: 1980, Phys. Rev. Lett. 45, pp.
    1307-1311.
Renzini, A.: 1979, Advanced Stages in Stellar Evolution, ed. Bovier, P.
    and Maeder, A., Geneva: Sauverny.
Sandage, A., and Yahil, A.: 1980, in Physical Cosmology, ed. Balian, R.,
    Audouze, J., and Schramm, D.: North Holland.
Sato, K. and Kobayashi, M.: 1977, Prog. Theo. Phys. 58, pp. 1775-1789.
Schramm, D.N., and Steigman, G.: 1981, Astrophys. J. 243, pp. 1-7.

Shvartsman, V.F.: 1969, Sov. Phys. JETP.Lett. 9, pp. 315-317.
Silk, J.: 1981, Proc. N.Y. Acad. Sci.,   (in press).
Steigman, G.: 1979, Ann.Rev. Nucl. Part. Sci. 29, pp. 313-338.
Szalay, A.S. and Marx, G.: 1976, Astron. Astrophys. 49, pp. 437-441.
Talent, D.L.: 1980 Ph.D. Thesis: Rice University, U.S.A.
Tytler, D.: 1981, Nature, 291, pp. 289-293.
Vaughn, M.T.: 1979, Zeit.. Phys. C 2, pp. 11-115.
Wagoner, R.V., and Steigman, G.: 1979, Phys. Rev. D. 20,pp. 825-829.
Weinberg, S.: 1972, Gravitation and Cosmology, New York: Wiley.
Zel'dovich, Y.B.: 1965, Adv. Astron. Astrophys. 3, pp. 241-397.
Zel'dovich, Y.B., Okun, L.B., and Pikel'ner,S.B.: 1966, Sov. Phys.
   Usp. 8, pp. 702-709.
Zel'dovich, Y.B., and Sunyaev, R.A.: 1981, Sov. Astron. Lett. 6,
   pp. 249-259.
Zimmerman, R.L., and Hellings, R.W.: 1980, Astrophys. J. 241, pp. 475-485.

BREAKING OF GAUGE SYMMETRIES IN THE COOLING UNIVERSE

Frans R. Klinkhamer
Sterrewacht, Postbus 9513, 2300 RA Leiden, The Netherlands

SUMMARY

The three observed particle interactions (electromagnetic, weak and strong) are thought to be low energy remnants of a unified theory (gauge symmetry group G) spontaneously broken at two energy scales $M_U \sim 10^{15}$ GeV and $M_{WS} \sim 10^2$ GeV:

$$G \xrightarrow{M_U} SU(3) \times \left[SU(2) \times U(1) \right]_{electroweak} \xrightarrow{M_{WS}} SU(3)_{colour} \times U(1)_{electromag}$$

Finite temperature calculations show that in a cooling Universe these symmetry levels become a temporal sequence, with two phase transitions at temperatures of order M_U and M_{WS}.

In this paper we review unified interactions and their phase transitions (PTs), which may considerably change the standard picture of the earliest Universe. Then we consider the impact of PTs on 1) the creation of the observed matter-antimatter asymmetry, 2) the origin of the density perturbations required for galaxy formation, and 3) the related problem of the creation of (too many) monopoles. It appears that 1) and perhaps 3) can be dealt with, provided the PTs, especially the first order ones, preserve the homogeneity of the Universe   But 2) probably arises from the unknown epoch of quantum gravity importance.

> *Je m'accommode assez, pour moi, des petits corps;*
> *Mais le vuide à souffrir me semble difficile,*
> *Et je goûte bien mieux la matière subtile.*
>
> *Molière, Les Femmes Savantes.*

1. INTRODUCTION

The expanding, homogeneous and isotropic Universe which is presently filled with low energy photons (T = 3K), is thought to have originated in a rapidly expanding and very hot phase: the Big

A. W. Wolfendale (ed.), Progress in Cosmology, 39–61.

Bang. The knowledge needed to describe the earliest epoch of the
Universe has been augmented significantly by recent advances in our
understanding of the interactions of elementary particles. The
essential idea is that of gauge theories, to which all four fundamental
interactions belong, namely gravity, electromagnetism and the weak
and strong forces. The latter three forces have been proved to be
renormalizable, i.e. quantum corrections are finite and calculable,
whereas a renormalizable quantum theory of gravity remains elusive.

Contrary to the case of the photon the mediating particles of
the weak interactions are massive, O(100 GeV), which is precisely
the reason why the interaction is so weak. This mass results from
the mechanism of spontaneous symmetry breaking (SSB), in which a non-
symmetric ground state occurs for a theory with symmetric interactions. More specifically the vacuum state is not invariant under
the gauge transformations while the Lagrangian is symmetric.

It is probable that these three observed forces are the low
energy remnants of a unified gauge theory, broken at several energy
scales. Because quarks and leptons are grouped together in the
representations, Grand Unified Theories (GUTs) have baryon number
(B) changing reactions, which, not being included in the electroweak
and colour interactions, are very weak at observable energies and
thus predict a large, but finite, decay time for the proton of
$\sim 10^{31 \pm 2} (M_X/6 \ 10^{14} \mathrm{GeV})^4$ years, where M_X is the mass of the relevant
boson mediating the B violating force. But in the very early Universe
at temperatures larger than or of the order of M_X the B violating
forces are quite effective and in fact they might well create a
small asymmetry between baryons and anti-baryons, which after the
annihilation of the pairs at much lower temperatures O(1 GeV) gives
the observed dominance of matter over antimatter.

After some introductionary work we will review the transitions
to lower gauge symmetry as the Universe cools down (sections 5 and 6).
These socalled phase transitions (PTs) might modify the expansion of
the Universe significantly, but we hope, of course, that they do
not invalidate the "standard" results of Heliumsynthesis and baryon
number creation (sections 2 and 4). Hopes that PTs might generate
the required density perturbations for galaxy formation are
tempered in §6.2.

Also we will briefly discuss in §6.3 the expected creation of
monopoles, i.e. localized, finite energy solutions of the equations
of motion with magnetic charge, which might be expected when the
spatial distribution of the symmetry breaking "directions" is non-
trivial.

Let us remark that the coverage of the references is only
indicative. We put $\hbar = K_{Boltzman} = c = 1$ and express   everything
in powers of GeV (roughly the proton mass), but in order to distinguish
gravity, which sets the stage only, we keep $M_{Pl} \gtrsim G^{-\frac{1}{2}} = 1.22 \ 10^{19}$ GeV.

Indices μ run 0,1,2,3; x^μ and ∂_μ denote a space-time point and derivative; and all indices occurring twice are summed over.

2.   THE BIG BANG   (cognoscenti may skip this section)

There are three basic cosmological observations (Weinberg, 1972).

1. On large enough scales ($\gtrsim$ 0 (100 Mpc)) the Universe is homogeneous and isotropic. Hubble found in 1929 a linear expansion: redshift $Z \equiv (\lambda_{observed} - \lambda_{emitted})/\lambda_{emitted} = H_o \times$ distance, with H_o a constant.

2. Penzias and Wilson discovered in 1965 the isotropic electromagnetic background radiation which has a Planck spectrum and temperature $\sim$ 3K.

3. There is a universal Helium abundance of $\sim$ 25% in mass and only a very low primordial abundance of heavier elements later to be enhanced by stellar evolution processes.

The theoretical model to understand these facts is very simple (Weinberg, 1972). From the Einstein equations of classical gravity and the restrictions on the metric tensor from homogeneity and isotropy (Robertson and Walker) we have an expansion equation

$$H^2 \equiv (\frac{\dot{R}}{R})^2 = \frac{8\pi G}{3} \rho - \frac{k}{R^2} , \tag{1}$$

where R(t) is the scale factor and k = -1,0,1 a curvature parameter. Also we have an energy conservation equation (adiabatic expansion)

$$\frac{d}{dR}(\rho R^3) = - 3pR^2. \tag{2}$$

For relativistic particles we have the energy density $\rho_{rel} = \frac{\pi^2}{30} N T^4$, with N the total effective number of helicity states (2 for the photon), and from entropy conservation (s $\propto T^3$; volume $\propto R^3$) $T \propto R^{-1}$. Thus $\rho \propto R^{-4}$ and the curvature term in Eq. (1) can be neglected for R $\rightarrow$ o. Eq. (1) then gives the Big Bang evolution

$$t = 2.4 \ 10^{-6} N^{-\frac{1}{2}} (\frac{T}{GeV})^{-2} \text{ seconds.} \tag{3}$$

Particles with mass M_Z will be roughly as abundant as the photons γ for T > M_Z, but for lower temperatures they will annihilate and reheat the other interacting particles somewhat. Note that they will be able to reach their equilibrium density n $\sim (M_Z T)^{3/2}$ exp $[- M_Z/T] \ll n_{rel} \sim T^3$ only if their interaction rate Γ_Z > H at T $\sim M_Z$. This point will be of importance in section 4.

We note that because of the finite age of the Universe the maximal

distance travelled by light, i.e. the causally connected region, is
limited. This socalled particle horizon is given by $d_H(t) =$
$R(t) \int_0^t R(\bar{t})^{-1} d\bar{t} = 2t$, as follows from the metric $ds^2 = dt^2 -$
$R(t)^2 [d_3s^2]$ and $R(t)$ given by Eq. (3).

Roughly the Big Bang scenario runs as follows. At very early
times the Universe is dominated by relativistic particles; precise
calculations show that at $t \sim 1$ minute some 25% He is synthesed and
practically no heavier elements; still later at $T \sim 4000$ K the
protons and electrons recombine and the photons expand freely
henceforth, no longer being Thomson scattered ($T \propto R^{-1}$; now $T_0 \sim 3K$);
at roughly the same epoch the energy density starts to be dominated
no longer by radiation but by non-relativistic matter, i.e.
hydrogen and Helium nuclei. In this last phase, the largest in
time (cf. Eq. 1), we can neglect the pressure and the precise
expansion solution depends on two constants only, which we take as
the present values of expansion rate and density ratio: $H_0 \equiv (\dot{R}/R)_0 =$
h 100 km s^{-1} Mpc^{-1} and $\Omega_0 \equiv (\rho/\rho_{cr})_0$, with $\rho_{cr} \equiv 3H_0^2/8\pi\rho = h^2 \, 2 \, 10^{-29}$
gcm^{-3}. According to present knowledge $h \sim 1/2$ to 1 and $\Omega_0 \sim 0.1$ to 1.
The age of the Universe is $t_0 = f(\Omega_0) \, H_0^{-1} \sim H_0^{-1} \sim h^{-1}$ 20 billion
years, but we will be discussing the very earliest relativistic
phase only, when $t \, (T = M_X) \sim 10^{-35}$ s.

3. PARTICLE INTERACTIONS : UNIFICATION

3.1 Immediately after the invention of quantum mechanics the
relativistic theory for electromagnetic interactions was sought and
the Dirac equation dates of 1928 already. But only in the late 40's
was the theory proven to be renormalizable: the infinite quantum co-
rections can be absorbed in a finite number of constants (charge and
mass of the electon for example) and the theory (Quantum Electro-
dynamics, QED) can be formulated using the observed finite masses
and charges only and finite quantum corrections. That this can be
done is highly non-trivial; infinities at each order in the
perturbation expansion might each require different counterterms in
the Lagrangian, whereas for QED all counterterms, say ($Z_1e + Z_2e^2 +$
--) $\bar{\psi}\psi$, are of a limited number of forms, say mass terms $m_0 \, \bar{\psi}\psi$, and
thus only a few constants can be redefined, say mass m_{renorm}. QED
has a very special form; it is a gauge theory, i.e. invariant under
transformations with local parameters (as Λ (x) in Eq. 4). The
global invariance (Λ = constant) leads to charge conservation.

What about the other two particle interactions? First let us
consider the weak interactions. Fermi gave in 1934 the theory for β
decay ($n \rightarrow p + e^- + \bar{\nu}$), in a form analogous to QED with a four
fermion interaction and dimensional coupling constant $G_F \sim 10^{-5}$ GeV^{-2}
Although very successful the theory could not be the final one:
it is not renormalizable and at high energies ($\gtrsim 300$ GeV) the cross-
sections violate the limits from unitarity (in simple terms: more out

than in; cf. Taylor, 1976). In the 50's and 60's two ingredients
towards a solution were put forward, but each had their problems.
One was the idea of spontaneous symmetry breaking (SSB) which may
give masses to the mediating bosons (see example below), but according
to the Goldstone theorem massless scalars are predicted to occur
which are not observed. The other idea is to extend the U(1)
symmetry of QED to a larger group, these are called Yang-Mills (YM)
theories. We then have more than one "photon", but they obviously
are massless, whereas the weak interactions are weak because the
mediating boson W is very heavy: $G_F = g^2/2 \; M_W^2$ with g the
dimensionless coupling constant (cf. e of QED). YM theories looked
promising for renormalization. Higgs showed that surprisingly a
spontaneously broken gauge theory only had the good properties:
massive gauge bosons but no massless scalars. Later 't Hooft showed
that the YM theories after the SSB remained renormalizable.

It will prove useful, for later work, to illustrate all this
for the simplest model, called the Abelian Higgs model (cf.
O'Raifeartaigh, 1979). The gauge symmetry group U(1) has one
parameter $\Lambda(x)$ and works on the gauge field A_μ and the complex
scalar field $\phi = \phi_1 + i \; \phi_2$ as follows, in infinitesimal form,

$$A_\mu(x) \rightarrow A_\mu(x) + \frac{1}{e} \partial_\mu \Lambda(x)$$

$$\phi(x) \rightarrow \left[1 - i\Lambda(x)\right] \phi(x). \qquad (4a)$$

The invariant Lagrange   density is

$$L = - \tfrac{1}{4} F_{\mu\nu} F^{\mu\nu} - \tfrac{1}{2} |D_\mu \phi|^2 - V(\phi), \qquad (5)$$

with field strength $F_{\mu\nu} = \partial_\mu A_\nu - \partial_\nu A_\mu$, gauge covariant derivative
$D_\mu \phi = \partial_\mu \phi + ieA_\mu \phi$, and the scalar potential

$$V(\phi) = -\mu^2 |\phi|^2 + \lambda |\phi|^4 \qquad (\lambda, \; \mu^2 > 0). \qquad (6)$$

The vacuum will be at the asymmetric minimum of V, which we choose
$\sigma \equiv < 0 \; |\phi| \; 0 > = <0 \; |\phi_1| \; 0> = (\mu^2/2\lambda)^{\frac{1}{2}}$.

Developing in fields around this minimum $\theta = \phi - \sigma$ we rewrite
(5) after a gauge rotation to eliminate θ_2:

$$L = -\tfrac{1}{4} F_{\mu\nu} F^{\mu\nu} -\tfrac{1}{2} M_A^2 A_\mu^2 - \tfrac{1}{2} |D_\mu \theta|^2 -$$

$$2\mu^2 \theta_1^2 + L_{interaction} \; (A_\mu, \; \theta_1), \qquad (7)$$

where the gauge boson has become massive $M_A^2 = e^2 \mu^2/2\lambda$. There is only
one massive scalar field θ_1 and $L_{interaction}$ retains the gauge
invariance. Realistic models require larger gauge groups and the
introduction of fermions, for completeness we give the general form.
Lie group G has generators t_r (r = 1 --- N, N = n^2 - 1 for G = SU(n)),
hence for $g \; \varepsilon \; G$: $g = \exp \left[\Lambda_r t_r\right]$. Define a Lie valued potential

$A_\mu = A_\mu^r \, t_r$. The finite gauge transformations are

$$A_\mu \to g \, A_\mu \, g^{-1} + \frac{1}{e} \, g \, \partial_\mu \, g^{-1}$$

$$\phi \to R \, (g) \, \phi,$$

(4b)

where $g = g(x)$, R a representation for the ϕ, and e the coupling constant.

The fields and covariant derivatives are

$$F_{\mu\nu} = \partial_\mu A_\nu - \partial_\nu A_\mu + e \, [A_\mu, A_\nu]$$

$$D_\mu \phi = \partial_\mu \phi + g \, R(A_\mu) \, \phi.$$

(8)

With the Cartan–Killing metric { , }, note $\{A_\mu, A^\mu\} = \sum_r A_\mu^r \, A^{\mu r}$, we

generalise Lagrangian (5) to

$$-L = \tfrac{1}{4} \, \{F_{\mu\nu}, \, F^{\mu\nu}\} + \tfrac{1}{2}\{D_\mu \phi, \, D^\mu \phi\} + \bar{\psi} \, \gamma^\mu D_\mu \, \psi +$$

$$+ \bar{\psi} \, m \, \psi + g \, \bar{\psi} \, \phi \, \psi + V(\phi).$$

(9)

The 3rd and 4th terms of the RHS are the generalised Dirac terms
for the fermions ψ in some representation R_ψ (replace R in D_μ
definition (8)), while the 5th term is a Yukawa interaction. Some
gauge fields A_μ^s acquire masses by SSB if $V(\phi)$ has non-trivial minima,
and if H is the remaining symmetry there are dim (H) fields A_μ^t
which remain massless.

With these ideas Glashow, Weinberg and Salam constructed the
electromagnetic-weak theory (table I; Taylor, 1976). The quarks u,
d and the leptons e^-, ν_e are put in left-handed doublets and right-
handed singlets (no ν_R), each generation is treated similarly. The
ratio $g_1/g_2 \equiv \tan \theta_w$ is determined experimentally : $\sin\theta_w^2 = 0.21$.
The SSB gives the weak bosons a mass and by the existence of the Z^o
predicted neutral currents ($e^- \, \nu_e \to e^- \nu_e$), as confirmed later by
CERN experiments. Also the prediction by Glashow-Iliopoulos–
Maiani of the charm (c) quantum number (in order to cancel sd $\to$
$W^+ \, W^- \to s\bar{d}$) has been verified, still later the beauty (b) quark was
found as $T = b\bar{b}$ at $\sim$ 10 GeV (truth (t) remains the current experimenters quest). Baryons are composed of 3 quarks and mesons of 2, with
the interquark forces described by a YM theory (called Chromodynamics,
with "colours" red, yellow, blue as charges, Table I). For high interaction energies q^2 the colour forces get weaker (cf. Eq. 10), probably
this explains the famous lepton scattering results on protons, which
apparently scattered on freely moving point particles (quarks). For
larger distances g_3 gets larger and this is related to the confinement
of quarks, i.e. no separate quarks exist, but rigorous calculations are
still lacking. Also the hadron production relative to $\mu\mu^+$ in $e^- + e^+$
collisions is beautifully explained by 5 quark flavours, each in 3

colour variants.

Table I : Low energy interactions

gauge group G	SU(2) x U(1)		SU(3)
coupling constants g,	g_2	g_1	g_3
strengths $\alpha = g^2/4\pi$ as observed	$\alpha_2 \sim 0.03$	$\alpha_1 \sim 0.01$	$\alpha_3 \sim 0.2$

fermion representations

$$\begin{pmatrix} u \\ d \end{pmatrix}_L \quad \begin{pmatrix} \nu_e \\ e^- \end{pmatrix}_L \qquad \begin{pmatrix} u_r & u_y & u_b \\ d_r & d_y & d_b \\ c_r & c_y & c_b \\ \cdot & \cdot & \cdot \\ \cdot & \cdot & \cdot \\ \cdot & \cdot & \cdot \end{pmatrix}$$

$$u_R, \ d_R, \ e_R. \qquad ^*$$

gauge fields	W^+, W^-, Z^0 mass $\sim g^\sigma$	8 gluons
	γ photon	

Higgs scalars

$$\begin{pmatrix} \phi^+ \\ \phi_o \end{pmatrix}, \qquad \text{none}$$

$$\text{with } <0|\phi|0>=\sigma\begin{pmatrix} 0 \\ 1 \end{pmatrix}$$

$$\sigma \sim 300 \text{ GeV}$$

*
similarly for the other two generations: read for (u, d, e, ν_e) respectively (c, s, μ, ν_μ) or (t, b, τ, ν_τ), where we neglect Cabbibo mixing of the down quarks.
L, R denote left- and right-handed components.

3.2   The three elementary particle interactions can thus be understood as the gauge theories summarized in Table I. But, as always, deeper questions remain unsolved:

(1)   there are still (too) many arbitrary parameters, e.g. 7 for the electroweak theory.

(2)   why is electric charge quantised and $|Q_{electron}| = |Q_{proton}|$?

(3) why do we have three so different forces?

(4) what is the relation between quarks and leptons, which on the

one hand are so different (quarks have strong interactions, leptons
not; quarks heavier than the leptons, per generation at least), but
on the other hand have some interrelations (notably the cancelling
of the triangle anomalies)?

(5) why is there a triplication of generations?

(6) are the Higgs scalars elementary particles or composites,
and what regulates the SSB precisely?

(7) what is the quantum theory of gravity?

The last three problems are still unsolved (perhaps 5 has something to do with the connection GUT-supergravity, see below), but the
first four might be tackled in the unification scheme (reviews:
Langacker, 1981; Nanopoulos, 1980: Ellis, 1980). The idea is to
have one simple gauge group G_U, hence one coupling constant g_u, and
quarks and leptons together in G_U representations. At a high energy
scale M_U the symmetry breaks spontaneously to the "observed" SU(3) x
SU(2) x U(1). But what about one coupling strength, whereas we
observe large differences (Table I)? The answer lies in the renormalization group equations (Georgi et al., 1974), which give the
effective coupling as a function of the interaction energies (q^2),
generally for G = SU(n)

$$\alpha_n(q^2) = \frac{12\pi}{(11n - 2F) \ln (q^2/\Lambda^2)} \qquad (10)$$

with F the number of quark flavours (6?) and Λ an energy scale. With
the observed α_3 and α_1 (with n = 1 in (10)) at $q^2 = (100$ GeV$)^2$ we
calculate equality at $M_U \sim 10^{15}$ GeV with $\alpha_U \sim 1/40$. But now we may
calculate $g_2(100$ GeV) and find: $\sin^2\theta_W \sim 0.20$! Thus the three
observed forces are low energy remnants of a unified interaction
broken at energies $\sim M_U$. Also questions (2) and (4) may be dealt
with (e.g. Nanopoulos, 1980). The smallest possible (rank 4) G_U is
SU(5) (Georgi and Glashow, 1974). There are stringent restrictions
on G which basically allow only 2 larger G, which will have more
fermions (cf. Barbieri, 1980). Table II gives the fermion representations and the required Higgs scalars for the SU(5) unification,
where again each generation is treated similarly.

The breaking by <0|H|0> in the Yukawa terms gives the down quark
and charged lepton equal masses. Thus for the third generation
$m_b \sim m_\tau$, which gets renormalized for the low observable energies
(using F = 6: a strong indication on the number of generations) to
$m_b \sim 3m_\tau$, which fits the experimental data (1.5 and $\sim$ 4.5 GeV). For
the first generation the observed ratio is quite different from 3,
see discussion in Ellis (1980).

Table II : Unified Interactions (Georgi and Glashow, 1979)

G_U :   SU(5)

α_U :   1/40

fermions* : $\bar{d}_r$ $0 \quad \bar{u}_b - \bar{u}_y - u_r - d_r$

$\bar{d}_y$ $0 \quad \bar{u}_r - u_y - d_y$

$\underline{5} = \quad \bar{d}_b \qquad \underline{10} = \qquad 0 - u_b - d_b$

e^- $- \text{etc} \quad 0 - e^+$

ν_e 0

L   L
                                     antisymm

scalars : $\phi:\underline{24}$, with $<0|\phi|0> \sim 10^{15}$ GeV diag $(1,1,1,-3/2,-3/2)$

H: $\underline{5}$, with $<0|H|0> \sim 10^2$ GeV $(0,0,0,0,1)^T$

* other generations similarly

But there is also something completely new in the unification hat. By the breaking through the ϕ scalars $24-(8+3+1)=12$ bosons X get a high mass (the ones of the SU(3) x SU(2) x U(1) remaining massless).   Some of these bosons mediate reactions between quarks and leptons (uu $\overset{X}{\to}$ e$^+\bar{d}$) which at our low energies leads to a very weak proton decay p = uud $\to$ e$^+$ + (π^0 = d$\bar{d}$) with a large lifetime due to the X propagator, but perhaps measurable. Another significant effect of the baryon number changing reactions might lie in the early Universe at temperatures T $\sim$ M_X.

4. CREATION OF THE MATTER-ANTIMATTER ASYMMETRY

All observations indicate the total absence of antimatter in the Universe (Steigman, 1976). The baryon number density n_B thus is non-zero and observations give the numerical value

$$\frac{n_B}{n_\gamma} = \frac{n_{baryon} - n_{antibaryon}}{n_\gamma} \sim \frac{n_{baryon}}{n_\gamma} \sim 10^{-10} \left(\frac{\Omega_{bo}}{0.01}\right) \qquad (11)$$

with Ω_{bo} the present baryon density relative to the critical density (section 2; cf, Olive et al. 1981). If baryon number would be totally conserved one might think the small number of Eq. (11) just to be given by initial conditions. But it would be more satisfactory if we could find a physical mechanism to generate this small baryon asymmetry. The basic idea is quite simple : the density ratios of

baryons : antibaryons : photons at very high temperatures is
something like 10^{10} : 10^{10} : 10^{10}, later the unknown mechanism
produces a small asymmetry $10^{10} + 1$: $10^{10} - 1$:10^{10}, which
after annihilation gives the final ratios 2:0 : $\sim 10^{10}$, as
observed. The key problem, of course, is the second step. Four
ingredients are required to generate a net baryon number density
(e.g. Weinberg, 1979):

1. Baryon number (B) non-conservation,
2. Charge conjugation (C) asymmetry,
3. C and parity (CP) asymmetry,
4. not thermal-equilibrium distribution functions of some of the
species involved.

      Conditions 2 and 3 are to avoid a cancelling of production by
particles and antiparticles (operation of C changes the sign of the
baryon number, because B (particle) = -B (antiparticle); P leaves
B unchanged). Also condition 4 is obvious; if anything first
stays in equilibrium no asymmetry can arise. As we have seen in
the previous section unified interactions will give B violating
reactions as well as C and CP asymmetries. Condition 4 is
implemented by the fast expansion of the Universe as given by Eq.
(1) (see also Appendix A of Kolb and Wolfram, 1980a). In order
to illustrate condition 4 Weinberg (1979) has proposed the delayed
decay scenario, which runs as follows. Assume complete equilibrium
at T_{p1} ($\gg M_X$); we have for the decay rate of the heavy X boson
(including a time dilution factor)

$$\Gamma_X \sim \alpha_X M_X^2 N \left[T^2 + M_X^2\right]^{-\frac{1}{2}} \tag{12}$$

and the expansion rate of the Universe (Eq. (1))

$$H \sim N^{\frac{1}{2}} T^2/M_{p1}. \tag{13}$$

The X and their antiparticles $\bar{X}$ will decay around a temperature T_d
when $\Gamma_X(T_d) \sim H(T_d)$. If the X and $\bar{X}$ decay at temperature

$$T_d < M_X \tag{14}$$

they will not be replenished by inverse decays (Boltzmann factor),
and the net baryon number to entropy ratio produced in the delayed
decay is simply proportional to their equilibrium density

$$\frac{n_B}{s} \sim \frac{45}{4\pi^4} \zeta(3) \left(\frac{N_X}{N}\right) \Delta B, \tag{15}$$

with s the specific entropy and ΔB the average net baryon number
from the decay of an X-$\bar{X}$ pair. Before discussing the estimates of
ΔB we must see whether or not condition (14) applies. Using Eqs.
(12) (13) we easily see that (14) implies

$$M_X > N^{\frac{1}{2}} \alpha_X M_{Pl}. \tag{16}$$

For gauge bosons ($\alpha \sim 1/40$) with a calculated mass of $M_X \sim 6 \ 10^{14}$GeV (Ellis et al., 1980a) we see that the delayed decay condition (14) does not hold. However Higgs bosons are believed to have very small Yukawa couplings (as in the Weinberg-Salam model, cf. Taylor, 1976) and the condition (14) may apply.

In order to find ΔB one calculates the different branching ratios of the X and $\overline{X}$ (cf. simple model below; Nanopoulos and Weinberg, 1979). For SU(5) with two 5's of Higgs and some honest estimates Yildiz and Cox (1980) calculated (15) to be of order of 10^{-10} (for further discussion of ΔB and the required CP breaking e.g. Ellis, 1980). Recently a direct connection was noticed (Ellis et al., 1981a) between the ΔB interference graphs and those leading to a finite renormalization ($\delta\theta_{GUT}$) of the θ parameter of the QCD (SU(3)) vacuum. [This θ defines the ground state when topologically distinct Yang-Mills vacua exist, analogous to the Bloch functions for periodic potentials; gauge transformations in topological equivalence class n implemented on the vacuum $|\theta>$ give $G_n |\theta> = e^{in\theta} |\theta>$ (for an introduction e.g. Crewter, 1978; Jackiw, 1980)].

Assuming $\theta = 0$ at a high energy, say M_{Planck}, the present observational limits on the electric dipole moment of the neutron d_n, which gets a dominant contribution of the CP violating term with parameter θ (1 GeV) $\gtrsim \delta\theta_{GUT}$, allow for practically no entropy generation after the n_B generation at unification energies (Ellis et al., 1981a,b). The reasoning goes as follows: $d_n \gtrsim 4.10^{-16}$ $\delta\theta_{GUT}$e-cm would violate the experimental upper limit (2.10^{-24} e-cm) if the generated $\{n_B/s\}_{GUT}$ had to be significantly larger than 10^{-10} to allow for later entropy generation. We remark that in comparing $\delta\theta_{GUT}$ and ΔB graphs moduli of typical unitary matrix elements U_{dm} connecting the different contributing Higgses, namely in decay and mass terms, were naturally assumed to be O(1). If these are substantially smaller they could alleviate the nearly conflicting theoretical ($\delta\theta > --- |U_{dm}|^2 n_B/s$) and observational ($d_n < ---$) limits (see Eq. (24) of Ellis et al., 1981a). Note that the above arguments do not hold if a global U(1) axial symmetry forces θ to be zero (Peccei and Quinn, 1977; Dine et al., 1981).

If condition (14) is not strongly satisfied the final result will be less than estimated in Eq. (15) and the rate equations of the B changing reactions have to be solved numerically.

Kolb and Wolfram (1980a) have introduced a simple model which incorporates the major physical ingredients. More detailed models are not quite relevant for the moment because of the lack of knowledge on the precise unification Lagrangian and the details of CP violation.

The model consists of two types of particles: nearly massless particles b and $\bar{b}$ carrying baryon numbers $B = \frac{1}{2}$ and $B = -\frac{1}{2}$ respectively, and massive bosons X and $\bar{X}$ mediating baryon-number violating reactions. The decay amplitudes M of these massive bosons are parameterized as

$$|M(X \to bb)|^2_2 = (1 + \eta)\frac{1}{2} |M_o|^2,$$

$$|M(X \to \overline{bb}|^2 = (1 - \eta)\frac{1}{2} |M_o|^2,$$

$$|M(\bar{X} \to \overline{bb})|^2 = (1 + \bar{\eta})\frac{1}{2} |M_o|^2,$$

$$|M(\bar{X} \to bb)|^2 = (1 - \bar{\eta})\frac{1}{2} |M_o|^2,$$

with $|M_o|^2$ of the order of a small coupling constant α. Because of unitarity and CPT invariance only two free parameters η, $\bar{\eta}$ are left, where $\varepsilon = \eta - \bar{\eta} = 0(\alpha)$ measures the amount of CP breaking. Thus a state initially containing an equal number of X and $\bar{X}$ ($n_X^o = n_{\bar{X}}^o$) will decay, inthe absence of back reactions, to a system with a net baryon number $n_B = (\eta - \bar{\eta})\frac{1}{2}(n_X^o + n_{\bar{X}}^o)$. For simplicity all particles are given only one spin degree of freedom, and obey Maxwell-Boltzmann distributions. Because in the expanding Universe all densities drop quickly, a convenient type of variable is

$$Y_A \equiv n_A/n_\gamma,$$

the relative number density of particle $A(= b, \bar{b}, X$ or $\bar{X})$ with respect to photons. The rate equations can be written as (Hut and Klinkhamer, 1981a)

$$\frac{dY_\Delta}{dx} = \frac{1}{2} x^2 K_1(x)$$

$$- \frac{\alpha}{4x_p} \{x \frac{K_1(x)}{K_2(x)} Y_\Delta(x) + \frac{1}{4} \varepsilon x^3 K_1(x) Y_B(x)\} \qquad (17a)$$

$$\frac{dY_B}{dx} = \frac{\alpha}{4x_p} \{ \varepsilon x \frac{K_1(x)}{K_2(x)} Y_\Delta(x) - x^3 K_1(x) Y_B(x)$$

$$\frac{-288\alpha}{\pi} x^{-4} Y_B(x)\} \qquad (17b)$$

with "time" parameter $x = M_X/T$, an effective expansion parameter $x_p = M_X/(7.5 \ 10^{18} \ N^{-\frac{1}{2}} \ GeV)$, $K_{1,2}$ modified Bessel functions, $Y_B = Y_b - Y_{\bar{b}}$, and the deviation from equilibrium parametrised as $Y_\Delta = Y_+ - Y_+^{eq}$, $Y_+ = \frac{1}{2}(Y_X + Y_{\bar{X}})$. In deriving Eq. (17) we put $\eta + \bar{\eta} = 0$, the final Y_B being quite insensitive to the exact value as long as $|\eta + \bar{\eta}| \sim 1$. In equation (17a) it is clear how the first RHS term determines the production of Y_Δ, independent of already existing Y_Δ and Y_B, as a function of temperature only ($T = M_X/x$). The second term destroys the deviation from equilibrium of the X's, and is therefore proportional to α/x_p or $\alpha G^{-\frac{1}{2}}$: the magnitude of the deviation is governed by a competition between particle reaction rates and Universe expansion. The third term is typically ten orders of magnitude smaller than the second one in our calculations ($\varepsilon = 10^{-6}$; $Y_B \lesssim \varepsilon$). Therefore the rate equation for Y_Δ is nearly completely Y_B - independent, and $Y_\Delta(x; x_o)$ is fixed by specifying the initial condition $Y_\Delta(x_o; x_o)$ independent of Y_B.

The rate equation for Y_B shows a production term $\propto Y_\Delta$, with the same reaction vs. expansion factor α/x_p, but also the small CP violation parameter ε. The next two terms $\propto Y_B$ determine the damping of Y_B, by means of all number changing processes, independent of ε, which already indicates that $Y_{B\ max} \leq \varepsilon$. The first of these two terms describes inverse decays of X, $\bar{X}$ (bb, $\to$ X, etc.), and drops off quickly for high x($K_1(x) \propto x^{-\frac{1}{2}} e^{-x}$ for x >> 1). The last term is the Fermi approximation for $2 \to 2$ scattering processes (bb $\to$ $\bar{b}\bar{b}$, etc.), which is the dominating, but still rather unimportant, term for x $\lesssim$ 20. For high energies this term is unimportant, since the large contribution to $2 \to 2$ scattering by the exchange of on-shell intermediate X's is already included in the previous term (see Kolb and Wolfram, 1980a, sect. 2.3.2). Therefore we just replaced x^{-4} by 1 for x < 1, since a detailed treatment of the $2 \to 2$ processes would be unnecessary in this regime.

Let us choose parameters $\alpha = 1/40$, $M_X = 10^{15}$ GeV, N = 100 and $\varepsilon = 10^{-6}$ (note that condition (16) is strongly violated). Starting from thermal equilibrium at $x_o = 0$ (or $x_o = M_X/M_{Pl} \sim 10^{-4}$) we find $Y_B(x)$ peaking around x $\sim$ 1 and then, mostly because of inverse decays, falling a factor 4 to the final value $Y_B(\infty, x_o = 0) = 1.43 \ 10^{-8}$ (this differs from Kolb and Wolfram (1980a) who find a somewhat larger drop, this difference could only be due to our treatment of the 2-2 scattering, but we overestimated their contribution somewhat at x $\sim$ 1). For other parameters Kolb and Wolfram (1980a) displayed similar damping of the Eq. (15) estimate in their figs. 3 and 4.

Finally we remark that because of the B violating processes operative at T = $0(M_X)$ any (not too large) primordial baryon number will be destroyed (second term RHS of Eq. (17b)), after which the final Y_B will be generated. The presently observed matter dominance thus sets via the required ΔB important constraints on the uni-fication theory and the CP violation.

5. PHASE TRANSITIONS

Up till now we have not included finite temperature effects in the particle theory. We are warned however by such drastic effects as super conductivity (another example of SSB, cf. section 3) vanishing for high enough temperatures: symmetry is restored by the thermal fluctuations. In the Abelian Higgs model we used the minimum σ of the classical (tree) potential: $\left[\dfrac{-dV^o}{d\phi}\right]_\sigma = 0$.

The vacuum state is determined by the <u>effective potential</u> (cf.
Coleman and Weinberg, 1973), where quantum corrections are included
(e.g. in the loop, i.e. $\hbar$, expansion). If one still requires a
non-trivial minimum of V_{eff} one finds that for 1-loop in A_μ the
quartic coupling must not be too small $\lambda > 3e^2/32\pi^2$ (cf. Linde,
1979). For finite temperature the effective potential is ideally
suited : usual field theory applies, but with the time components
of the momenta discrete and the relevant integrals replaced by
sums (de Fetter and Walecka, 1971). The expectation value of
operator O is $<O> = \text{Tr}\left[O \exp(-H/T)\right]/\text{Tr}\left[\exp(-H/T)\right]$, with H the
Hamiltonian and T the equilibrium temperature. For the T = 0
potential $V^o = -\frac{1}{2}\mu^2 \phi^2 + \frac{1}{4}\lambda \phi^4$ one finds the equation for the
minimum σ modified: $\sigma(\lambda\sigma^2 - \mu^2 + \frac{1}{4}\lambda T^2) = 0$. The minimum $\sigma(T)$ shifts
continuously from $\sigma(T = 0) = (\mu^2/\lambda)^{\frac{1}{2}}$ to $\sigma(T_c) = 0$ at a critical
temperature $T_c = 2\mu/\lambda^{\frac{1}{2}}$; this is called a <u>second order phase transition</u>
(2PT). For $T > T_c$ the theory is symmetric. Strictly speaking
quantum corrections will always introduce a weak discontinuity in
$\sigma(T)$. For the Abelian Higgs model one finds analogously:
$\sigma(\lambda \sigma^2 - \mu^2 + 1/12(4\lambda + 3e^2)T^2) = 0$. And for $\lambda < e^4$ we see that
although there is a non-trivial minimum σ' at $T_c^- \sim (15\lambda/2\pi^2)^{\frac{1}{4}}\mu$
with $V(\sigma = 0) = V(\sigma')$, there is a barrier in between. The transition
from $\sigma = 0$ to $\sigma' \neq 0$ will be discontinuous and occurs at a
temperature different from T_c : <u>a first order phase transition</u> (1PT).
For completeness we remark that for a temperature range $|T - T_c|$
$\overset{<}{\sim} e^2 T_c$ perturbation theory breaks down, the $m_{Higgs} \to 0$ and infrared divergencies occur (Weinberg, 1974). For more details we refer
to the reviews of Kirznits and Linde (1976) and Linde (1979). Also
for more complicated models (GUTs) it holds true that 1PTs occur
if the Higgs particles H are light compared to the gauge bosons G
(cf. Eq. 7). Alas our understanding of the Higgs sector in GUTs
is little. Generally one might say that if the scalars H are
fermion composites $H = F\bar{F}$ one expects $M_H \sim M_F \sim M_G$; but if the
breaking is through radiative corrections (Coleman and Weinberg,
1973) we have typically $M_H \sim \alpha^{\frac{1}{2}} M_G$. Perhaps there are two
indications that the breaking is radiative (at least at the unification scale[2]). The unexplained hierarchy of energy scales $M_{WS} \sim$
10^2 GeV $<<< M_U \sim 10^{15}$ GeV $<< M_{Pl} \sim 10^{19}$ GeV could originate as
follows (Ellis et al. 1980b; Weinberg, 1979b): if the quartic
couplings λ of the 2 Higgs systems (with bare masses zero) needed
for the breakings are of the same strength $\sim g^2$ at M_{Pl}, the λ of the
24 will diminish much more rapidly at lower energies than those of
the 5, and the first will give a breaking at $\sim 10^{-4} M_{Pl}$ when $\lambda \sim 0$,
while the second will be ~ 0 at very much lower energies. Secondly
Ellis et al. (1980c) have looked for the GUT which can be incorporated in the extended supergravity theory with composite
states, and find only G = SU(5), fermion representations $3(\underline{10} + \bar{5})$
(exactly the 3 generations as observed) and probably massless $\overline{24}$
and $\underline{5}$ scalars. Alas the argument involves a lot of speculations,
but some have a field theoretical backing in 2 dimensional models.

6. PHASE TRANSITIONS FROM UNIFICATION IN THE UNIVERSE

6.1 After all the preparations of the previous sections we now get to business[3]. At very high temperatures ($T > T_c = 0(10^{15}$ GeV)) the vacuum is in the symmetric state and the gauge bosons and fermions (the observed ones get their masses only at the 10^2 GeV breaking) are massless and forces long range. If the vacuum gravitates normally (cf. weak equivalence principle, Weinberg 1972) the symmetric minimum at $T > T_c$ has a large energy density $\rho_v = 0(\sigma^4)$, because the present cosmological constant is zero or small $\lesssim (10^{-2}$ eV)^4 (cf. Kolb and Wolfram 1980b). This just amounts to setting the zero energy level of the potential: $V(\sigma, T = 0) = 0$. For $T > T_c$ this $\rho_v \sim \sigma^4 \sim T^4$ will be negligible compared to the particles $\rho \sim NT^4$. For a 2PT the vacuum shifts at T_c to the broken state so that never in the history of the Universe was ρ_v dominant. But for a 1PT the transition is blocked for $T < T_c$ and the constant energy density ρ_v will soon dominate over that of the particles which is redshifted by the expansion. From Eq. (1) and $\rho_v \sim \sigma^4$ we then have a very rapid expansion approaching the de-Sitter solution.

$$R(t) \sim R(t_c) \exp (\frac{t - t_c}{\tau}) \qquad (18)$$

$$\tau \sim (8\pi/3)^{-\frac{1}{2}} M_{P1}/\sigma^2$$

which stretches the particle horizon to

$$d_H(t) \sim \tau \exp (\frac{t - t_c}{\tau}) \quad . \qquad (19)$$

The particle temperature T still goes as R^{-1}, so there results a large supercooling if the symmetric vacuum is blocked from the transition (hence called the false vacuum) for times quite larger than $t_c \sim N^{-\frac{1}{2}} T_{P1} T_c^{-2}$ (Eq. (13)).

Because we presently are in the broken state we know that the transition must have occurred. As in the usual first order phase transitions the transition goes through nucleation. The bubbles of true vacuum will have a minimum size for which the gained energy $\propto \rho_v r_{min}^3$ compensates the surface energy $\propto - r_{min}^2$. There are two ways the bubbles may originate: by thermal excitations, or by barrier penetration. The nucleation rates are, $\Gamma_T \propto \exp[-E_3/T]$ and $\Gamma_Q \propto \exp \cdot [-S_4]$, where E_3 is the free energy of the critical solution and S_4 the action of the solution in 4-dimensional Euclidean space (but see Cant, 1981). Γ_T peaks at $T \sim \frac{1}{4}T_c$ (Guth and Weinberg, 1980), whereas Γ_Q is constant but small, barrier penetration being a weak effect, of course (Coleman, 1977). There are basically three alternatives to end the supercooling at

T_{end}:

1. By thermally nucleated bubbles, but then T_{end} is not far below T_c, otherwise the few bubbles cannot catch up the rest of the exponentially expanding Universe (Sato, 1981).

2. By the tunneled bubbles filling the Universe.

3. The false vacuum becoming unstable.

If the barrier is large the thermally nucleated bubbles will be too sparse, so we expect strong supercooling either terminated by 2 or 3. But alternative 2 is full of problems, to name a few (for monopoles see § 6.3):

(a) Strictly speaking the transition never ends, Γ_Q is constant but a bubble will cover only a finite region in co-moving co-ordinates (cf. Eqs. 18, 19), hence there always remains some false vacuum. Guth (1981) also noted that bubbles true vacuum probably do not percolate.

(b) Neglecting point (a) the problem remains to thermalise (how?) the released heat, which is put in the kinetic energy of the bubble walls, of the largest bubbles, when there is only 1 second before Helium synthesis starts.

(c) If the bubbles give rise to strong inhomogeneities it is not clear whether or not we reheat to T_R of Eq. (20) below, which is important for the baryon number creation (§ 6.2).

It thus appears that an instability of the false vacuum at T_1, is the cleanest way to end the supercooling period. At T_1, the vacuum at every space point shifts to the broken state. The "latent heat" $\rho_v \sim \sigma^4$ reheats the Universe to

$$T_R = \sigma\left[\frac{30}{N_R \pi^2} + \frac{N_1}{N_R}\left(\frac{T_1}{\sigma}\right)^4\right]^{1/4} \sim 0.4\sigma , \qquad (20)$$

as follows from energy conservation with N_1 and N_R the number of states before and after reheating (cf. Einhorn and Sato, 1981).

After these generalities we must be more specific on what does occur at $T \lesssim 10^{15}$ GeV. For G = SU(5) Daniel and Vayonakis (1981) and Guth and Weinberg (1981) calculated the phase diagrams. What transition occurs depends on the coupling constants in the Lagrangian, but a not too strong 1PT is typical. For certain parameters the false vacuum destabilizes (region d of Guth and Weinberg, 1981; for Abelian Higgs $3e^4/16\pi^2 < \lambda < e^4$). If the breaking is of the Coleman-Weinberg type (CW; see section 5) Daniel (1981) and Billoire and Tamvakis (1981) found a very large supercooling $T_{end} = O(1 \text{ GeV})$

when at last the nucleation rate Γ_Q equals the expansion H. The potential for $\phi \ll T \ll \sigma$ and with the adjoint scalars Φ on the critical orbit $\Phi = \phi$ U^+ diag $(1,1,1,^{-3}/2, ^{-3}/2)U$ (arbitrary U) is to 1-loop (cf. Abbott, 1981)

$$V_{eff}^1 = \frac{5}{8} g^2 T^2 \phi^2 + B\phi^4 (\ln \frac{\phi^2}{\sigma^2} - \frac{1}{2}) + \frac{B}{2} \sigma^4$$

$$B = 8 \ 10^{-4} \tag{21}$$

$$\sigma \sim 10^{15} \ \text{GeV}.$$

This potential has a barrier whose width shrinks for lower temperatures and thus there always will be a T_{end} when the nucleation rate $\Gamma_Q(T_{end})$ is large enough. This need not be the case for certain barriers originating from quantum corrections, which are more or less constant as is the expansion rate of the false Universe $H \sim \frac{1}{T}$, (Eq. 18). Recently Sher (1981) noted the importance of the running SU(5) coupling constant g(T), cf. Eq. (10), because we are dealing with exponentials. A more detailed investigation (Tamvakis and Vayonakis, 1981) shows a destabilization by a non-perturbative $<F_{\mu\nu}^2>$ term, calculated for instantons, which will become important at T = 0 ($\Lambda_{SU(5)}$), with $\Lambda_{SU(5)} \sim 2 \ 10^6$ GeV, cf. Eq. (10).

Another reason for the premature ending of the supercooling might lie in the small gravitational effects (cf. the chiral effects noted by Witten (1981a) in the supercooling at the Weinberg-Salam scale, inducing a transition at O(100 MeV); but see our note 2). Abbott (1981; see also Fujii, 1981) considered some small coupling terms $\pm R\phi^2$ added to the potential of Eq. (21). For curvature determined by ρ_V, $R = -32\pi G(\frac{B}{2}\sigma^4)$ we have two completely different histories:-) the false vacuum destabilizes at O(10^{10}GeV), or +) the barrier always is too large (cf. remark above) and the transition is never completed. Hut and Klinkhamer (1981b) noted another possible gravitational effect: for T $\le$ O(10^{11} GeV) the wavelengths associated with the barrier are larger than the event horizon $D_H = (3/8\pi)^{\frac{1}{2}} M_{Pl}$ $\rho_V^{-\frac{1}{2}}$, probably invalidating the usual flat space-time calculations and global gravitational effects in analogous to the Hawking radiation might induce a shift to the vacuum.

These three different arguments thus indicate that the super-cooling for a CW SU(5) potential is ended at much higher temperatures and more smoothly than first thought.

6.2   In this paragraph we will see whether or not the simple ideas on baryon number creation of section 4 survive the complications from phase transitions (Hut and Klinkhamer, 1980a). For this purpose we will use the simple model of Kolb and Wolfram (1980a). If the vacuum is in the disordered state ($<\phi>$ = 0) the theory is symmetric and massless, so clearly no net B can be generated, the out-of-

equilibrium driving mechanism is lacking. For a second order PT
this occurs at $T > T_c$. At $T \sim T_c$ the theory is broken and the
baryon number producing processes start their build up. We thus
solve the equations (17) starting from thermal equilibrium at $x_o = M_X/T_c \sim g\sigma/T_c = 0(1)$ and find, not surprisingly, the same n_B/s as
the standard calculations with $x_o = 0$. Larger differences might be
expected for first order PTs. The baryon number will be generated
after reheating, which must be quite smooth, or in other words the
thermalisation must be effective, if we want to preserve the homogeneity of the Universe. This will be the case if the vacuum
becomes unstable at T_1 (10^6 - 10^{10} GeV?), because when the barrier
vanishes the bubbles from the last flash of nucleation probably will
have sizes and thermalisation times $0(\sigma^{-1})$. But before, in the
supercooling period, we have an important bonus: the washing out of
any truly primoridal n_B/s will be much stronger than in the standard
scenario (cf. Kolb and Wolfram, 1980a). For example the B changing
collisions will have $\sigma_{2-2} \sim \alpha_{U}^2/T^2$ (Ellis et al., 1980d) so that the
damping factor is

$$\exp\left[-M_X/x_p \int_{T_{P1}}^{T_1} <v\sigma_{2-2}>\ dT\right] \sim \exp\left[-\alpha_U^2\ N^{-\frac{1}{2}}T_{P1}/T_1\right] \tag{22}$$

which is $\stackrel{<}{\sim} 10^{-40}$ for $T_1 \stackrel{<}{\sim} 10^{-2}T_c$. This exponential damping follows
from the rate equation $dY_B/dt \sim -2Y_B n\gamma <v\sigma_{2-2}>$ (cf. the third term
on the RHS of Eq. 17b, where we had inserted the Fermi approximation
valid for $T << M_X$; see § 4 of Kolb and Wolfram, 1980a).

With the possibly huge damping we must now consider the n_B/s
generated after smooth reheating. Again this is simply starting
from equilibrium $(Y_\Delta(x_o) = Y_B(x_o) = 0)$ at $x_o \sim M_X/T_R \sim g\sigma/0.4\sigma \sim 0(1)$,
which gives the $[n_B/s]_{final}$ equal to the standard value (this holds
generally for $x_o < 5$). Also the numerical solutions show that the
n_B/s destroying processes are still effective enough at $x > x_o$ to wash
out a possible baryon number density $(Y_B(x_o) \neq 0)$ from the thermalisation
of the bubbles themselves. But this contribution to $[n_B/s]_{final}$ came
from the gauge bosons, whereas lighter Higgs bosons are present.
These will be the most important for the n_B/s created, since 1) their
CP violating diagrams are of lower order than for gauge bosons,
typically (remember from Eq.(17b)$Y_{B\ final}$ roughly $\propto \epsilon$), and 2) the
dilution will be less (cf. Kolb and Wolfram, 1980a, fig. 4).

We thus conclude that the generated baryon number density after
a second order PT or after the smooth reheating in a first order PT
probably will be the same as in the usual calculations neglecting
PTs. Finally some remarks (Hut and Klinkhamer, 1980a, Appendix B)
on the role of 1PTs for galaxy formation, where the major question
is the origin of the small density perturbations, which grow under
self-gravity into the observed bound systems, galaxies up to clusters.
Strong 1PTs provide two interesting ingredients: 1) after reheating
the particle horizon has been stretched by a factor $\sim N^{\frac{1}{2}}\ T_c/T_1$

relative to the standard horizon $\sim 2\ t_c$, as follows easily from
Eqs. (18) and (19), and 2) nucleation probably will lead to density
perturbations. But point 2) includes all unsolved problems mentioned
earlier and whereas the wanted density perturbations on galaxy
scales perhaps could be made, this arises not at all naturally nor
is it clear how to avoid unwanted aspects of the perturbation
spectrum, such as unobserved strong metric perturbations (primordial
black holes). Probably the density perturbations already exist
before the baryon number creation epoch. This would then result
in density perturbations of the adiabatic type ($\delta\rho_b/\rho_b = \delta\rho_\gamma/\rho_\gamma$, or
n_B/s constant): in each region of size d_H in the huge galaxy-sized
density enhancement (of very small amplitude) the same n_B/s is
made at $T \sim M_X$, depending on physical parameters only (ε, M_X, α etc.)
albeit at different eigentimes. The only reasonable way to have
isothermal perturbations ($\delta\rho_\gamma = 0$) later, appears to be from large-
scale shear at $T \sim M_X$ (Bond et al., 1981), but here we are already
deviating from the charming simplicity of the standard model
(section 2).

6.3 Another aspect of the breaking of gauge symmetries in the early
Universe is the possible creation of monopoles, which depends on
the "directions" of the breaking. From topological arguments
monopoles are expected to occur if the breaking has a stage with
$G \to H_j$ where $\pi_2(G/H_j) \neq \{0\}$, which always is the case if G is
simply connected and H_j contains a U(1) factor, because then $\pi_2(G/H_j)$
$= \pi_1(H_j)$ which is certainly non-trivial (remember: $\pi_n(H)$ is the nth
homotopy group with as elements the different equivalence classes of
the mapping n-sphere S_n in H). Hence in the unification scheme
monopoles are to be expected. The actual 't Hooft-Polyakov solution
for G = SU(2) and a triplet Higgs ϕ^a (reviews: Actor, 1979, Prasad,
1980) illustrates this. We are looking for classical[4] and
stationary solutions with finite energy, hence ϕ at infinity must
be on the orbit of minima of the potential $V = -\frac{1}{2}\ \mu^2\ \phi^2 + \frac{1}{4}\ \lambda\ \phi^4$.
The exact 't Hooft-Polyakov solution, which asymptotically goes
as $\phi^a \xrightarrow[\infty]{r} \hat{r}_a\ [\mu\ \lambda^{-\frac{1}{2}} + (\text{const}/g\ r)\ \exp(-\sqrt{2}\ m\ r)] \sim \hat{r}_a\ \mu\ \lambda^{-\frac{1}{2}}$, cannot
be deformed continuously to a fixed direction in group space $\hat{n}^a$,
and thus is not in equivalence class n = 0 but in n = 1 (remember:
SU(2) $\simeq S_2$). More detailed analysis shows that this solution is a
magnetic monopole with magnetic charge g_m = n/e and mass M = C
$(4\pi/g^2)\ g\ \mu\ \lambda^{-\frac{1}{2}}$, with a constant $C(\lambda/g^2) = O(1)$. Recently great
progress has been made with multipmonopole solutions (Ward, 1981;
Prasad, 1981, Prasad and Rossi, 1981; Jaffe and Taubes, 1980). For
larger G the analysis can be extended.

We must thus seriously consider the creation of monopoles (non-
trivial distribution of "directions" of ϕ^a) during phase transitions
("magnitudes") at unification temperatures. Alas a correct calculation taking care of gauge subtleties has not even been attempted,
but the creation of vortices in superconduction experiments is
indicative of the reality of the effect. In the following we
assume the monopoles are not confined (cf. the suggestion of Linde,

1980). Naively one expects the ϕ^a directions to be uncorrelated over separations > d_H. which indicates O(1) monopole per volume element O(d_H^3) (cf. Kibble, 1976; Einhorn, 1980). Because of the smallness of d_H, the large monopole mass O(100 M_X) and the slow annihilation their energy density would completely destroy the standard Helium synthesis result (Presskill, 1979). It was soon realised (Guth and Tye, 1980; Einhorn et al., 1980) that the stretching of d_H in a 1PT might be relevant. If the supercooling period ends by a shift at T_1, the directions in the many small bubbles being uncorrelated, some rough arguments on the correlation length seem to require $T_1 \lesssim O(10^8$ GeV) (Einhorn and Sato, 1981), not too far from the values discussed in § 6.1. If there is a 2 PT, or in other words $M_{Higgs} \gtrsim M_{gauge}$, perhaps thermal fluctuations reduce the monopole density enough (Bais and Rudaz, 1980).

7. CONCLUSION

Theories of the unification of the separate gauge theories for the three types of elementary particle interactions (electro-magnetic, weak and strong) have direct relevance for the earliest phases of the Universe. The history of the Universe is described by the Hot Big Bang model, which gives the epoch of important unified interactions at times $t \sim 10^{-36}$s and temperatures and typical energies of $\sim 10^{15}$ GeV $\sim 10^{38}$K. This theoretical hubris is rewarded with an explanation of the presently observed matter-anti-matter asymmetry. We have reviewed in some detail the finite temperature effects on the field theory, namely the phase transitions (PTs) at temperatures of order 10^{15} GeV (and 10^2 GeV) between the different gauge symmetries. These PTs may have dramatic effects on the history of the expansion of the Universe, but we showed that the final "standard" results on the matter-antimatter asymmetry are hardly effected. Also we briefly discussed the expected creation of monopoles at these transitions. Naive estimates indicate much too high monopole densities, invalidating the synthesis of Helium with abundance $\sim 25\%$, as observed. Perhaps strong supercooling in first order PTs reduces the monopole density. We emphasized that in order to preserve the baryon number creation and Helium synthesis phase transitions are not allowed to create strong inhomogeneities. This might be a problem for first order transitions and the best way to end a 1PT appears to be a destabilisation of the false vacuum at temperatures of order 10^{10} - 10^6 GeV, for which there are indications that this indeed occurs.

This is the appropriate place to acknowledge discussions with A. Guth, D.V. Nanopoulos, K. Tamvakis, E. Weinberg and especially P. Hut, who also commented on the manuscript as did C. Norman. It is a pleasure to thank the organizers of this symposium.

Notes

[1] Ideas that the value of Eq. (11) might hold locally from imperfect (statistical) annihilation do not seem to work and also give typically $n_B/n_\gamma \stackrel{<}{\sim} 10^{-18}$ (cf. Steigman, 1976).

[2] If the Weinberg-Salam breaking is radiative, Witten (1981a) has calculated an entropy generation of $10^5 - 10^6$ from the 1PT, which appears not to be allowed by the $\delta\theta$ arguments mentioned in section 4. More than one Higgs doublet, as seems required for ΔB, might reduce this supercooling (Flores and Sher, 1981).

[3] For completeness we remark that all our discussion of phase transitions might be completely changed if recent speculations on a global super-symmetry hold true. The major problem of GUTs is a natural explanation of the hierarchy $M_{WS}/M_U \sim 10^{-13}$, or why do some scalars remain massless in the breaking at unification energies M_U? The heuristic scenario runs as follows (Witten, 1981b):

The gauge symmetry is broken to SU(3) x SU(2) x U(1) at energies M_U at the tree level, without breaking a global supersymmetry (GSS); the GSS remains unbroken up to all finite orders in perturbation theory; at low energies the GSS (and SU(2) x U(1)) is broken by a non-perturbative mechanism. The last step is still uncertain, though there are some analogies in lower dimensions, and, of course, how to reconcile GSS with finite temperature?

[4] There is the possibility that quantum corrections invalidate these classical results. This might indeed happen for the classical instanton solution, where the boundary condition of pure gauge at infinity ($A_\mu = \frac{1}{e} G \partial_\mu G^{-1}$ for a particular $G(\theta, \phi)$, cf. Equation 4b) need not hold because of large quantum fluctuations (cf. Witten, 1979, Pagels and Tomboulis, 1978). Also for monopoles the boundary condition at infinity being the vacuum state might be far from $|\phi| = \mu\lambda^{-\frac{1}{2}}$.

References

Actor, A. (1979) Rev. Mod. Phys. 51, 461.

Abbott, L.F. (1981) Nucl. Phys. B185, 233.

Bais, F.A., Rudaz, S. (1980) Nucl. Phys. B170, 507.

Barbieri, R. (1980) in International School of Physics E. Fermi, Varenna (preprint CERN-TH 2935).

Billoire, A., Tamvakis, K. (1981) preprint CERN-TH 3019.

Bond, J.R., Kolb, E.W., Silk, J. (1981) preprint.

Cant, R.J., (1981) Phys. Lett. 104B, 121.

Coleman, S. (1977) Phys. Rev. D15, 2929 (also D16, 1762 and D21, 3305).

Coleman, S., Weinberg, E. (1973) Phys. Rev. D7, 1888.

Crewter, R.J. (1978) Acta Physica Austriaca Suppl. XIX, 47.

Daniel, M. (1981) Phys. Lett. 98B, 371.

Daniel, M., Vayonakis, C.E. (1981) Nucl. Phys. B180, 301.

Dine, M., Fisschler, W., Serednicki, M., (1981) Phys. Lett. 104B, 199.

Einhorn, M.B. (1980) in Unification of Fundamental Particle Interactions, eds. Ferrara, S., Ellis, J., Nieuwenhuizen, P. van, New York : Plenum.

Einhorn, M.B., Sato, K. (1981) Nucl. Phys. B180, 385.

Ellis, J. (1980) in Gauge theories and experiments at high energy, eds. Bowler, K.S., Sutherland, D.G.

Ellis, J., Gaillard, M.K., Nanopoulos, D.V., Rudaz, S. (1980a) Nucl. Phys. B176, 61.

Ellis, J., Gaillard, M.K., Peterman, A., Sachrajda, C.T. (1980b) Nucl. Phys. B164, 253.

Ellis, J., Gaillard, M.K., Zumino, B. (1980c) Phys. Lett. 94B, 343.

Ellis, J., Gaillard, M.K., Nanopoulos, D.V. (1980d), in Unification of Fundamental Particle Interactions, eds. Ferrara, S., Ellis, J., Nieuwenhuizen, P. van, New York : Plenum.

Ellis, J., Gaillard, M.K., Nanopoulos, D.V., Rudaz, S. (1981a) Phys. Lett. 99B, 101.

Ellis, J., Gaillard, M.K., Nanopoulos, D.V., Rudaz, S. (1981b) Nature 293, 41.

Fetter, A.L., Walecka, J.D. (1971) Quantum Theory of Many Particle Systems, New York : McGraw-Hill.

Flores, R.A., Sher, M. (1981) Phys. Lett. 103B, 445.

Fujii, Y. (1981) Phys. Lett. 103B, 29.

Georgi, H., Glashow, S.L. (1974) Phys. Rev. Lett. 32, 438.

Georgi, H., Quin, H.R., Weinberg, S. (1974) Phys. Rev. Lett. 33, 451

Guth, A.H. (1981) Phys. Rev. D23, 347.

Guth, A.H., Tye, S.H. (1981) Phys. Rev. Lett. 44, 631, 963.

Guth, A.H., Weinberg, E. (1981) Phys. Rev. D23, 876.

Hut, P., Klinkhamer, F.R. (1981a) submitted Astron. Astrophys. (preprint 3 June 1981).

Hut, P., Klinkhamer, F.R. (1981b), Phys. Lett. 104B, 439.

Jackiw, R. (1980) Rev. Mod. Phys. 52, 661.

Jaffe, A., Taubes, C. (1980) Vortices and Monopoles, Boston: Birkhauser,

Kibble, T.W.B. (1976) J. Phys. A9, 1387.

Kirzhnits, D.A., Linde, A.D. (1976) Ann. Phys. 101, 195.

Kolb, E.W., Wolfram, S. (1980a) Nucl. Phys. B172, 224.

Kolb, E.W., Wolfram, S. (1980b) Astrophys. J. 239, 428.

Langacker, P. 1981, Phys. Rep. 72, 185.

Linde, A.D. (1979) Rep. Prog. Phys. 42, 389.

Linde, A.D. (1980) Phys. Lett. 96B, 293.

Nanopoulos. D.V. (1980) in XVieme Rencontre de Moriond (preprint
   CERN-TH 2896).

Nanopoulos, D.V., Weinberg, S. (1979) Phys. Rev. D20, 2484.

Olive, K., Schramm, D.N., Steigman, G., Turner, M.S., Yang, Y. (1981)

O'Raifeartaigh, L. (1979) Rep. Prog. Phys. 42, 159.

Pagels, H., Tomboulis, E. (1978) Nucl. Phys. B143, 485.

Peccei, R.D., Quinn, H.R. (1977) Phys. Rev. D15, 1791.

Prasad, M.K. (1980) Physica 1D, 167.

Prasad, M.K. (1981) Comm. Math. Phys. 80, 137.

Prasad, M.K., Rossi, P. (1980), Preprint MIT-CTP-903 (also Phys.
   Rev. Lett. 46, 806).

Presskill, J.P. (1979) Phys. Rev. Lett. 43, 1365.

Sato, K. (1981) Monthly Notices Roy. Astron. Soc. 195, 467.

Sher, M. (1981) Phys. Rev. D24, 1699.

Steigman, G. (1976) Ann. Rev. Astron. Astrophys. 14, 339.

Taylor, J.C. (1976) Gauge Theories of Weak Interactions Cambridge :
   Cambridge UP.

Tamvakis, K., Vayonakis, C.E. (1981) preprint CERN-TH 3108.

Ward, R.S. (1981) Comm. Math. Phys. 79, 319.

Weinberg, S. (1972) Gravitation and Cosmology, New York : Wiley.

Weinberg, S. (1974) Phys. Rev. D9, 3357.

Weinberg, S. (1979a) Phys. Rev. Lett. 42, 850.

Weinberg, S. (1979b) Phys. Lett. 82B, 387.

Witten, E. (1979) Nucl. Phys. B149, 285.

Witten, E. (1981a) Nucl. Phys. B177, 477.

Witten, E. (1981b) Nucl. Phys. B188, 513.

Yildiz, A., Cox, P.H. (1980) Phys. Rev. D21, 906.

PARTICLE PHYSICS ASPECTS IN COSMOLOGY[+]

Harald Fritzsch
Sektion Physik
Universität München
and
Max-Planck-Institut für Physik, München
Germany

1. Introduction

The new theories developed by the particle physicists during the last decade about the structure of matter at small distances are important for astrophysics in several respects. These theories give us an insight into the dynamics of matter at extremely high densities and temperatures, i.e. in situations which are encountered in the collapse of massive stars and in the first moments after the birth of the universe (provided, the current theories about the "hot big bang" are correct). The unified theories of quarks and leptons provide us with a possibility to understand the generation of baryons shortly after the big bang. Furthermore it might turn out that the gravitational space-time dynamics of the universe is dominated by huge "clouds" of massive neutrinos.

On the other hand it may turn out in the future that the universe as a whole can be used as a testing ground for high energy physics. The high temperatures and densities achieved during the first stages of the universe implied collisions of particles at energies much higher than the energies reached in our earth- based accelerators. In those collisions all sorts of particles have been produced, probably including particles nobody has seen yet or thought of. Some of these particles may be stable or have life times of billions of years, and they might be lying around on earth waiting for their detection.

Surely, the field of cosmology and in particular the interplay between cosmology and high energy physics is a rather speculative field; the chances of making severe errors are large. Nevertheless it seems clear that we are dealing here with a rapidly developing field of physics. I expect that the present-day love affair between particle physics and cosmology will develop into a lasting marriage.

[+]Invited talk

A. W. Wolfendale (ed.), Progress in Cosmology, 63–74.
Copyright © 1982 by D. Reidel Publishing Company.

In this talk I shall discuss those topics of particle physics which at the present time are relevant for astrophysics and cosmology. With respect to the specific applications for cosmology, the reader is advised to consult the other contributions to these proceedings.

2. Hadrons and QCD

During the last few year the most impressive **success** in high energy physics has been the check of the predictions of quantum chromodynamics for hadronic physics (for recent reviews see e.g. Buras (1981), Fritzsch (1980), Mueller (1981)). It has become clear that baryons consist of three quarks, and mesons of a quark and an antiquark. QCD has developed to a realistic theory of the hadrons and their interactions. The quarks u and d relevant for the stable matter in the universe are essentially massless. In QCD the hadronic mass scales, e.g. the proton mass, have nothing to do with the intrinsic masses of the quarks, but with the mass scale entering into the theory in a non-perturbative way via the QCD coupling constant. The strong interaction coupling constant g has been determined recently in the lepton- hadron scattering experiments and in e^+e^--annihilation with a relatively good accuracy. One finds at an energy of about 30 GeV: $g_s^2/4\pi = \alpha_s \approx 0.15$, i.e. $g_s \approx 1.4$. Both in the in the lepton- hadron scattering experiments and in e^+e^--annihilation one observes effects (e.g. scaling violations) due to gluon radiation off quarks, which are the QCD analogs of the well-known QED bremsstrahlung. Furthermore one has found evidence for the production of quark -antiquark pairs by gluons (the QCD analog of the Bethe - Heitler process), e.g. by observing the production of charmed or b- flavored particles in hadronic collisions. In addition there are indications that gluons couple directly to gluons, as predicted by the non-Abelian nature of the QCD gauge theory. One can draw this conclusion by investigating the change of the gluon distribution function of the proton at increasing energies.

We conclude: the agreement between experiment and the theoretical predictions based on perturbation QCD is excellent. According to the theory the QCD coupling constant α_s at high energies behaves like

$$\frac{1}{\alpha_s(q^2)} = \frac{1}{\alpha_s(\mu^2)} + \frac{11 - \frac{2}{3}n_f}{4\pi} \ln\left(\frac{q^2}{\mu^2}\right) = \left(11 - \frac{2}{3}n_f\right)\ln\left(\frac{q^2}{\Lambda^2}\right)$$

(μ: arbitrary renormalization scale,

n_f: number of quark flavors).

The experiments give $\Lambda \approx 160$ MeV (see e.g. Buras (1981).

It is remarkable that the numerical value of $\wedge$ is close to the inverse radius of the nucleon (e.g. the (charge radius)^2 of the proton is 0.7 fm^2). This suggests that $\wedge^{-1}$ is related in a simple way to the confinement length of the theory, i.e. to the length at which the force between the quarks becomes strong. Furthermore the proton mass must be directly related to $\wedge$: $M_p = K \cdot \wedge$, where K is a numerical constant which is in principle calculable from QCD and which is of the order of 6 (perhaps K = 2π ?). Unfortunately the progress in our understanding of nonperturbative effects in QCD is rather slow (for a recent review see e.g. t'Hooft (1981), however some insight has been gained within the lattice approach to QCD (see e.g. Hasenfratz (1981)). Especially it has been possible to relate the slope of the Regge trajectories (a measure of the hadronic scale like the proton mass) to the $\wedge$-parameter; one finds $\wedge \approx$ 200 MeV, in good agreement with observation. Nevertheless one can say that the breakthrough needed in order to calculate nonperturbative effects in QCD has not occurred yet, and one has to wait.

As far as astrophysics and cosmology is concerned, we can say that the hadronic matter at high densities and temperatures looks much simpler than previously thought. In a good approximation it is simply a free gas of spin 1/2 quarks and spin 1 massless gluons.

The question which remains completely open within the framework of QCD is the relation between the QCD mass scale $\wedge$ (or, what is the same, the proton mass) and the gravity scale. What determines the weight of the proton? Presumably an answer can be found only within a unified theory of QCD and general relativity.

3. Electroweak Forces

The standard SU(2) x U(1) theory (for a recent review see Fritzsch and Minkowski (1981)) implies that the weak interactions are mediated by W and Z bosons, which are very massive. According to the experimental determinations of the neutral current coupling parameters one expects M_W = 78 GeV and M_Z = 89 GeV (uncertainty about 1.5 GeV). The scale of the weak interactions, which reflects itself in the masses of W and Z, is generated quite differently than the hadronic mass scale of QCD, namely by the scalar Higgs field φ assuming a nonvanishing vacuum expectation value. The latter is directly related to the Fermi constant G:

$$\frac{G}{\sqrt{2}} = \frac{1}{2<\varphi>^2} \quad (<\varphi> = 246 \text{ GeV}).$$

At the present time we have tested the SU(2) x U(1) theory only in the low energy region, where we can describe the weak interactions by an effective current x current Hamiltonian, involving both the charged and the neutral current. The agreement between experiment and theory is excellent.

The present experimental limits on M_W and M_Z are M_W, $M_Z \gtrsim 40$ GeV. There is a good chance to find the W and Z bosons fairly soon in the upcoming experiments at the proton- antiproton collider in CERN

4. Unified strong and electroweak interactions

Particle physics and cosmology get in close touch in the field of unified interactions, involving both the strong and electroweak inter- actions. The idea is well-known: the aim is to embed both the QCD gauge group $SU(3)^C$ and the electroweak gauge group $SU(2) \times U(1)$ in a larger group G:

$$SU(3)^C \times SU(2) \times U(1) \subset G.$$

The "observed" quarks and leptons can be divided into three generations:

$$\begin{pmatrix} \nu_e & \vdots & u \\ e^- & \vdots & d \end{pmatrix} \quad \begin{pmatrix} \nu_\mu & \vdots & c \\ \mu^- & \vdots & s \end{pmatrix} \quad \begin{pmatrix} \nu_\tau & \vdots & t \\ \tau^- & \vdots & b \end{pmatrix}$$

(We note that the t quark has not been observed yet; its mass must be larger than ~ 18 GeV).

If we disregard the family problem and consider only the first family (all lefthanded fermions and antifermions, the corresponding righthanded fermions are provided by CPT) we are dealing with the re- presentation

$$(1,2, -\tfrac{1}{2}) + (1,1,1) + (3,2, \tfrac{1}{6}) + (\bar{3},1, \tfrac{1}{3}) + (\bar{3},1, -\tfrac{2}{3})$$

$$= (\nu_e, e^-)_L + (e_L^+) + (u,d)_L + (\bar{d})_L + (\bar{u})_L$$

under the group $SU(3)^C \times SU(2) \times U(1)$. Fifteen lefthanded fermions are needed. One may add one further fermion, the lefthanded antineutrino $\bar{\nu}_{eL}$, in which case one obtains sixteen fermions. The maximal symmetry of the free fermion Lagrangian is SU(15) or SU(16) respectively (see e.g. Fritzsch and Minkowski (1975)). However these groups cannot be used as gauge groups, due to the existence of anomalies which spoil the re- normalizability of the theory. One may ask the question whether there exists a group such that

a) it is a subgroup of SU(15) or SU(16)

b) the 15 or 16 fermions form a representation of this group which has the desired transformation property under $SU(3)^C \times SU(2) \times U(1)$

c) there exist no anomalies.

The solutions to this problem are well-known. In the 15-fermion case one finds SU(5) (Georgi and Glashow, 1974), while in the 16-fermion case the solution is SO(10) (Fritzsch and Minkowski (1975)).In both schemes the colorgroup SU(3) and the electroweak group SU(2) x U(1) is embedded in the larger group SU(5) or SO(10). In particular the U(1)-generator is represented by one of the generators of SU(5) or SO(10), implying that the electric charges of the fundamental fermions are quantized. Furthermore the coupling constants g_3, g_2, and g_1 are equal (up to group factors of order 1) to the fundamental gauge coupling g. This implies in the symmetry limit:

a) The value of the SU(2) x U(1) mixing angle is given by
$\sin^2\theta_W = 3/8 = 0.375$.

b) $\alpha_s = \dfrac{g_3^2}{4\pi} = 8/3 \cdot \alpha \approx 1/50$.

Of course, these values are in disagreement with the values of the coupling constant observed in the experiments. The only way to achieve consistency is to assume that the energy at which the symmetry limit is obtained is far away from the region studied in the present experiments. As emphasized in 1975 (Georgi, Quinn and Weinberg, 1974), the renormalization effects increase both $\sin^2\theta_W$ and α_s, and a good agreement with experiment is achieved, if the unified symmetry sets in at about 10^{15} GeV (for a recent review see Langacker (1980)).

In the SU(5) scheme the 15 fermions are described by the reducible representation $\bar{5}$ + 10, which decompose under the subgroup SU(3) x SU(2) x U(1) exactly as needed:

$$\bar{5} = (\bar{3}, 1, \tfrac{1}{3}) + (1, 2, -\tfrac{1}{2}) = \bar{d}_L + (\nu_e, e^-)_L$$

$$10 = (\bar{3}, 1, -\tfrac{2}{3}) + (3, 2, \tfrac{1}{6}) + (1, 1, 1) = \bar{u}_L + (u, d)_L + e^+_L.$$

With respect to the SU(5) gauge interactions both the ($\bar{5}$)representation and the (10)- representation of SU(5) cause anomalies. Those are exactly equal in magnitude, but opposite in sign. Thus in the sum they cancel, as required by renormalizability.

In the SO(10) scheme the 16 fermions are described within the irreducible 16 - dimensional spinor representation of SO(10) (for a general discussion of orthogonal group see Weyl (1946)). Under the subgroup SU(3) x SU(2) x U(1) one finds the decomposition:

$$16 = (1, 2 - \tfrac{1}{2}) + (1, 1, 1) + (3, 2, \tfrac{1}{6}) + (\bar{3}, 1, \tfrac{1}{3})$$

$$+ (\bar{3}, 1, -\tfrac{2}{3}) + (1, 1, -1).$$

The group SU(5) is a subgroup of SO(10), and one can decompose the fermions with respect to SU(5) as follows:

$$16 = \bar{5} + 10 + 1.$$

i.e. one obtains the fermions of the SU(5) scheme and a SU(5) - singlet, the righthanded neutrino. It is interesting to observe that the SO(10) group is able to accommodate the left-right symmetric gauge theory of the weak interactions:

$$SO(10) \supset SU(3)^C \times SU(2)_L \times SU(2)_R \times U(1)$$

With respect to $SU(3)^C \times SU(2)_L \times SU(2)_R$ one finds:

$$16 = (3, 2, 1) + (1, 2, 1) + (\bar{3}, 1, 2) + (1, 1, 2)$$

$$= (u, d)_L + (\nu_e, e^-)_L + (d^c, - u^c)_L + (e^+, - \nu^c)_L.$$

In the standard SU(2) x U(1) theory the U(1) generator is a fairly complicated mixture of T_3 and Q_e. However in the left-right symmetric theory this is no longer true; the U(1) charge is simply the charge B-L (baryon number minus lepton number).

Both in the SU(5) and the SO(10) schemes quarks, leptons and antiquarks appear in the same representation. Hence baryon number is not conserved and the proton is unstable.

In the SU(5) scheme the charge (B-L) is exactly conserved (for a discussion see Langacker (1980)). Thus the proton decays mostly via the modes $p \to e^+ + (\pi^0, \rho^0, \omega, ...)$ and $p \to \bar{\nu}_e + (\pi^+, \rho^+, ...)$. The lifetime of the proton can be estimated to be

$$\tau(\text{proton}) = 4 \cdot 10^{31 \pm 1.3} \left(\frac{\Lambda}{0.16 \text{ GeV}} \right)^4 \quad \text{years}$$

(see e.g. Langacker ((1980)). The best value of the QCD Λ-parameter is about 0.16 GeV, hence the proton lifetime is expected to be about 10^{31} years.

In the SO(10) scheme the decay of the proton proceeds essentially along the same lines as in the SU(5) scheme, if the SO(10) symmetry is broken at $\sim 10^{15}$ GeV down to $SU(3)^C \times SU(2)_L \times U(1)$. The life time of the proton is equal to the one predicted within the SU(5) scheme. However the situation is different if the left-right symmetric electroweak theory $SU(2)_L \times SU(2)_R \times U(1)$ is valid down to much smaller energies, say a few hundred GeV. In this case the proton life time is typically larger than 10^{31} years and may well be above 10^{33} years.

In the SO(10) scheme (B-L) is not conserved and in general one expects that the neutron mass term contains a (small) Majorana term, which leads to n - n̄ oscillations (see e.g.: Marshak and Mohapatra ((1980)).

The n-n̄ oscillation time must be more than $\sim 10^5$ sec, otherwise there are problems with the stability of nuclei, implying the n-n̄ transition mass term to be less than 10^{-20} eV. However in most models one finds that δm is highly suppressed, i.e. much smaller than 10^{-20} eV. From a theoretical point of view, the prospects of observing n-n̄ oscillation in dedicated reactor experiments are not good, since one expects $\tau(n-\bar{n}) >> 10^5$ sec. Nevertheless they may be there, in which case one finds interesting applications for cosmic ray physics. In particular the observed large flux of antiprotons, observed in the primary cosmic rays, may be due to n-n̄ oscillations.

Probably the most interesting application of the idea of a unified interaction becoming relevant at a large energy scale (> 10^{15} GeV) is the idea of spontaneous baryon number generation (for a recent review see Langacker (1980)).

In particular the decays of superheavy gauge bosons or superheavy scalars can produce a net baryon number. The idea is that soon after the "big bang" the baryon number violating interactions came into equilibrium, and any initial baryon asymmetry (if present) was washed out. If the expansion rate of the universe is fast compared to the decay rate of the superheavy particles, the latter drop out of equilibrium, and a net baryon number is generated.

In the SU(5) model including several 5-representations of scalars of net baryon number generated by the decay of superheavy scalars is close to the observed one (Nanopoulos and Weinberg (1979)). In the SO(10) scheme the baryon number depends heavily on the breaking of $SU(2)_L \times SU(2)_R \times U(1)$. This group is P and C invariant. Thus the baryon number will be essentially zero, unless $SU(2)_R$ is broken at a very large mass scale ($\sim 10^{15}$ GeV) (Kurzmin and Shaposhnikov (1980)). This is a strong argument in favor of the assumption that parity is violated already at the unification mass scale of the order of 10^{15} GeV.

Recently one has observed an unexpectedly large flux of antiprotons in the primary cosmic rays (Buffington et al., 1981), wich has stimulated interest in the baryon symmetric domain cosmology (Stecker (1978), Sato (1981)). In tese theories C and CP are broken spontaneously, in which case one may obtain a domain structure of the universe. In some areas matter is dominated by quarks, in other areas causally disconnected from the previous areas by antiquarks. (Of course, one could redefine the antiquarks to be quarks, since baryon number is not conserved. However the weak interactions violate parity maximally. Hence the different domains of the universe differ in the signs of parity violating amplitudes.) It remains to be seen if the idea of a domain structure of the universe is correct.

5. Massive neutrinos

In the standard electroweak theory neutrinos are massles since

a) no righthanded neutrino fields are introduced,

b) lepton number is conserved (hence a Majorana mass term is forbidden).

Recently many papers have been written about massive neutrinos. Most of these papers have been stimulated by the development of unified gauge models such as the SO(10) model discussed previously in which right-handed neutrino fields must be present. In such a case it would be most natural to have a neutrino mass term which may be a Dirac type mass term, a Majorana type mass term or a mixture of both.

If neutrinos are massive, there is no reason to expect the mass term to be diagonal in flavor space. Hence neutrinos oscillate. No convincing evidence has been found so far that they do in the laboratory (for a recent review see e.g. von Feilitzsch (1981)). An experimental group reports to have found a positive signal for the $\bar{\nu}_e$ - mass: 14 eV $\lesssim m(\bar{\nu}_e) \lesssim$ 46 eV (see Lubimov (1980)). Whether the electron neutrino has a mass or not, remains to be seen.

If neutrinos are massive and stable, they could provide a substantial fraction of the mass density in the universe. If the sum of the masses of stable neutrinos is more than about 10 eV, the neutrinos will provide the dominant part of the mass density in the universe (for a recent review see: Rees (1981)).

What can we expect within our theoretical ideas about mass generation for the neutrino masses? First of all I would like to stress that there exists no satisfactory theoretical framework in which one is able to calculate the fermion masses (quark masses, lepton masses, including the neutrino masses). Hence all what I say must be considered to be rather preliminary.

As mentioned above, in the conventional scheme and within the SU(5) theory the neutrino masses do vanish. In the SO(10) theory the neutrino masses are not zero in general, but they depend strongly on specific details of the dynamical symmetry breaking. In the simplest scheme one predicts something which is unacceptable: the neutrinos are Fermi- Dirac particles, and their masses are equal to the corresponding quark masses: $m(\nu_e) = m_u$, $m(\nu\mu) = m_c$, $m(\nu\tau) = m_t$. However this problem can easily be avoided if the (B-L) quantum number is broken at a superheavy mass scale M, which could be of the same order of magnitude as the mass scale describing the onset of the ground unification. In that case the right-handed neutrino (the singlet of the SU(5) subgroup of SO(10)) aquires a large mass M, and the neutrino mass matrix can be written as follows (for a recent review see Langacker (1981)):

$$(\bar{\nu}_L \quad \bar{\nu}_L{}^c) \quad \begin{pmatrix} 0 & m_q/2 \\ m_q/2 & \mu \end{pmatrix} \quad \begin{pmatrix} \nu_R{}^c \\ \nu_R \end{pmatrix}$$

($\nu_R{}^c$ denotes the charge conjugated neutrino field m_q stands for m_u, m_c, m_t respectively). This mass matrix implies that the lefthanded neutrino has a vanishing Majorana mass, the righthanded one has a Majorana mass μ, and the neutrino has in addition a Fermi- Dirac mass which is equal to the corresponding quark mass. Diagonalizing the neutrino mass matrix, one finds that the lefthanded neutrino aquires a neutrino mass $m_q^2 / 4 \mu$. This mass is very small; for $m_q = m_c = 1$ GeV one finds e.g. $m(\nu_\mu) \approx 10^{-5}$ eV. However the Majorana mass μ may well be several orders of magnitude smaller than the 10^{15} GeV; for example radiative corrections induce a mass μ of the order of $(\alpha/\pi)^2$ M $\approx 10^{10}$ GeV. For $\mu = 10^{10}$ GeV one finds e.g. $m(\nu_e) = 6 \cdot 10^{-7}$ eV, $m(\nu_\mu) = 0.04$ eV, $m(\nu_\tau) = 10$ eV.

If this pattern of neutrino masses deduced from the SO(10) scheme has anything to do with the real neutrino masses, it is implied that only the τ- neutrino can be the one which is relevant for astrophysics. Both the muon and the electron neutrino are too light to play a significant rôle in astrophysics.

In the same approach the flavor mixing angles in the neutrino sector are related to the corresponding ones in the quark sector (Cabibbo-like angles). The latter ones are observed to be rather small. Hence it will be difficult to observe neutrino oscillations. In any case neutrino oscillations could not be made responsible for the fact that the neutrino flux of the sun is observed to be less than predicted.

If neutrinos have a mass, they would exhibit several properties which are otherwise not expected. For example they have a magnetic moment, which however depends severely on specific details of the underlying weak interaction theory. In many models the typical scale is $\mu(\nu) \sim 10^{-19} m(\nu) \cdot \mu_B$ (μ_B: Bohr magneton). Of course, it would be extremely difficult to detect such a small magnetic moment. However the presence of a magnetic moment would imply that the "black body" neutrino background in the universe contains a large number of righthanded neutrinos (the estimate of $\mu(\nu)$ given above is made for Dirac-type neutrinos).

Furthermore one might expect that neutrinos decay, e.g. $\nu_\tau \rightarrow \nu_\mu \bar{\nu}_e \nu_e$, or $\nu_\tau \rightarrow \nu_\mu + \gamma$. Due to the small neutrino masses the phase space for these decays is extremely small. In order to see what the typical scale for the neutrino decay might be, we assume that the decay $\nu_\tau \rightarrow \nu_\mu \bar{\nu}_e \nu_e$ has the same strength (in amplitude) as the μ-decay. In this case one expects

$$\frac{\tau(\nu)}{\tau(\mu)} \;=\; \left(\frac{m(\mu)}{m(\nu)}\right)^{5}$$

This gives $\tau(\nu) \approx 10^{+29}$ s, i.e. a lifetime too large to be observable in laboratory experiments. In addition decays like the ones noted above are flavor changing processes which in the case of the charged fermions are observed to be suppressed in amplitude. Nevertheless one cannot exclude the existence of new types of interactions involving e.g. Higgs scalars, in which case the life times of the neutrinos are in the vicinity of 10^{29} s. In that case one might be able to observe e.g. the decay $\nu \rightarrow \nu' + \gamma$, by finding evidence for the emission of photons in the neutrino decays (see e.g. Sciama (1982) and Stecker (1981)).

Finally I would like to mention the possibility to discover the effects of a neutrino Majorana mass by observing the neutrinoless double β-decay. Recent investigations indicate that the Majorana mass of the electron neutrino cannot be much larger than 15 eV (Haxton et al. (1981)). Refined measurements may improve this limit by an order of magnitude. Of course, no such limit exists for a Dirac mass of ν_e. Furthermore no information can be obtained this way about the masses of the other neutrinos.

6. Final comments

Despite the rather deep understanding of the strong, weak and electromagnetic interactions achieved during the last ten years, one must say that essential clues are still missing. For example not much progress has been made in getting an insight into the mechanism which generates the masses (especially the lepton and quark masses). The relationship between the particle physics mass scales and the Planck mass is still mysterious. Perhaps the missing clues can be found in the existence of new interactions which have not been observed yet and which are associated with mass scales above 100 GeV. It is quite possible that the new p - p̄ collider going in operation in the fall of 1981 at CERN will give us information about such new mass scales. Those could in particular be related to the bound state structure of leptons and quarks. The leptons are observed to be pointlike down to a distance of 10^{-17} cm, i.e. a distance which is of the order of the length scale given by the Fermi constant. It may well be that many surprises are found once the energy region between 100 and 1000 GeV is explored. No doubt, such new insights would have severe consequences for astrophysics and cosmology.

References

1. Buffington, A., et.al., Ap. J. in press (1981).

2. Buras, A. (1981), proceedings of the Int. Symposium on Lepton Interactions (Bonn, 1981), in press.

3. Feilitzsch, F.v. (1981), proceedings of the Int. Conference on High Energy Physics (Lisbon, 1981), in press.

4. Fritzsch, H. and Minkowski, P. (1975), Annals of Physics 93, 193 - 266.

5. Fritzsch, H. and Minkowski, P. (1981). Physics Reports 73, 67 - 173.

6. Fritzsch, H. (1980), Physica Scripta 24, 847 - 860.

7. Georgi, H. and Glashow, S. (1974), Phys. Rev. Lett. 32, 438 - 441.

8. Georgi, H., Quinn, H., and Weinberg, S., Phys. Rev. Lett. 33, 451 - 454.

9. Hasenfratz, P. (1981), proceedings of the Int. Conf. on High Energy Physics (Lisbon, 1981), in press.

10. Haxton, W. (1981), Phys. Rev. Lett. 47, 153 - 156.

11. t'Hooft, G. (1981), proceedings of the Int. Conf on High Energy Physics (Lisbon, 1981), in press.

12. Kuzmin, V. and Shaposhnikov, M. (1980), Phys. Lett. 92 B, 115 - 118.

13. Langacker, P. (1980), SLAC-preprint SLAC-PUB-2544, to appear in Physics Reports (1981), in press.

14. Lubimov, V. et al. (1980), Phys. Lett. 94 B, 226 - 268.

15. Marshak, R. and Mohapatra, R. (1980), Phys. Lett. 94 B, 183 - 186.

16. Mueller, A. (1981), Physics Reports 73, 237 - 368.

17. Nanopoulos, D. and Weinberg, S. (1979), Phys. Rev. D 20, 2484 - 2493.

18. Sato, K. (1981), Phys. Lett. 99 B, 66 - 70.

19. Sciama, D. (1982), these proceedings, 75.

20. Stecker, F., proceedings of the Int. Conference on
     Neutrino Physics (Wailea, Hawaii, 1981), to appear.

21. Stecker, F.W. (1978) Nature 273, 493

22. Weyl, H., "The classical groups" (Princeton, 1946)

MASSIVE NEUTRINOS AND ULTRA-VIOLET ASTRONOMY

D.W. Sciama
Department of Astrophysics, Oxford University.
Department of Physics, University of Texas at Austin.

Two astronomical ionisation processes are tentatively attributed to photons emitted by massive neutrinos. In the first, the intergalactic medium is assumed to be ionised by photons emitted by a cosmological distribution of neutrinos. The observations then require a neutrino mass lying between 50 and 110 electron volts and a radiative lifetime of order 10^{27} seconds. These values are compatible with cosmological constraints and with particle physics expectations if the GIM mechanism is not operating.

The second ionisation process involves the production of CIV and SiIV high up in the galactic halo. This is attributed to photons emitted by neutrinos dominating the halo. It would require a neutrino mass exceeding 96ev and a lifetime of order 10^{27} seconds. These values are compatible with those required by the intergalactic hypothesis.

INTRODUCTION

Recent developments in elementary particle physics have led to the suggestion that neutrinos may possess a non-zero rest-mass, possibly lying in the range 10^{-3}-100 electron volts. As is well-known, if the hot big bang picture of the origin of the universe is correct, there would then exist a cosmological distribution of such neutrinos, with a present concentration about one quarter of that of the photons in the 3 K background, that is, about 100cm^{-3}. A rest-mass of 100ev would then imply a contribution to the density of the universe of the critical order required to close it. Indeed, the estimated age of the universe would lead to an upper limit on the sum of the masses of the various neutrino types (lying in the mass range $\ll$1 Mev) of about 200ev, unless there is a non-zero cosmical constant.

It is also well known that the further suggestion has been made that the "missing matter" in clusters of galaxies and in individual

75

A. W. Wolfendale (ed.), Progress in Cosmology, 75–85.

galaxies, including the Milky Way, may be in the form of massive
neutrinos. These neutrinos would then constitute about 90 per cent
of the total mass in clusters and in single galaxies. Whether such
objects could form in the expanding universe is now under debate, and
no consensus has yet been reached. One condition must certainly be
satisfied, however, and this is a phase space constraint which leads
to a lower limit on the relevant neutrino masses. For the Milky Way
this limit is (Peebles 1980)

$$m_\nu \gtrsim 30 \text{ eV.}$$

This limit is determined solely by the observed characteristics of our
Galaxy and by the value of Planck's constant. It is therefore encouraging that it brings us into the regime for which the neutrinos
would have a dominant importance in cosmology, and yet is compatible
with the upper limit on m_ν set by the age of the universe without
the need to invoke a cosmical constant.

The existence of missing mass in galaxies is deduced mainly from
their flat rotation curves which are observed to extend out to several
tens of kiloparsecs. If the missing mass forms a spherical halo its
density would then have to decrease outwards as the inverse square
of the distance from the centre of the galaxy. For the Milky Way,
the neutrino density at the sun's location would have to be given by

$$n_\nu \sim 10^7 \left(\frac{100 \text{ eV}}{m_\nu} \right) \text{ cm}^{-3}$$

if the neutrinos are to account for the missing matter. Thus for
neutrinos in the relevant mass range their concentration is enhanced
over that in the universe as a whole by a factor of order 10^5.

It would be highly desirable to test these ideas by detecting the
neutrinos in a manner independent of the gravitational field which
they produce. The most promising possibility is to search for the
photons which the neutrinos would be expected to emit, according to
some modern particle theories. These photons would probably lie in
the ultra-violet (de Rujula and Glashow 1980). This possibility is
the subject of the present article.

Photons from Decaying Massive Neutrinos

The decay process involved would be

$$\nu_1 \rightarrow \nu_2 + \gamma.$$

If the parent neutrino ν_1 is at rest relative to the observer, then
the photon energy E_γ would be given by

$$E_\gamma = \frac{m_1^2 - m_2^2}{2 m_1}.$$

If $m_2 \ll m_1$ this simplifies to

$$E_\gamma \sim \tfrac{1}{2} m_1 .$$

For purposes of exposition we shall adopt this approximation throughout. Then for m_1 in the range 10-100ev, E_γ would lie in the range 5-50ev, that is, in the ultra-violet.

The flux of photons produced by decaying neutrinos depends, of course, on the radiative lifetime involved. This question has been discussed by de Rujula and Glashow. The lifetime depends critically on whether a process known as GIM suppression is operating. If it is, then the lifetime τ typically ranges from

$$\tau \sim \frac{10^{36}}{\sin^2 2\beta} \left(\frac{30\,eV}{m_1} \right)^5 sec$$

to ten times this quantity (Pal and Wolfenstein 1981). Here β is the mixing angle between ν_1 and ν_2, which in some models is large (Goldman and Stephenson 1981).

GIM suppression can itself be suppressed if, for example, one makes wider assumptions about the number of neutrino types (and associated Higgs particles). The lifetime is then drastically reduced and one obtains (de Rujula and Glashow)

$$\tau \sim \frac{1.5 \times 10^{30}}{\sin^2 2\beta_1} \left(\frac{30\,eV}{m_1} \right)^5 sec$$

or possibly less (Pal and Wolfenstein).

Observational Lower Limits on the Neutrino Lifetime

We note first the coincidence that the photon flux at the Earth coming from galactic neutrinos has the same order of magnitude as that coming from cosmological neutrinos (ignoring absorption effects for the moment). This follows from the fact that the galactic enhancement in the neutrino concentration for $m_\nu \sim 100$ev is of the same order ($\sim 10^5$) as the ratio of the radius of the Universe to the scale-height of the galactic halo (~ 30 kpc). The main difference is that the galactic flux is nearly monochromatic (since the velocity-dispersion of the galactic neutrinos ~ 200 km.sec^{-1} $\ll c$), whereas the cosmological flux would be drawn out into a continuous spectrum by the differential red shift associated with the expansion of the Universe.

In fact this spectrum would have the form

$$I_\lambda \atop (\lambda \geqslant \lambda_0) = \frac{c\, n_\nu(z=0)}{H_0\, 4\pi \tau} \frac{\lambda_0^{3/2}}{\lambda^{5/2}} \left(1 + (2q_0 - 1)\left(1 - \frac{\lambda_0}{\lambda}\right) \right)^{-1/2} ,$$

where λ_0 is the rest wavelength of the decay photon, λ the observed wavelength, τ the neutrino lifetime and q_0 the deceleration parameter. Any absorption effects would have to be added to this relation. (There is a numerical error of $\sim 10^4$ in equation (7c) of de Rujula and Glashow (1980) which has been corrected by Kimble, Bowyer and Jakobsen (1981)).

It has been pointed out by Stecker (1980) and Kimble, Bowyer and Jacobsen (1981) that one can use this spectrum to obtain lower limits on τ from the observed u-v background even if λ_0 corresponds to a wavelength at which the Galaxy is opaque (e.g. $\lambda_0 < 912$Å). Since the observed background is probably due to other sources, it has been used to limit τ rather than to determine it. From an observed background $\sim$ 200-300 photons cm^{-2} sec^{-1} ster^{-1} Å^{-1}, they deduce that

$$\tau > 10^{22} - 10^{23} \ sec \qquad 10 \ ev < \frac{hc}{\lambda_0} < 50 \ ev$$

if the intergalactic medium is transparent out to the largest redshifts involved (z $\sim$ 6). We return to the question of this transparency later.

Stecker also discussed the possibility that a reported increase in the background at 1700Å might represent a photon flux from galactic neutrinos and he suggested that $\tau \sim 3 \times 10^{24}$ sec. According to Kimble, Bowyer and Jakobsen, this increase has not been confirmed and is best treated as an upper limit to the actual intensity.

These authors also derived a lower limit on τ for galactic neutrinos, from the general observed background in the 30-50ev range (which corresponds to the mass range in which neutrinos of cosmological origin could dominate the Galaxy). Their limit depends on the uncertain opacity of the Galaxy at these photon energies, and on the photon energy itself, but lies in the range

$$\tau > 10^{20} - 10^{22} \ sec \qquad 30 \ ev < E_\gamma < 50 \ ev$$

More stringent limits have been derived by Shipman and Cowsik (1981) and Henry and Feldman (1981) from optical and ultra-violet observations of the Virgo and Coma clusters of galaxies. If these clusters are dominated by neutrinos of appropriate mass the derived limits are

$$\tau > 10^{23} - 10^{25} \ sec \qquad 1 \ ev < E_\gamma < 10 \ ev$$

Shipman and Cowsik consider that with existing or proposed instruments one could improve these limits up to the range $10^{26} - 10^{27}$ sec.

All these limits are derived from direct observations of photon fluxes. One can also use arguments derived from the ionising effects of the photons, as pointed out by Melott and Sciama (1981). These

differ from the previous arguments in that they can lead to
limits on (or evaluations of) photon fluxes at positions distant
from the galactic plane, where the effects of absorption by neutral
hydrogen or dust will be different (and in general less). For
example, Melott and Sciama (1981) demanded that these photon fluxes
should not destroy the High Velocity Clouds. The clouds are neutral
hydrogen features observed at 21 cm. to be predominantly approaching
us with velocities of a few hundred kilometres per second. Various
arguments suggest that some clouds lie at least a kiloparsec above
the galactic plane, while a recent estimate puts them at tens of
kiloparsecs away. In fact the limits obtained are valid so long as the
clouds lie within the Local Group of galaxies, but not within the
galactic plane. One finds in this way that

$$\tau > 10^{24} \text{ sec} \qquad E_\gamma > 13 \cdot 6 \text{ eV}$$

There is no suggestion in any of these arguments (with the exception of Stecker's) that one is close to observing a real effect which
could be attributed to photons from neutrino decay. Moreover, the
theoretically expected value of τ is many orders of magnitude greater
than even the largest of the lower limits so far quoted. The question
therefore arises whether a more sensitive ionisation process can be
discovered which would lead either to an actual effect being identified, or at least to a much more stringest limit on τ in the relevant photon energy range.

The Ionisation of the Intergalactic Medium

We here examine the hypothesis that photons emitted by a cosmological distribution of massive neutrinos are mainly responsible
for the ionisation of the intergalactic medium to a level compatible
with the absence of absorption troughs in the spectra of quasars
(Gunn-Peterson (1965) effect). This ionisation is usually attributed
to photons emitted by the quasars themselves (Sherman 1981). Such a
mechanism is plausible but is hard to quantify, especially because of
uncertainties in the emissivity of quasars in the hard ultra-violet
and in their spatial distribution at large red shifts. In particular,
the IGM seems to be highly ionised at least out to a red shift
$\sim$ 3.5, where the quasar distribution may have a sharp cut-off
(Osmer 1981). It is therefore of interest to examine the alternative
hypothesis that the ionising photons are emitted by a cosmological
distribution of decaying neutrinos.

The absence of absorption troughs due to neutral hydrogen and
helium in the IGM (Green et al 1980, Wampler et al 1973, Ulrich et al
1980) leads to the following upper limits on the concentration of
neutral hydrogen and helium:

$$n_{HI} < 2 \times 10^{-12} \ cm^{-3} \qquad z = 0$$

$$n_{HI} < 10^{-11} \ cm^{-3} \qquad z \sim 2$$

$$n_{HI} < 5 \times 10^{-11} \ cm^{-3} \qquad z \sim 3.5$$

$$n_{HeI} < 10^{-11} \ cm^{-3} \qquad z \sim 2$$

However, recent observations with IUE (Gondhalekar et al 1981) have indicated that there may be a cut-off in the continuum emission of a quasar with a red shift of 3.2 at wavelength shorter than HeII 304 A in the quasar rest-frame. This has been tentatively attributed by those observers to a positive Gunn-Peterson effect due to singly ionised helium. If this is correct it would mean (a) that an IGM has been detected for the first time, (b) that it contains helium and (c) that the helium is singly ionised but not doubly ionised. The observers derive a lower limit to the optical depth for scattering HeII 304A photons of 3, from which they deduce that the density of helium in the IGM at $z \sim 3$ satisfies

$$n_{He}(3) > 10^{-9} \ cm^{-3}.$$

There is a large uncertainty set by the noise in their spectrum, and the true density could be much greater than this lower limit.

If the helium/hydrogen ratio in the IGM takes its "cosmical" value of 0.1, as would be expected in the canonical hot big bang picture, then we would have for the density of (ionised) hydrogen in the IGM at $z \sim 3$

$$n_{H}(3) > 10^{-8} \ cm^{-3}.$$

Our ionisation hypothesis leads to stringent restrictions on the possible mass m_{ν} and radiative lifetime τ of the decaying neutrinos. Since helium in the IGM is singly ionised we require $m_{\nu} \gtrsim$ 50ev. This would be compatible with the upper limit on m_{ν} ($\sim$ 200ev) derived from lower limits on the age of the Universe. If helium in the IGM is indeed not doubly ionised we require $m_{\nu} \lesssim$ 110ev, which is close to the cosmological upper limit. Thus the relevant neutrino type must have a mass lying in the range

$$50 \ ev \lesssim m_{\nu} \lesssim 110 \ ev.$$

We can place limits on the radiative lifetime τ by requiring that we satisfy the observational constraints given above. For simplicity we shall adopt an Einstein-de Sitter model with a Hubble constant of 50 km sec^{-1} Mpc^{-1} and an age t of 1.3×10^{10} years. Our results do not depend critically on this assumption. At large red shifts the universe would be dense enough to be opaque to photons in the range 25-50ev, and the emission rate per unit volume of these photons would be $n_\nu(0)/\tau \cdot (1+z)^3$, where $n_{\nu_3}(0)$ is the present concentration of decaying neutrinos (~ 100 cm^{-3}). The total number of photons emitted per unit volume up to an epoch corresponding to a red shift z would be

$$\frac{n_\nu(0)\, t\,(1+z)^{3/2}}{\tau}.$$

The hydrogen density at that red shift would be

$$n_H(0)\,(1+z)^3.$$

The hydrogen will become highly ionised at a critical red shift z_i for which there is one ionising photon available per hydrogen atom. (It can be shown that the ionization cross-section is large enough for the ionisation time to be less than the expansion time for all z, so this is a good approximation). This gives

$$n_H(0)\,\tau = n_\nu(0)\,t\Big/(1+z_i)^{3/2}$$

Since we require $z_i \geqslant 3.5$ we must have

$$n_H(0)\,\tau \leqslant 5 \times 10^{18}\ cm^{-3} sec$$

It is possible that the quasar cut-off at $z \sim 3.5$ is due to absorption by a neutral IGM at $z > 3.5$, rather than, say, to quasars being first formed at this red shift. In that case we would have $z_i \sim 3.5$ and

$$n_H(0)\,\tau \sim 5 \times 10^{18}\ cm^{-3} sec.$$

We must next ensure that the high ionisation levels implied by the observational constraints can be achieved. At the critical red shift z_i we would have

$$\frac{n_{HI}}{n_e} = \frac{\alpha n_e}{\sigma n_\gamma c} = \frac{\alpha}{\sigma c},$$

since we have $n_e \sim n_H$ and $n_\gamma \sim n_H$ at z_i. Here α is the recombination coefficient and σ the photoionisation cross-section. The recombination rate depends on the electron temperature, which in our model is itself mainly determined by the ionisation process. This problem has been discussed by Rees and Setti (1970). Following their methods we find that, at $z \sim z_i$, $T \sim 2.5 \times 10^4\ {}^\bullet K$. At this temperature $\alpha \sim 2 \times 10^{-13}$ cm^{3}sec^{-1}. For a photon energy ~ 50ev (see later) we have $\sigma \sim 10^{-19}$ cm^2. Introducing a further factor 2/3 to allow for the fact that in the transparent regime n_γ is determined by an integration over the past light cone, we have

$$n_{HI} \sim 10^{-4} \, n_e.$$

For purpose of illustration we take $z_i \sim 3.5$. We must ensure that at this red shift $n_{HI} \lesssim 5 \times 10^{-11} \, cm^{-3}$. Since $n_e(z_i) \sim n_H(z_i)$,

we require

$$n_H(0) \lesssim 5 \times 10^{-9} \, cm^{-3}.$$

This upper limit is a few per cent of the mean density due to baryons in galaxies if all dynamically determined invisible mass has a non-baryonic form, e.g. as massive neutrinos. This would mean that galaxy formation is a highly efficient process. While this is not unreasonable, we prefer to be conservative at this point by adopting $n_H(0) \sim 5 \times 10^{-9} \, cm^{-3}$. Accordingly we require that

$$\tau \sim 10^{27} \, sec.$$

A similar calculation shows that the ionisation level for hydrogen at other red shifts, and for helium, would also then satisfy the observational constraints. This value for τ is compatible with the lower limit of 10^{24} seconds derived by Melott and Sciama (1981) for hydrogen-ionising photons if the massive neutrinos also dominate the halo of our Galaxy.

Our limits on m_ν and τ have relevance for particle physics. Various Grand Unified Theories (Langacker 1981, Fukugita, Yanagida and Yoshimura 1981) suggest that the largest neutrino mass could be as great as 100ev. The calculated lifetime in the absence of GIM suppression agrees well with our cosmological estimate if our value of 100ev for m_ν is adopted, unless the mixing angle β is very small. In view of the sensitive dependence of τ on m_ν $(\propto m_\nu^{-5})$ we regard this agreement as encouraging. Further encouragement follows from our second ionisation hypothesis which, as we shall see, leads to similar values for m_ν and τ .

The Ionisation of the Upper Galactic Halo

Recent observations with IUE have shown that highly ionised carbon (CIV) and silicon (SiIV) exist in regions ten to fifteen kiloparsecs from the plane of the Galaxy with column densities typically $\sim 10^{14} \, cm^{-2}$ and $3 \times 10^{13} \, cm^{-2}$ respectively. Savage and de Boer (1979), Savage and de Boer (1981), Ulrich (1980), Bromage and Sciama (1980), Bromage (1981). The existence of these highly ionised stages can be plausibly attributed either to collisional ionisation in a hot halo gas with $T \sim 10^5$ °K, or to ionisation by photons from hot stars or the soft x-ray background. As with the ionisation of the IGM by quasars, these processes are hard to quantify, and we therefore think it worthwhile to investigate the possibility that the ionisation arises instead from photons emitted by massive neutrinos dominating the galactic halo.

The ionisation potentials of the precursor stages CIII and SiIII are 47.9ev and 33.5ev respectively. Thus to produce CIV by our proposed mechanism we require

$$m_{\nu} \geqslant 95 \cdot 8 \text{ ev.}$$

As we have seen, when this inequality is satisfied neutrinos could dominate galactic halos as we are assuming. If we combine this with our previous inequalities on m_{ν} from the ionisation of the IGM, we find that the neutrino mass must lie in the narrow range given by

$$95 \cdot 8 \text{ ev} \leqslant m_{\nu} \leqslant 110 \text{ ev.}$$

We now consider the value of the radiative lifetime τ implied by our hypothesis. The most detailed data relating to ionised elements in the galactic halo have been obtained by Savage and de Boer (1981), who also give an extensive discussion of their results. The ions are detected from their absorption lines observed by IUE in the spectra of distant stars e.g. in the Magellanic Clouds. One uses velocity data to determine the location of the absorbing material, on the provisional assumption that the halo is co-rotating with the Galaxy. One determines in this way that much of the material lies at distances from the galactic plane of ten to fifteen kiloparsecs.

The derived CIV number density $\sim 10^{-9} \text{cm}^{-3}$ and the density ratios of SiII, SiIV, CII, CIV and NV are roughly 10:1:120:4: $<$ 1 (unobserved), with localised variations. In a photionised gas these ratios are set by charge exchange with HI, HII and HeII as well as by direct photionisation and recombination with free electrons. The rates of the processes which we need to know have all been given in the literature (Aldovandi and Pequignot (1973), Butler and Dalgarno (1980)), and it is straightforward to determine whether a monochromatic flux of ~ 50ev photons can reproduce both the observed ionisation ratios and the absolute values of the ion densities. We find that with an assumed ambient gas density $\sim 10^{-4} \text{cm}^{-3}$ and a cosmic abundance of He, C, Si and N, we can satisfy all the observations (including the non-detection of NV) with a 50ev photon flux $\sim 600 \text{ cm}^{-2} \text{ sec}^{-1}$. With this value of the photon flux the steady state ionisation balance which we are assuming to hold would be established in a time less than the age of the Galaxy.

The photon flux produced by halo neutrinos has been estimated by de Rujula and Glashow. Since this flux could be comparable with the flux from extragalactic neutrinos we must consider the opacity of halo material for 50ev photons. The mean free path ℓ for such a photon is given by

$$\ell \sim \frac{1}{n(HI)\sigma}.$$

If the halo is opaque the column density of HI within one mean free path is

$$n(HI)\,\ell \sim 10^{19}\,cm^{-2}.$$

If $n(H) \sim 10^{-4}$ cm^{-3} and $n(HI) \sim 1/3n(H)$ we have $\ell \sim 100$ kpc, which is of the same order as the size of the halo. The observed column density of HI at high galactic latitudes, including the disk contribution, $\sim 2 \times 10^{20}$ cm^{-2}, comfortably larger than the value we are taking for the halo alone. We conclude that it is likely that at a height of 15 kpc the extragalactic flux is shielded out, but that in computing the internal flux the opacity is not an important factor.

If we insert into de Rujula and Glashow's estimate the neutrino density referred to earlier we find that the required photon flux I_γ is given by

$$I_\gamma \sim \frac{6 \times 10^{29}}{\tau}\ cm^{-2}\,sec^{-1}.$$

Since the flux required to account for the observed column-density of CIV and SiIV ~ 600 cm^{-2} sec^{-1}, we conclude that

$$\tau \sim 10^{27}\ sec.$$

This agrees well with our estimate from the ionisation of the IGM, and the implications of this estimate for particle physics given there hold for our present discussion as well. This encourages us to make the further hypothesis that the halos of other galaxies are also ionised by photons emitted by their ambient neutrino distribution. This would be relevant to quasar absoption line spectra of intermediate red shift, which usually include highly ionised stages such as CIV and SiIV (but often not NV) and are thought to arise from intervening galactic halos.

Of course further observations are required before the radical ideas proposed here can be taken seriously. A good test would be to search above the atmosphere for a faint narrow emission line ($\Delta\lambda \sim 10^{-3}\lambda$) at high galactic latitudes with a photon energy lying between 47.9 and 54.4ev.

ACKNOWLEDGEMENTS

I am grateful to A.L. Melott for convincing me of the potential importance of massive neutrino for cosmology, and for collaborating with me on some of the ideas presented here. I am also grateful to G.E. Bromage, A.C. Edwards, G. Feinberg, P. Gondhalekar and L.W. Wolfestein for helpful discussions and correspondence.

REFERENCES

S.M.V. Aldrovandi and D. Pequignot, Astron. and Astrophys. 25, 137, 1973.

S.L. Baliunas and S.E. Butler, Ap.J.Lett. 235, L45, 1980.

G.E. Bromage, A.H. Gabriel and D.W. Sciama, Proc.2nd European IUE Conf. 26-28 March 1980 (ESA SP-157, April 1980).

G.E. Bromage in Nature, 289, 123, 1981.

S.E. Butler, T.G. Heil and A. Dalgerno, Ap.J. 241, 442, 1980.

S.E. Butler and A. Dalgerno, Ap.J. 241, 838, 1980.

M. Fukugita, T. Yanagida and M. Yoshimura, to be published, 1981.

S. Goldman and T. Stephenson, Phys. Rev D. 24, 236, 1981.

P.M. Gondhalehar, D.F. Malin, A. Boggess, R. Wilson and C-C. Wu, to be published 1981.

R.F. Green, J.R. Pier, M. Schmidt, F.B. Estabrook, A.L. Land and H.D. Wahlquist, Ap.J. 239, 483, 1980.

J.E. Gunn and B.A. Peterson, Ap.J. 142, 1633, 1965.

R.C. Henry and P.D. Feldman, Phys.Rev.Lett. 47, 618, 1981.

R. Kimble, S. Bowyer and P. Jakobsen, Phys.Rev.Lett. 46, 80, 1981.

P. Langacker. Physics Report 1981.

A.L. Melott and D.W. Sciama, Phys.Rev.Lett. 46, 1369, 1981.

D.E. Osterbrock, The Astrophysics of Gaseous Nebulae p. 34 (Freeman, San Francisco 1974).

P.B. Pal and L. Wolfenstein, to be published 1981.

P.J.E. peebles, in Physical Cosmology (ed.R. Balian, J. Audouze and D.N. Schramm) les Houches Summer School 2-27 July 1979, p. 265 (North-Holland, Amsterdam 1980).

A. de Rujula and S.L. Glashow, Phys.Rev.Lett. 45, 942, 1980.

B.D. Savage and K. de Boer, Ap.J.Lett. 230, L77, 1979.

B.D. Savage and K de Boer, Ap.J. 243, 460, 1981.

R.D. Sherman, Ap.J. 246, 365, 1981.

H.L. Shipman and R. Cowsik, Ap.J.Lett. 247, L111, 1981.

F.W. Stecker, Phys.Rev.Lett., Phys.Rev.Lett. 45, 1460, 1980.

M.H. Ulrich et al. Mon.Not.Roy.Astr.Soc. 192, 561, 1980.

E.J. Wampler, L.B. Robinson, J.A. Baldwin, and E.M. Burbidge, Nature 243, 336, 1973.

W.D. Watson and R.B. Christensen, Ap.J. 231, 627, 1979.

TIME-ASYMMETRY, COSMOLOGICAL UNIFORMITY AND SPACE-TIME SINGULARITIES

Roger Penrose
Mathematical Institute, University of Oxford, U.K.

We know from the "singularity theorems" of general relativity that any universe model which is compatible with the standard interpretation of background radiation and radio source counts, and which satisfies Einstein's equations with physically reasonable assumptions about the positivity of matter density and absence of closed timelike curves, must be singular. Thus the big bang is "stable" in the sense that perturbations away from exact symmetry cannot remove the singularity. Likewise the singularities of gravitational collapse are stable in the same sense. Thus, physics as we presently understand it is limited in its scope and some new physical ideas are needed in order that the regions of the universe referred to as "space-time singularities" can be treated according to the methods of science.

There is, however, some very strong evidence that there are, indeed, powerful laws governing the structure of singularities. This evidence comes from two apparently independent sources. On the one hand there is the observed large-scale uniformity of the universe, which implies that the big-bang singularity was, at least to a high degree of approximation, of a special highly symmetrical type. On the other hand there is indirect evidence which comes from the second law of thermodynamics. The second law in effect implies that the final singularities which arise in collapse must be unconstrained whereas the big bang must be highly constrained. The Bekenstein-Hawking formula for black-hole entropy enables an estimate to be made of the "degree of specialness" of the big bang. In a closed universe containing 10^{80} baryons, for example, this degree of specialness works out as something of the general order of one part in $10^{10^{123}}$.

This suggests that there must indeed have been very precise physical laws in operation at the big bang itself. The new physics involved is necessarily time-asymmetric. A certain "thought-experiment" appears to imply that such time-asymmetric physics should also have relevance to the "observation" problem in quantum mechanics.

A. W. Wolfendale (ed.), Progress in Cosmology, 87–88.

The figure 10^{80}, for the baryon number of the universe,
is chosen here for convenience only, in order that some calculation
can be made. (The popularity of this particular figure seems
to stem from the state of the art of cosmology in Eddington's day !)
For a universe with B baryons its "specialness" would be about
one in $10^{10^3 B^{3/2}}$. This figure is not affected if one supposes the
present baryon number B to arise from baryon non-conserving
processes in an initially baryon-antibaryon-symmetric universe,
since it refers to the possible entropy that can be gained by
collapse into black holes, and this entropy easily swamps
all other entropy production processes.

While it may be the case that the existence of a neutrino
mass of a few tens of eV's will "close the universe", it can be
argued that even so, the figure of B = 10^{80} is an "improbable"
one - roughly for the reason that this figure leads to a
timescale for recollapse which is not far from m_p^{-3} in absolute
units (G = c = $\hbar$ = 1), but which is related in no simple way to
the neutrino mass (the neutrino being taken as the dominant
contributor to the mass of the universe). The present apparent
universe age is related to the lifetimes of main sequence stars
(since it is this that determines "our present epoch"), which leads
to a universe lifetime (so far) of the general order of m_p^{-3} in
absolute units. In the absence of a theoretical reason that
the ultimate recollapse time (if k = +1) is also approximately
of this order, one is led to guess that the final timescale is
$\gg m_p^{-3}$ and probably unobservable directly.

References

Carter, B. (1974) In Confrontation of Cosmological Theories
   with Observational Data (IAU Symp. 63) ed. M.S. Longair.
   Reidel: Boston.

Penrose, R. (1979) In General Relativity, An Einstein
   Centenary Survey (ed. S.W. Hawking & W. Israel.) Cambridge
   University Press.

Penrose, R. (1981) In Quantum Gravity 2: A Second Oxford
   Symposium (ed. C.J. Isham, R. Penrose & D.W. Sciama)
   Oxford University Press.

ANTIMATTER IN COSMIC RAYS

Péter KIRÁLY
Central Research Institute for Physics
H-1525 Budapest POB 49, Hungary

1. Introduction

Electromagnetic waves, the traditional carriers of astrophysical information, are unable to distinguish between matter and antimatter. Until neutrino astronomy can really set off, cosmic rays should give the only direct information about any antimatter existing in our astrophysical environment. After the pioneering work of Aizu and co-workers /1961/ and of Brooke and Wolfendale /1964/ the search continued with negative results, but with improving upper limits /for a survey see Steigman 1976/, until Golden and co-workers /1979/ identified the first cosmic antiprotons /p̄/ at several GeV in a balloon-borne superconducting-magnet spectrometer. At about the same time, Bogomolov and co-workers /1979/ also reported on p̄ observations at similar energies /2 events/. Finally, Buffington and co-workers /1981 a,b,c/, covering a much lower energy range, where p̄ production by collisions is much less efficient, also found a substantial p̄ flux.

Of course, the detection of p̄ in cosmic rays does not automatically imply the existence of any macroscopic accumulation of antimatter /such as stars, galaxies, or clusters of galaxies/. In fact, fairly stringent upper limits exist for the antimatter content of our environment, see Steigman /1976/. The important question is, whether high-energy collisions between cosmic rays and interstellar gas in a conventional model can explain the observations. If not /which appears to be the case/, alternative models of cosmic ray propagation and acceleration have to be checked for consistency with all the observational data. Since cosmic ray physics is still a mixture of folklore and hard facts, a considerable margin of uncertainty exists, which may be wide enough to accommodate the p̄ fluxes observed at several GeV. Buffington's data, on the other hand, are so far from

A. W. Wolfendale (ed.), Progress in Cosmology, 89–99.
Copyright © 1982 by D. Reidel Publishing Company.

conventional expectation that - as will be discussed later -
even exotic origin models /like those based on a baryon-
symmetric Universe/ have to be stretched somewhat in order
to get some semblance of consistency.

Although the recent $\bar{p}$ observations have been planned,
carried out and evaluated with great care and ingenuity,
one should remember that experimenters are not infallible
either. Some causes for worry will be mentioned later.
Thus, in view of the huge potential importance of the results, new and even more sophisticated $\bar{p}$ searches in several
energy regions should be carried out in the near future. A
better upper limit /or positive identification/ of the $\bar{\alpha}$
flux would also be very useful for clarifying the situation.
Until then exotic interpretations are worth examining, but
the present evidence is not yet compelling enough to dis-
card conventional views.

Part of the present review is based on a paper published in Nature /Király et al. 1981/.

2. Observations

The main problem of $\bar{p}$ flux measurements is background
elimination. All observations so far have been done by
balloon-borne equipment, under a residual atmosphere of
5-20 gcm^{-2}. Because of the residual atmosphere, the background contains a variety of secondary particles in addition to the flux of primary protons and nuclei. Part of
these are negatively charged and fairly massive /μ^-, π^-, K^-/,
and in that respect similar to the $\bar{p}$ component to be measured. Nuclear interactions and fragmentations in the equipment can also simulate $\bar{p}$ behaviour in many respects. In
view of all these difficulties, fairly complex equipments
have to be used and a detailed record of any interesting
event has to be kept for further analysis. Mountain-top
calibrations and computer simulations are unavoidable.
Because of the huge number of primary protons, records of
all events cannot be kept, therefore some trigger requirements are also needed to eliminate most of the uninteresting
events.

The main components of the $\bar{p}$ detectors used so far
were scintillators, multiwire proportional counters, Cherenkov detector, and superconducting magnet /Golden et al.
1979/, optically viewed spark chambers, Cherenkov scintillation counter telescope, and magnet /Bogomolov et al. 1979/,
scintillators, spark chambers giving a detailed visualiza-
tion of events, and Cherenkov counter /Buffington et al.
1981 b/. In the $\bar{p}$ detectors applying magnets positive back-

ground particles are eliminated by magnetic deflection,
negative mesons by a combination of magnetic deflection
and Cherenkov signal, electrons by detecting cascades in
the multilayer detectors. In the equipment of Buffingon and
co-workers $\bar{p}$ annihilation stars were looked for by studying
their multiprong topology. 'High-energy' protons /E $\gg$ 300
MeV/ were eliminated by including the Cherenkov-counter
into the trigger requirement. 'Low-energy' /E $\lesssim$ 400 MeV/
protons were mostly eliminated by ionisation losses in the
chamber /a coincidence of the top and bottom scintillators
was also needed for the trigger/. The selective trigger
provides a 1000-fold reduction in the proton background,
while it accepts $\bar{p}$ annihilations with about 8% probability,
since pions emerging from the annihilation have about that
probability to reach the bottom scintillator.

The equipment of Buffington and co-workers /1981 b,c/
was flown in June 1980 from Canada, at an average atmospheric depth of 11 gcm^{-2}. The $\bar{p}$ energy range covered was
130 to 320 MeV. Altogether about $2 \cdot 10^4$ triggers were recorded. That number was first reduced to 1500 events, containing an incident track each which connects to three or
more prongs. Most of these contained more than one energetic
track or some other uncertainty. All topological requirements defined by the experimenters were satisfied by 64
events, 28 of which were found to have an incident particle
arriving from the side. 19 of the remainder proved to be
due to fragmenting He nuclei. One of the remainder was also
discarded as He-induced and two due to the presence of accompanying particles. The experiment was thus considered
to have observed 14 $\bar{p}$ events. Several crosschecks were then
carried out in reasonable agreement with Monte Carlo predictions for simulated $\bar{p}$ events.

While the above procedure is probably the best under
the circumstances, many would agree that alternative processes may have contributed to the events classified as $\bar{p}$
and that this important result needs confirmation by independent experiments. There is also some cause for concern
at higher energies because of small statistics /Bogomolov
et al.1979/ and because the data of Golden and co-workers
do not appear to reproduce the normally accepted flux of
protons /Szabelski et al. 1981/.

The fluxes inferred from the three $\bar{p}$ measurements are
shown in Fig.1. along with a conventional and a less conventional prediction to be discussed later. While the observations corrected to the top of the atmosphere should be
close to the local interstellar values at the two higher
energies, solar modulation effects are important for Buffington's point. That observation was made at a time near

to the maximum solar activity and to the polarity reversal
of the interplanetary magnetic field, making the modulation
correction both large and somewhat uncertain. A good approximate description of solar modulation appears to be an
average adiabatic energy loss of the interstellar flux, the
magnitude of which changes with the phase of the solar cycle.
Following our previous, more detailed treatment /Király et
al. 1981/, the correction was done with three different
values of the adiabatic loss /400, 600 and 900 MeV/. It is
important to note that adiabatic energy changes conserve
the density in phase space, which is proportional to p^{-2}
times the differential flux, where p is the momentum. Demodulation thus transfers the observed flux 'upwards and
to the right' in Fig.1.

The $\bar{p}/p$ flux ratios were found as $/5.2\pm1.5/ \times 10^{-4}$
/Golden et al. 1979/, $/6+4/ \times 10^{-4}$ /Bogomolov et al. 1979/,
and $/2.2\pm0.6/ \times 10^{-4}$ /Buffington et al. 1981 b/. In Buffington's result, however, there is some uncertainty connected with possible charge-dependent modulation effects
and with the normalization of the proton flux /Buffington
et al. 1981 b/, thus no significant energy dependence is
seen in the flux ratios.

As a secondary objective of their experiment, Buffington and co-workers /1981 b/ have found a new upper limit
on the $\bar{\alpha}/\alpha$ ratio $/2.2 \times 10^{-5}/$, which is the most stringent
upper limit so far. No $\bar{\alpha}$ was observed. The fact that this
upper limit is about an order of magnitude smaller than
the $\bar{p}/p$ ratios observed, is a strong argument against considering the high $\bar{p}$ fluxes as indicators of the existence
of bulk antimatter in the Universe.

3. Conventional predictions

The most conventional, i.e. simplest and most often
used model of cosmic ray propagation is the 'leaky box'
model. Particles and their secondaries are assumed to be
in equilibrium, both escaping from the Galaxy after many
collisions with its boundary, and after penetrating the
same average amount of matter $/\lambda \sim 5$ gcm^{-2} at a few GeV/.
It is an approximation of the more complicated diffusion
models, and it is expected to be valid for quantities
which depend on the amount of matter traversed, but not
necessarily for quantities depending on cosmic ray age
/Protheroe 1981/. Based on this model, and on accepted values for the cosmic ray spectra and for interstellar gas
densities and composition, and using accelerator data for
inclusive $\bar{p}$ production cross sections, the calculation of
the $\bar{p}$ flux is straightforward. Although there are some

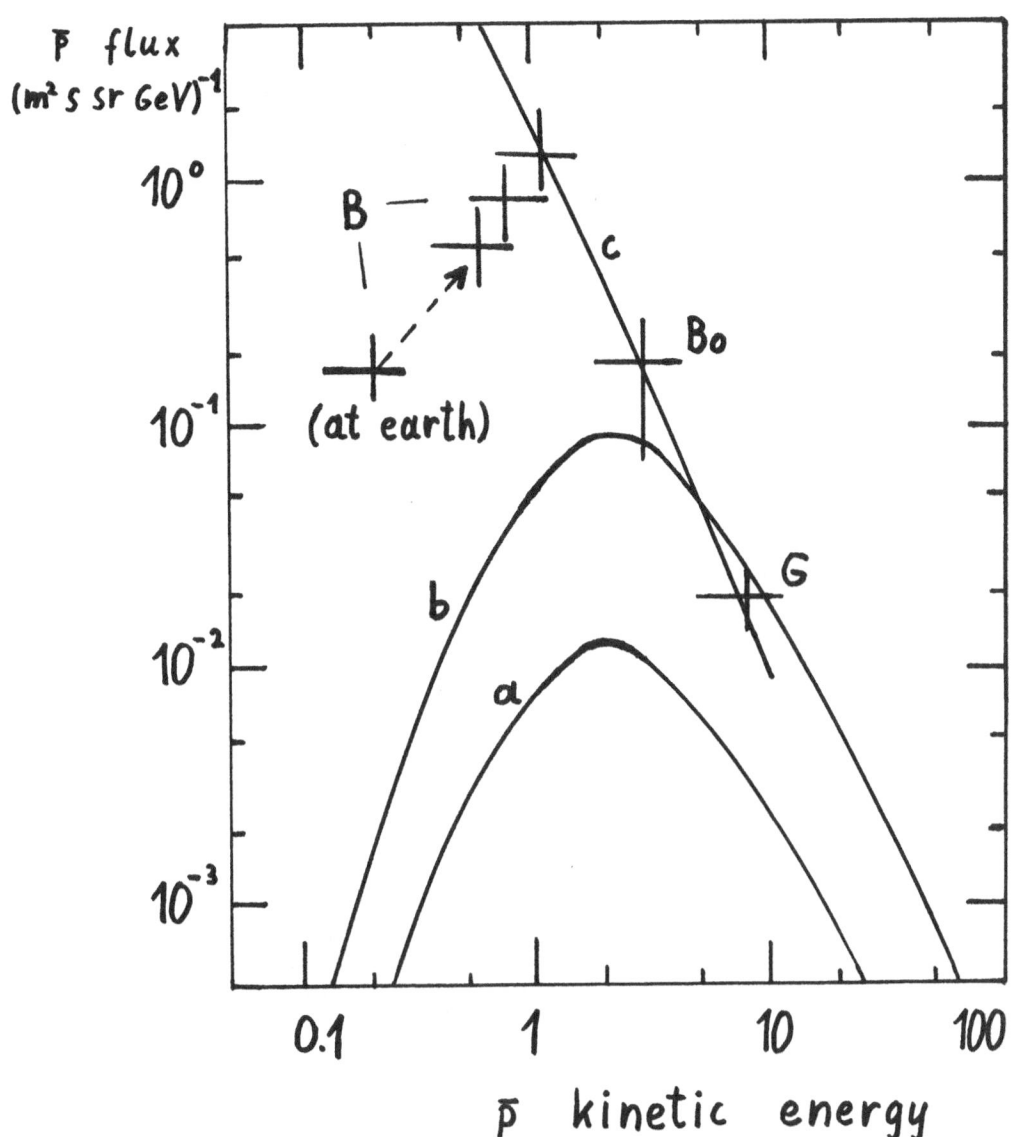

Fig.1. Measured and predicted $\bar{p}$ flux values. 'a' leaky box model with λ = 5 gcm^{-2}; 'b' closed Galaxy model; 'c' proton spectrum scaled down by a factor of 2000. B denotes Buffington's observed point at earth and the corresponding interstellar fluxes obtained for three possible values of the adiabatic energy loss parameter /400, 600 and 900 MeV/. Bo denotes Bogomolov's, G Golden's point.

differnces among the values accepted by various groups,
the most recent results of their calculations do not deviate by more than 60% /Király et al. 1981/, and the overall
uncertainty should be definitely less than a factor of 2.
As seen in curve 'a' of Fig.1, there is a definite disagreement with $\bar{p}$ observations, ranging from a factor of 5 at
the highest energies to a factor of 100 at Buffington's
point.

4. Attempts at 'mild' modifications

These attempts were motivated by the $\bar{p}$ observations
at higher energies, where the discrepancy with 'leaky box'
predictions was much smaller than is the case for the very
recent data of Buffington and co-workers. Also, due to some
numerical errors in the calculations, the discrepancy
appeared to be smaller than it really is. In these models
the $\bar{p}$ flux is secondary as before, but the effective amount
of gas penetrated /the 'grammage'/ was assumed to be bigger
than the conventional 5 gcm^{-2}. With the current value of
the predicted $\bar{p}/p$ ratio at 10 GeV /Szabelski et al. 1981/
the grammage needed turns out to be about $80 gcm^{-2}$, very
much in excess of the values suggested earlier /20 gcm^{-2}/.
This new value is near to that suggested by the closed
Galaxy model of Rasmussen and Peters /1975/, thus the modification to the leaky box model is not very mild, after all.
Predictions based on the closed Galaxy model are shown in
curve 'b' of Fig.1. As can be seen, Buffington's point is
still too high. There are also other objections to the
closed Galaxy model, the most important one being that it
predicts 4 to 9 times too many secondary positrons at about
3 GeV /Giler et al. 1977/. A somewhat milder modification
proposed by Cowsik and Gaisser /1981/ postulates an important contribution from bright gamma-sources shrouded in
thick clouds; at 10 GeV, an enhancement of 3 in the $\bar{p}/p$
ratio and a consistency with positron observations is
claimed. Stephens /1981/ proposed a modified closed Galaxy
model, in which 50% of the observed nucleons are old, the
remaining ones coming from a so-called 'nested leaky box'
confinement. Neither of these models can reproduce Buffington's point. In fact, a spectrum proportional to that of
protons, but scaled down by a factor of 2000 /curve 'c' in
Fig.1/ gives a much better approximation to the observations than any of the previous models. This might indicate
that the $\bar{p}$ component is of primary origin.

Earlier suggestions by Szabelski et al. /1980/ for a
possible role of interstellar deceleration in increasing
the $\bar{p}/p$ ratio at low energies were taken up by Buffington
and co-workers /1981 a/ for explaining their observed anti-

protons as secondaries. Because of the difficulty of substantially increasing the density in phase space by deceleration, this explanation appears very unlikely. An even
more important argument is that simply there are not enough
high-energy antiprotons to supply enough decelerated ones
into the energy range below a few GeV.

5. Exotic suggestions

The models discussed so far tried to achieve agreement
with antiproton observations by changing the scenario of
cosmic ray propagation, but no fundamentally new physical
or astrophysical phenomena had to be invoked. Their failure
means that exotic possibilities have to be considered seriously.

Baryon-symmetric cosmology as a solution was favoured
by Stecker /1981/ and was also discussed by him in this
Symposium. Matter and antimatter are supposed to be separated in the Universe at e.g. the supercluster level, but
high-energy particles might be able to find their way into
neighbouring regions and leak into galaxies, thus giving
rise to an antimatter component of cosmic rays. The shape
of the expected $\bar{p}$ spectrum might be not very unlike curve
'c' in Fig. 1, if intergalactic propagation and galactic
modulation do not cause deformations. The stringent observational upper limit on the $\bar{\alpha}/\alpha$ ratio is a difficulty of
the model. It is in principle possible that the majority of
$\bar{\alpha}$ particles are destroyed in the active antigalaxies
contributing most to the local $\bar{p}$ flux, but the expected
gamma-radiation from those galaxies would then be above the
observed 'isotropic' gamma-background /Király et al. 1981/.
It is also highly uncertain, whether 1 GeV particles can
propagate in intergalactic fields fast enough to reach
other superclusters in an expanding Universe. Winds flowing
out of at least some galaxies and clusters also tend to
keep low-energy particles and antiparticles apart. The scale
of separation of matter and antimatter regions is of course
also hypothetical, and the scale corresponding to the distance of neighbouring superclusters is about the smallest
compatible with the gamma-ray background. In spite of these
shortcomings, the model can not be completely rejected. A
much more stringent upper limit for $\bar{\alpha}/\alpha$ /say by a factor
of 10/ would, however, probably eliminate this interpretation.

The high-energy p-$\bar{p}$ component of intergalactic gas
might be baryon-symmetric even if dominant low-energy matter
is not. Some unspecified exotic source population, which is
not concentrated in galaxies, might directly emit protons

and antiprotons with a power-spectrum, or else a power-spectrum might arise by subsequent intergalactic acceleration. Such a component could simulate antigalaxy-produced antiprotons, without necessarily containing antinuclei.

Turning now to the Galaxy, the Galactic centre explosion model of Khazan and Ptuskin /1977/ can be mentioned as a possible scenario for $\bar{p}$ production. The astrophysical background of that model is, however, too vague to allow any quantitative prediction.

An interesting possibility for $\bar{p}$ production in the Galaxy was pointed out by Sawada and co-workers /1980/. The idea was based on the recent suggestion by Mohapatra and Marshak /1980/ that baryon-number non-conserving neutron-antineutron oscillations can be expected in certain unification models of particle interactions. Limits on the oscillation time can be inferred either indirectly, from the lower limit of the decay time of normal matter /which should be much longer than the oscillation time because of the difference in the energy levels of n and $\bar{n}$ in a nucleus/, or from direct reactor observation of free n transforming into $\bar{n}$. Up to now, only the first method has been used /with inferred lower limits on the oscillation time ranging from 10^5 to 10^8 seconds, see Marshak, 1981/, but several experiments are in progress to use the second method as well. If such oscillations really take place, cosmic neutrons created in nuclear fragmentation reactions could turn into $\bar{n}$ and then decay into $\bar{p}$. Using the above indirect estimates for the oscillation time, too few $\bar{p}$ would be expected, but the estimates are rather uncertain. Another difficulty is the presence of the interstellar magnetic fields, which should split the energy levels of both n and $\bar{n}$ and would suppress n - $\bar{n}$ oscillations to some extent. This suppression effect would be more important at higher energies, where the magnetic field felt by the n is increasing with the Lorentz-factor. Arafune and Fukugita /1981/ estimate that Buffington's data could be explained if the oscillation time was as low as 10^4 seconds /see also Marshak 1981/. Although we have not seen their detailed calculations, it appears from the phenomenology described by Mohapatra and Marshak /1980/ that the $\bar{p}$ flux would be too small even with such a low oscillation time. Also, a sharp decrease of the $\bar{p}$/p ratio with increasing energy is expected, which is contrary to observations. Thus probably a closed Galaxy model should be invoked as well to explain the points at higher energies /as arising from cosmic ray - gas collisions/ and to supply a sufficient number of fragmentation neutrons. The direct n - $\bar{n}$ oscillation experiments are expected to decide in the near future whether an important contribution to the observed $\bar{p}$ flux is feasible.

The evaporation of primordial black holes /Hawking 1974, Page and Hawking 1976, Carr 1976/ could also contribute to the observed $\bar{p}$ flux. Black holes on a variety of mass scales may have been created in the early Universe with a number spectrum depending only slightly on the equation of state /Carr 1976/. They are expected to be concentrated in galactic halos, and in first approximation to radiate like black bodies of temperature $10^{26}m^{-1}$ K, where m is the black hole mass in grams. The power of black hole radiation increases with decreasing mass as m^{-2}, thus less massive holes evaporate earlier. Primordial black holes with original masses below about 5×10^{14} grams have already evaporated, but that section of the spectrum is continually repopulated as the mass of more massive holes decreases. Including only those particles which can be directly emitted by black holes as a thermal radiation, the time-integrated energy distribution above a limit depending on particle mass is proportional to E^{-3}, where E is total particle energy. By using production rates for various particles as given in Carr /1976/ we have examined a scenario in which $\bar{p}$ emission by black holes with masses below about 5×10^{13} grams account for the observed flux /Király et al. 1981/. The spectrum to be expected was found to be very near curve 'c' in Fig.1. Of course, the number of black holes in the Galaxy had to be inferred from the normalisation of that curve.

One should remark, however, that the $\bar{p}$ flux from a black hole might be much lower than one would expect from that simplified picture. The radii of the black holes which could contribute to the $\bar{p}$ flux are very small, comparable to the size of nucleons. If in such circumstances quarks and gluons are directly emitted from the black holes, and not entire nucleons, then the resulting jets fragment preferentially into mesons, particularly at low invariant masses /see e.g. Duinker 1981, describing $e^{+}e^{-}$ annihilation results/. If black hole evaporation follows that pattern, too many mesons /and their decay products/ would be produced and the model could be excluded.

6. Conclusions

There is no completely satisfactory explanation of the recently observed high cosmic $\bar{p}$ flux values. Further checks and new, independent observations are needed in order to reach an understanding of this very interesting new area of cosmic ray research.

Acknowledgements

The author is grateful for discussions with Prof. A.
W. Wolfendale, Prof. S.W. Hawking, Prof. J. Wdowczyk, Dr.
G.W. Gibbons, Mr. J. Szabelski, Dr. B.J. Carr and Dr. G.
Erdős.

References

Aizu,H., Fujimoto,V., Hasegawa,S., Koshiba,M., Mito,I.,
    Nishimura,J., Yokooi,K., Schein,M. /1961/ Phys. Rev.
    121, pp. 1206-18
Arafune,J. and Fukugita,M. /1981/, Submitted to the Proc.
    of the 17th Int. Cosmic Ray Conf., Paris
Bogomolov, E.A., Lubyanaya,N.D., Romanov,V.A., Stepanov,S.
    V., Shulakova,M.S. /1979/ Proc. 16th Int. Cosmic Ray
    Conf. Kyoto 1, pp. 330-5
Brooke,G. and Wolfendale,A.W. /1964/ Nature 202, pp. 480-1
Buffington,A. and Schindler,S.M. /1981a/ Ap.J. 247, pp.
    L105-9
Buffington,A., Schindler,S.M., Pennypacker,C.R. /1981b/
    Ap.J., to be published
Buffington,A. and Schindler,S.M. /1981c/ Proc. 17th Int.
    Cosmic Ray Conf. Paris 2, pp. 98-101
Carr,B.J. /1976/ Ap.J. 206, pp. 8-25.
Cowsik,R. and Gaisser,T.K. /1981/ Proc. 17th Int. Cosmic
    Ray Conf. Paris 2, pp. 218-21
Duinker,P. /1981/ DESY report 81-012
Giler,M., Wdowczyk,J., Wolfendale,A.W. /1977/ J.Phys. A
    10, 843-59
Golden,R.L., Horan,S., Mauger,B.G., Badhwar,G.D., Lacy,J.L.,
    Stephens,S.A., Daniel,R.R., Zipse,J.E. /1979/ Phys.
    Rev.Lett. 43, pp. 1196-9
Hawking,S.W. /1974/ Nature 248, pp. 30-1
Khazan,Y.M. and Ptuskin,V.S. /1977/ Proc. 15th Int. Cosmic
    Ray Conf. Plovdiv 2, pp. 4-8
Király,P., Szabelski,J., Wdowczyk,J., Wolfendale,A.W.
    /1981/ Nature 293, pp. 120-2
Marshak,R.E. /1981/, VPI-HEP 81/3 report
Mohapatra,R.N. and Marshak,R.E. /1980/ Phys.Lett. 94B,
    pp. 183-6
Page,D.F. and Hawking,S.W. /1976/ Ap.J. 206, pp. 1-7
Protheroe,R.J. /1981/ Proc. 17th Int. Cosmic Ray Conf.,
    Paris 2, pp. 198-201
Rasmussen,I.L. and Peters,B. /1975/ Nature 258, pp. 412-3
Sawada,O., Fukugita,M., Arafune,J. /1980/ KEK-TH 19 report
Stecker,F.W. /1981/ NASA Tech. Mem. 82083
Steigman,G. /1976/ Ann.Rev.Astron.Astrophys. 14, pp. 339-72
Stephens,S.A. /1981/ Proc. 17th Int. Cosmic Ray Conf. Paris
    2, pp. 214-7

Szabelski,J., Wdowczyk,J., Wolfendale,A.W. /1980/ Nature
      285, pp. 386-8
Szabelski,J., Wdowczyk,J., Wolfendale,A.W. /1981/ Proc.
      17th Int. Cosmic Ray Conference, Paris 2, pp. 206-9

POPULATION III AND THE MICROWAVE BACKGROUND

M.Rowan-Robinson and P.Tarbet

Dept of Applied Mathematics
Queen Mary College
Mile End Rd, London E.1.

Abstract

A natural consequence of the isothermal density fluctuation
picture is the possibility of a pregalactic generation of
stars, Population III, where "stars" includes objects of up
to 10^5 or 10^6 solar masses. The most dramatic role proposed
for Population III is the explanation of the whole microwave background, provided the problem of thermalisation can
be overcome. A less revolutionary consequence of pregalactic
star-formation and nucleosynthesis would be the distortion
of the spectrum of the microwave background at wavelengths
shortward of 2 mm, in a manner which turns out to be surprisingly similar to that observed by the Berkeley group.
The third and most conventional application of Population
III, and virtually indistinguishable from the Prompt Initial
Enrichment (PIE) scenario, is the production of the first
heavy elements in galaxies.

These three possibilities have been explored in a quantitative way, with plausible assumptions about the mass spectrum
of the stars, the helium and heavy element yields of stars
of different masses (in the range $10 - 10^6$ solar masses)
and the remnants left behind. Viable Population III scenarios can certainly be found, though to make any noticeable
contribution to the microwave background without violating
constraints on the total helium and heavy element yields,
Population III light output must be dominated by accreting
black holes.

I. INTRODUCTION

Soon after the hot Big Bang picture had been correctly
formulated and had been confirmed by the discovery of the
microwave background radiation, Doroshkevich et al(1967)

A. W. Wolfendale (ed.), Progress in Cosmology, 101–117.

and Peebles and Dicke(1968) recognised the possibility that
a pregalactic population of objects might form soon after
the decoupling of radiation and matter. As the Jeans mass
at decoupling is about 10^6 solar masses, Peebles and Dicke
suggested that globular clusters might be this population.
The work of Eggen et al(1962) had shown however that the
formation of old Population II stars in the Galaxy probably
took place during the collapse phase of the protogalaxy,
an event now associated with an epoch corresponding to
redshift ~ 10 (with an uncertainty of a factor of 3 or so).
Doroshkevich et al, on the other hand, supposed that the
primary condensations of 10^6 $M_\odot$ formed unstable protostars
which exploded and heated up the remaining gas. Neither of
these proposals attracted much attention, perhaps because
Silk(1967) showed that adiabatic density fluctuations on a
scale smaller than $\sim 10^{12}$ $M_\odot$ would suffer damping during
the radiation dominated era.

Even so the possibility of an Atlantean pregalactic population of objects remains a fascinating one and it is surprising that this idea was neglected for so long (but see
Rees 1978 for references to work on Population III in the
intervening period). Within the framework of the isothermal
density fluctuation picture it is natural, perhaps inevitable, that fluctuations should exist on mass scales down
to the Jeans mass at decoupling. The work of Peebles and
others (see Peebles 1980) on the covariance function shows
that the isothermal picture provides a natural explanation
of the observed clustering of galaxies and that the amplitude of the fluctuations increases as mass decreases. Indeed,
extrapolation of the relationship obtained from the covariance function suggests that the density fluctuations on
the scale of $\sim 10^6$ $M_\odot$ would have an amplitude of order
unity in $\Delta \rho / \rho$ and would therefore collapse rapidly after
the decoupling epoch (in about 10^6 years).

The main objection to the isothermal density fluctuation
picture is that recent GUT scenarios to unify particle
physics and cosmology lead naturally to adiabatic fluctuations (but see, eg, Barrow and Turner 1981). Although this
is a pity, we might be entitled to take the view that the
theory and observation of galaxy clustering is on a more
solid basis than GUTs are at the moment. We shall therefore
assume that the isothermal density fluctuation picture is
still alive and that some kind of pregalactic population
of objects, Population III, is a serious possibility. What
are the observable consequences of this ?

If objects in the range $10^6 - 10^7$ $M_\odot$ collapsed and fragmented into "stars" at epoch z_f , where $1000 > z_f \gg 1$,
where "star" is understood to mean self-gravitating, self-

luminous objects with masses $< 10^7$ $M_\odot$, there are three main types of consequence. Firstly we may expect light to be generated and this may contribute to the background radiation today. The most interesting possibility here is that part (conceivably all) of the microwave background may have been generated by Population III. Secondly, nucleosynthesis in these objects could contribute part (conceivably all) of the helium and heavy elements found in the oldest stars in the Galaxy. One of the most plausible ways of explaining the deficiency of stars in the Galaxy with low metal abundance ($Z < 10^{-4}$) is through Prompt Initial Enrichment (PIE, cf Audouze and Tinsley 1976). This involves an early generation of stars of mass $\gg 1$ $M_\odot$, normally assumed to have formed during the collapse phase of the Galaxy. However it is equally plausible that this population could be pregalactic (Truran and Cameron 1974, Hartquist and Cameron 1977). Indeed, it is hard to draw the distinction between these alternatives, since in the standard hierarchical clustering picture Population III objects would already be situated in density perturbations of larger scale but lower amplitude (i.e. the protogalaxies). Finally, the dark remnants of Population III will contribute to the baryonic density of the universe and may help to solve the "missing light" problem.

It was this latter aspect that White and Rees(1978) concentrated on when they revived the idea of Population III. They proposed that 90% of the matter in the universe formed Population III objects which evolved rapidly to dark remnants and clustered together. The remaining 10% of the matter then condensed in the potential wells defined by the dark matter, to give galaxies with massive dark haloes. Rees (1978) took this scenario further by proposing that the radiation from these Population III objects could provide the whole microwave background, with the thermalisation being provided by dust, molecules or free-free absorption. Rowan-Robinson et al (1979) showed that for dust to provide adaquate thermalisation at longer wavelengths the abundance of heavy elements in the form of dust would have to be $\gtrsim 10^{-3}$ relative to hydrogen (by mass). A rather contrived explanation would then be required to explain the stars of lower abundance seen in our Galaxy.

2. DISTORTION OF THE MICROWAVE BACKGROUND SPECTRUM DUE TO POPULATION III RADIATION ABSORBED BY DUST

Instead, Rowan-Robinson et al(1979) suggested that the distortion observed by Woody and Richards(1979) could be due to radiation from Population III, absorbed and re-emitted by silicate dust condensed from material ejected by the stars

at a redshift of about 200. Although Woody and Richards
(1981) have revised the significance of their claimed distortion from a blackbody spectrum from 5 to 2.7 σ , there
still appears to be a 1 - 2 mm excess to explain, particularly when compared with the most accurate of the longer
wavelength observations. Negroponte et al(1981) have presented more detailed and accurate calculations of the Rowan-
Robinson et al model. Figure 1 shows the grain properties
investigated by Negroponte et al (1981). Figure 2 shows

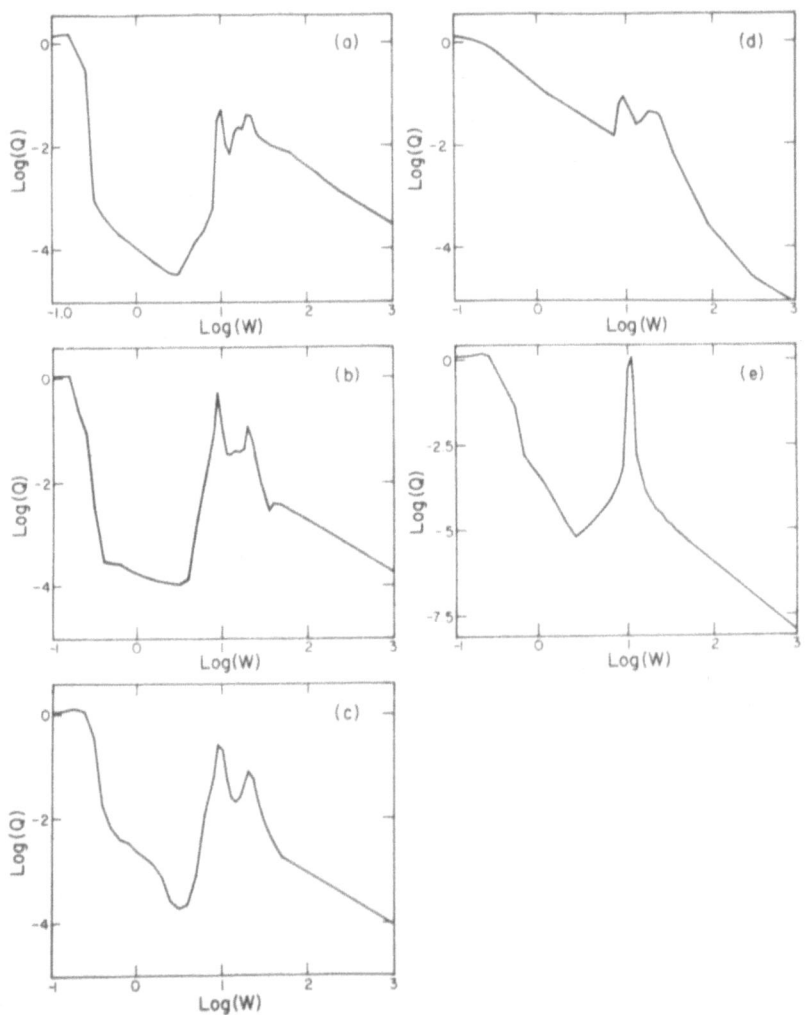

Figure 1. Grain absorption efficiency Q as a function of wavelength in microns W : (a) amorphous silicates, (b) obsidian, (c) basaltic glass, (d) dirty
silicates, (e) silicon carbide.

the grain temperature history for the models whose parameters are given in Table I and Figure 3 shows the corresponding background spectrum, shown in terms of the thermodynamic temperature, T_{th} , where $B_{\nu}(T_{th}) = I_{\nu}$, the intensity
of the background. Several of these models are consistent
with the Woody and Richards data. In these models the
grains are assumed to be formed after the stars have ceased
to radiate, when the energy density has dropped sufficiently
for the grains to condense. If the stars continue to radiate

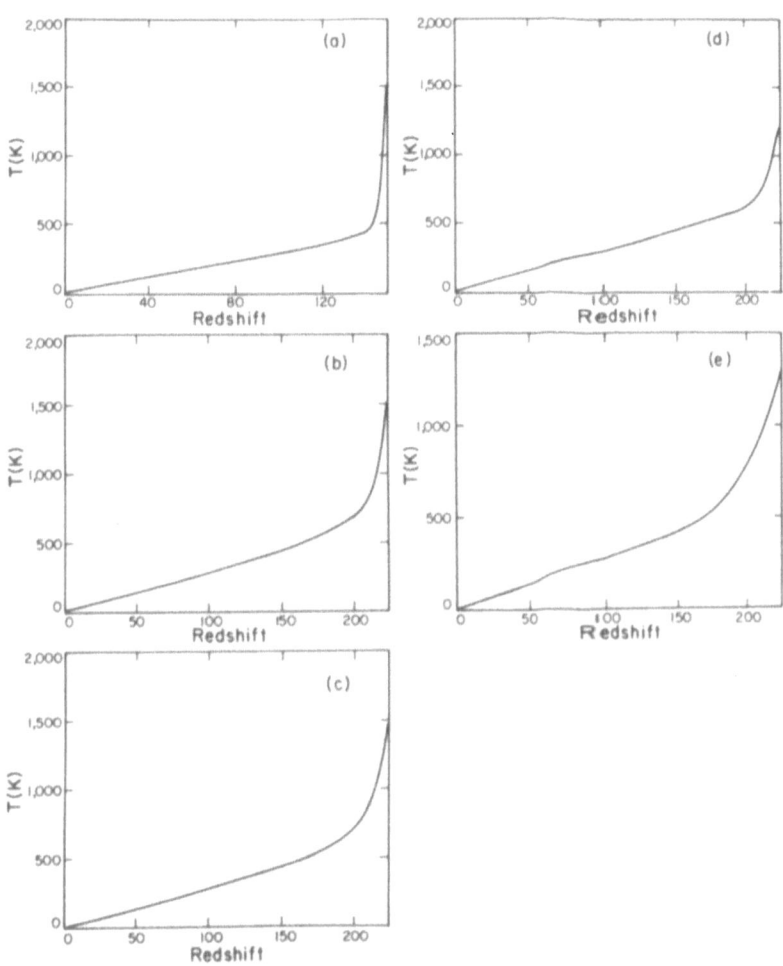

Figure 2. Grain temperature as a function of
redshift for the models (a)-(e) of Table I.

Table I.   Parameters of Negroponte et al models.

grain type	$1 + z_f$	Z	β
(a) amorphous silicate	150	7.5×10^{-6}	0.19
(b) obsidian	225	5.0×10^{-6}	0.42
(c) basaltic glass	225	2.5×10^{-6}	0.26
(d) dirty silicate	225	2.5×10^{-6}	0.40
(e) silicon carbide	225	5.0×10^{-7}	0.24

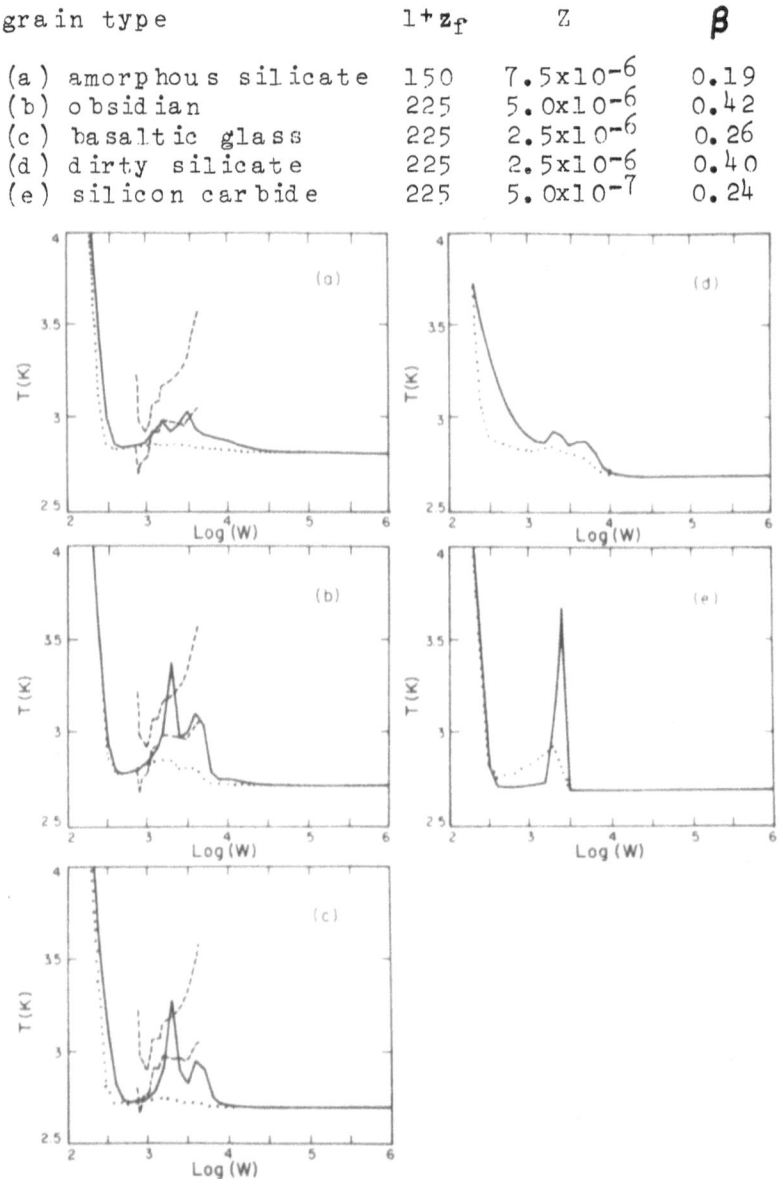

Figure 3.   Present background spectrum: thermodynamic temperature as a function of the wavelength in microns, W, for the models of Table I.   Dashed lines are $\pm 1\sigma$ points from the data of Woody and Richards($\overline{1}$979).   The dotted lines give the spectrum if the dust temperature is extrapolated linearly from the present to z_f .

for a time t_s after the grains have formed, the predicted
distortion tends to be smeared out, as shown in Figure 4
for $t_s = 10^6$ and 10^7 years. For $t_s \gtrsim 10^8$ years the distortion is smeared out too much to resemble the Woody and Richards data. Negroponte et al.(1981) also investigate the
variant of Puget and Heyaerts(1980), in which the formation
and heating of the grains takes place at $z_f \sim 10 - 20$.
This does not give such a satisfactory fit to the Woody and
Richards spectrum and the abundance in heavy elements
required exceeds the lowest values seen in our Galaxy by an
order of magnitude.

Carr(1981) has taken the Rowan-Robinson et al(1979) model
a stage further by suggesting that the 2.7 K blackbody radiation assumed by them to be primordial might actually be
radiation from earlier Population III objects (specifically,
accreting black holes) thermalised by free-free absorption.
While the details of this thermalisation remain to be demonstrated, we shall assume for the moment that this part of
Carr's hypothesis can be verified.

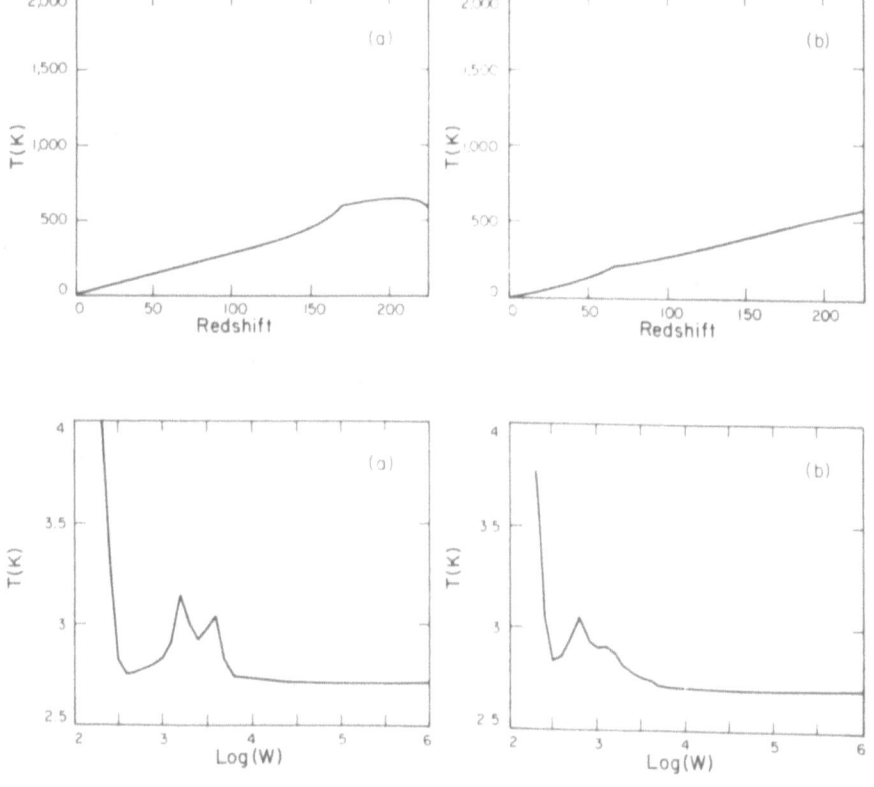

Figure 4. Dust temperature distribution (upper panels) and background spectrum (lower panels) for
obsidian models with $t_s = 10^6$ years(a), 10^7 years(b).

3. THE HELIUM AND HEAVY ELEMENT YIELDS OF POPULATION III

To investigate the viability of the different Population
III scenarios mentioned above, it is necessary to consider
the yields of helium and heavy elements from Population III.
Tarbet and Rowan-Robinson(1982) have investigated these in
detail for 3 possible roles for Population III:
        (a) the generation of the Woody and Richards dist-
        ortion,
        (b) the production of the entire microwave back-
        ground and of the observed "primordial"helium
        abundance,
        (c) the production of the first metals in galaxies.

The mean mass-density of the universe at the present epoch
Ω_o, can be considered to have 3 components: Ω_g, the contribution of visible matter; Ω_r, that due to dark matter,
eg black holes and neutron stars, which we will assume are
predominantly the remnants of Population III; and Ω_ν , due
to non-baryonic matter, eg neutrinos with mass.

$$ \text{Thus} \quad \Omega_o = \Omega_g + \Omega_r + \Omega_\nu = \Omega_b + \Omega_\nu \quad . $$

where Ω_b is the total baryonic contribution.

Assume that at some epoch in the early universe a population of stars was born and subsequently ceased to produce
radiation at a later epoch z_f (in the case of black holes
this would be the time when they stop accreting significantly). If Ω_{ej} is the contribution to the present visible
matter that was cycled in pregalactic stars and then eject-
ed, then $\Omega_{III} = \Omega_r' + \Omega_{ej}$ would be the density contained in these objects today if they had not ejected any of
their mass, where Ω_r' is the contribution of remnants
prior to any accretion.

$$ \text{Then} \quad \Omega_{III} = \frac{8\pi G}{3 H_o^2} \quad (1+z_f)^{-3} \; A \int_{M_1}^{M_2} M \; N(M) \; dM, \quad (1) $$

where A N(M) dM is the number-density of stars with masses
in the range M to M+dM at epoch z_f , and we shall take
$N(M) = M^{-\alpha}$.

If the cosmic (pre-Population III) helium abundance is Y_p
and Y(M), Z(M), are the fraction of a star's mass ejected
in the form of helium and heavy elements, respectively, then
the fraction of visible matter in the form of helium and
heavy elements at z_f is:

$$Y = (1 - \Omega_{ej}/\Omega_g) \, Y_p + \frac{8 \pi G}{3 \, H_o^2 \Omega_g} (1+z_f)^{-3} \, A \, .$$

$$\cdot \int_{M_1}^{M_2} Y(M) \, M \, N(M) \, dM \, , \qquad (2)$$

$$Z = \frac{8 \pi G}{3 \, H_o^2 \Omega_g} (1+z_f)^{-3} \, A \int_{M_1}^{M_2} Z(M) \, M \, N(M) \, dM. \qquad (3)$$

Similarly if R(M) is the fraction of a star's mass left as a dark remnant, then

$$\Omega_r = \frac{8 \pi G}{3 \, H_o^2} (1+z_f)^{-3} \, A \, (\int_{M_1}^{M_*} R(M) \, N(M) \, M \, dM$$

$$+ \, \mu_B \int_{M_*}^{M_2} R(M) \, N(M) \, M \, dM \,) , \qquad (4)$$

where it is assumed that the remnant is a black hole for $M > M_*$ and μ_B is the ratio of the mass accreted by a black hole to the mass of the initial black hole.

The constant A can be related to β, the factor by which the energy density of the cosmic background is enhanced due to Population III:

$$\beta \, \alpha \, T_o^4 (1+z_f)^4 = A \int_{M_1}^{M_2} \epsilon(M) \, M \, N(M) \, dM, \qquad (5)$$

where $\epsilon(M)$ is the fraction of a star's mass converted into radiation and $T_o(1+z_f)$ is the temperature of the primordial cosmic background at epoch z_f. If the efficiency of producing radiation in the conversion of H to He is ϵ_Y (~ 0.007) and of He to heavy elements is ϵ_Z (~ 0.003) then

$$\epsilon(M) = \epsilon_Y (Y(M)+Z(M)-Y_p) + \epsilon_Z \, Z(M), \text{ for } M \leq M_*,$$

$$= \epsilon_B \, \mu_B \, R(M), \text{ for } M > M_*, \qquad (6)$$

where ϵ_B is the efficiency with which material accreted by a black hole is converted to radiation. Tarbet and Rowan-Robinson(1982) take $\epsilon_B = 0.1$ and consider $\mu_B = 1$, 5 (cf Carr 1981).

Schramm and Steigman(1981) have given a firm lower limit

for Ω_g of $(0.9-3.5)\times10^{-3}$ h_o^{-1}, where $h_o = H_o/100$. We have
taken 0.2 as the upper limit for Ω_b for cases (a) and (c)
since above this value, primordial helium production becomes
excessive.

4. THE FUNCTIONS N(M), Y(M), Z(M), R(M)

(i) N(M)
We have assumed N(M) to have the form $M^{-\alpha}$. For the solar
neighbourhood the Salpeter value $\alpha = 2.35$ applies and
$M_1 = 0.01 - 0.1$ $M_\odot$, $M_2 = 60 - 100$ $M_\odot$. In a medium with
zero metal abundance, cooling will be less efficient and
fragmentation will be inhibited (Silk 1977). We may expect
that M_1 will be increased (Silk argues that $M_1 \sim 20$ M_O)
and α may be reduced. For $\alpha < 2$, M_1 is not a very important
parameter and we have adopted $M_1 = 12$ $M_\odot$. The cocoon star
mechanism that limits M_2 in the solar neighbourhood (Kahn
1974) also does not apply in the metal-free primordial
environment, and we assume that M_2 can extend up to the
scale of the Jeans mass ($10^6 - 10^7$ $M_\odot$).

(ii) Y(M), Z(M), R(M)
In the mass range $12 \lesssim M/M_O \lesssim 100$, stars are believed to
be able to burn all of their nuclear fuels through to iron
without mishap (Weaver et al 1978). The core then undergoes
hydrodynamic collapse but the implosion is reversed when
nuclear densities are reached and a large amplitude bounce
reflects the incoming kinetic energy, blowing off all layers
external to the core in a supernova explosion (Arnett 1978a,
van Riper 1979). The exact details of core bounce are not
yet known and the complex reaction networks of static and
dynamic burning make yield determinations uncertain. Here
the helium yields of Arnett (1978b), scaled for an initial
helium content Y_p, are used, the formula of Weaver and
Woosley (1978) is used for the metals, and the neutron star
remnants are taken to be 1.4 $M_\odot$.

Massive stars ($M > 60$ $M_\odot$) are prone to mass loss because
they are mainly supported by radiation pressure ($\Gamma \sim 4/3$)
and because they are prone to nuclear energised pulsations
(Schwarzschild and Harm 1959, Stothers and Simon 1970,
Stothers 1972). The fact that mass loss may be significant
coupled with these stars being almost completely convective
makes them sites of high helium and low metal production.
Studies by, for example, Appenzeller (1970a,b), Talbot
(1971a,b) and Papaloizou (1973), imply that the nuclear
driven pulsations do not completely disrupt the stars, but
for those of initial mass > 300 $M_\odot$, a substantial fraction
of the star's mass will be lost on a time-scale much shorter
than the hydrogen-burning phase. In the range 100-200 $M_\odot$

the stars lose mass at about 10^{-5} $M_{\odot}$ yr^{-1}, i.e. they survive their main-sequence lifetimes without copious loss, and
those below 100 $M_{\odot}$ suffer little from this instability.
However other mechanisms, eg radiatively driven winds, may
be important and the observations of Humphreys and Davidson
(1979) seem to point to the possibility that mass loss severely restricts the evolution of stars with initial mass $>$
60 $M_{\odot}$ to cooler parts of the HR diagram. Rowan-Robinson
(1982) has suggested that OH-IR II sources with deep silicate
absorption may be the remnants of massive stars which have
suffered substantial mass-loss.

Talbot and Arnett(1971) looked at helium production from
massive stars with zero initial helium and suffering mass-
loss due to nuclear pulsations and concluded that $>$ 17% of
the star could be shed in the form of helium, with a net
yield of about 0.5 for stars $>$ 300 $M_{\odot}$.

Stars with initial mass greater than about 100 $M_{\odot}$ do not
burn fuels all the way through to iron but encounter the
electron-positron pair instability and undergo core collapse (Barkat et al 1967). The question is, can subsequent
fast nuclear burning reverse this collapse or is the ultimate fate of the star a black hole ? Woosley and Weaver
(1982) and Arnett et al (1982) calculate that stars of
initial mass $>$ 300 $M_{\odot}$, evolving without substantial mass-loss,
will be unable to reverse this collapse, whilst those in the
range $\sim$ 100 - 300 $M_{\odot}$ suffer complete disruption.

Bearing in mind the uncertainties in mass-loss and in the
upper mass limit for complete disruption, the following
simple approach was taken. Stars of mass $>$ 300 $M_{\odot}$ lose half
their mass due to winds and so, scaling for a non-zero initial helium content, shed 0.275 of their mass as helium. The
remnant then contributes to the yield after disruption to
give a total of 0.575 as helium and 0.2 as metals. Stars
in the range 500 $M_{\odot}$ up to 10^{6} $M_{\odot}$ return the wind component
(0.275 helium yield) but the remnant collapses to a black
hole. For stars in the range 50 - 300 $M_{\odot}$ we simply interpol-
ated $Y(M)$ and $Z(M)$ linearly between the values adopted at
50 and 300 $M_{\odot}$. This is consistent with assuming the mass
lost through winds ranges from zero at 50 $M_{\odot}$ to 50% at 300
$M_{\odot}$.

Objects exceeding $\sim$ 10^{5} $M_{\odot}$ encounter a general relativistic
instability during quasi-static collapse and if they contain
no elements heavier than helium, nuclear burning is not
switched on at a significant rate before catastrophic collapse (Fricke 1973). Thus virtually the whole star collapses
to a black hole on a short time-scale ($\sim 2.4 \times 10^{8}$ $\mu^{-2}(M/M_{0})^{-1}$
years, where   is the mean molecular weight).

The adopted forms for Y(M), Z(M), R(M), are summarised in Figure 5.

5. DISTORTION OF THE MICROWAVE BACKGROUND SPECTRUM

Negroponte et al (1981) find that to produce the distortion observed by Woody and Richards(1979), it is necessary to take z_f in the range 150-225, β in the range 0.2-0.4, and a metal abundance relative to the total mass of $2.5-7.5 \times 10^{-6}$ for $\Omega_0 = 1$ and $1-3 \times 10^{-5}$ for $\Omega_0 = 0.1$ ($H_0 = 100$), depending on the grain material. To test this type of model we take $z_f = 200$, $\beta = 0.25$ and look for values of the parameters Ω_g, H_0, M_2, α, M_B, which can give a heavy element yield in the right range. Allowing for the uncertainties in the grain material and the observed distortion, we take this to be $\Omega_g Z = 5 - 50 \times 10^{-7}$. We have taken $H_0 = 50$ or 100, $\lg_{10} M_2^g = 2(1)7$, $\alpha = 0.5(0.5)2.5$, $M_B = 1.5$, and in each case varied Ω_g between 0.001 and 0.2. The corresponding value of Ω_r is then determined by equations (4)-(6). The

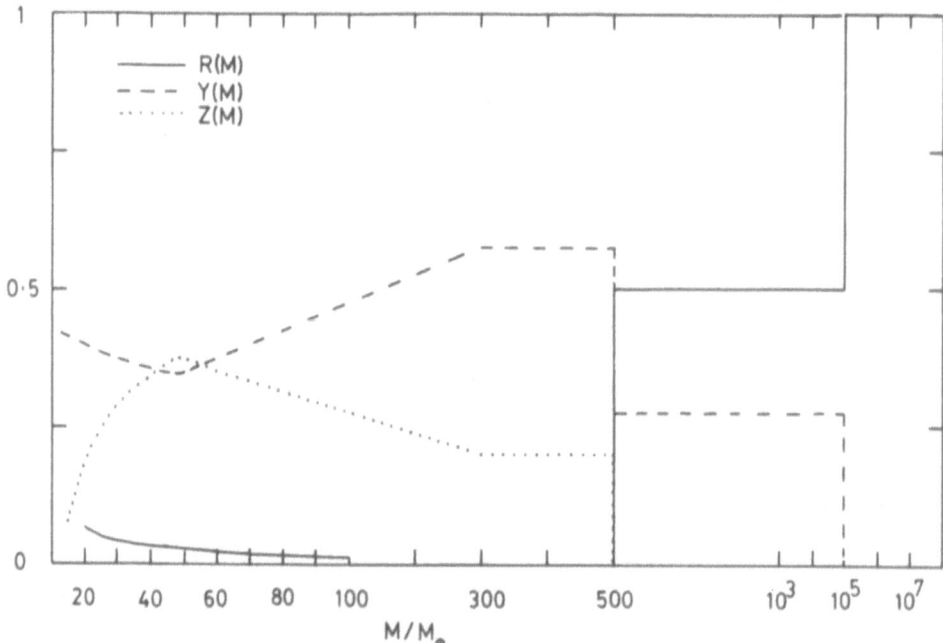

Figure 5. The fraction of a star's mass ejected as helium (broken curve) and heavy elements(dotted curve), and left as a remnant (solid curve), as a function of the star's mass. Note that the mass scale changes at 100, 500 and 1000 $M_\odot$.
        (from Tarbet and Rowan-Robinson 1982)

characteristic run of Y (and Y_p) against Z is shown in Figure 6 for the model with $M_2 = 10^6 M_\odot$, $\alpha = 1$, $\mu_B = 5$, $H_0 = 100$. At low values of Ω_g the metal and helium yields from Population III are quite high: at higher values of Ω_g, the higher value of Ω_b leads to a higher primordial helium yield Y_p but Population III yields only small amounts of helium and metals. The primordial helium yield has been taken from the work of Yang et al (1979), assuming $T_0 = 2.7$ K, in turn based on the code of Wagoner (1973). The model illustrated in Figure 6 gives slightly too low a metal yield. A better value is obtained by changing μ_B to 1, in which case the minimum total helium yield is $Y = 0.268$ at $Z = 3.65 \times 10^{-5}$, $\Omega_g = 0.042$, $\Omega_b = 0.066$. This is slightly higher than recent observed values: $Y = 0.216 \pm 0.015$ (French 1980), $Y = 0.228 \pm 0.014$ (Lequeux et al 1979). This is a problem shared by the standard big bang model unless $\Omega_b \leq 0.03$ (Olive et al 1981). However these low values are disputed by Kunth (1981), who derives $Y = 0.243 \pm 0.010$. As discussed by Olive et al (1981), Y_p could be reduced by taking a lower value for the neutron half-life. Bondarenko et al (1978) have determined a value of 10.13 ± 0.09 min., compared with the value of 10.7 used by Wagoner (1973) and the value 10.56 ± 0.09 recommended in the review by Wilkinson (1980). This could reduce Y_p by about 0.01. Alternatively if the τ-neutrino has a mass $\gg 1$ Mev/c^2, then the 2-neutrino case rather than the 3-neutrino case should be used in the helium nucleosynthesis calculation. The result is illustrated by the dotted curve in Figure 6. However this case leads to implausibly high values of Ω_g unless the mass of the τ-neutrino is greater than 2 Gev/c^2 and this is inconsistent with the experimental upper limit of 250 Mev/c^2 (see the review by Cowsik 1981). The corresponding deuterium yield is 2×10^{-5}, almost identical to the primordial value and consistent with observations.

We now consider the effect of varying the other parameters of our model. Changing H_0 to 50 leads to almost identical Y-Z curves to those shown in Figure 6, but the value of Ω_g Z is increased by a factor of 4 (the required value is increased by a factor of 2). With $M_2 = 10^5 M_\odot$ we need $\alpha = 0.5$, $\mu_B = 1$; with $M_2 = 10^7 M_\odot$ we need $\alpha = 1.5$, $\mu_B = 5$. No other acceptable sets of parameters were found. To summarise, for these distortion models to give acceptable values of Y and Z it is essential that the mass-function be flat and that it extend to very massive objects, so that most of the light is generated by accreting black holes.

6. GENERATION OF THE WHOLE MICROWAVE BACKGROUND

In the scenario of Carr (1981), we now need to increase
to correspond to the whole background, i.e. β = 1. We take
T_o = 2.9 K so as to give a total energy density equal to
that observed by Woody and Richards (1979,1981). The helium
yield Y now has to be generated entirely by Population III
stars. If we are still trying to fit the Woody and Richards
distortion with the Rowan-Robinson et al (1979) model then
we require metal yields as in section 5, and we take z_f = 200.

Varying the parameters of the model as before, the highest
helium yield consistent with the limits on Z is found for
M_2 = 10^5 $M_\odot$, α = 0.5, μ_B = 5, H_o = 100, in which case Y=0.224,
Z = 9.3x10^{-5} (corresponding to Ω_g = 0.023, Ω_r = 0.076).
The Carr scenario can therefore simultaneously account for
the whole background, explain the Woody-Richards distortion
with the Rowan-Robinson et al mechanism and give the correct
helium yield and extreme Population II metals. However in
this model deuterium has to be produced by some ad hoc astrophysical mechanism, for example that of Epstein (1977). It
would perhaps be worthwhile to investigate the predicted

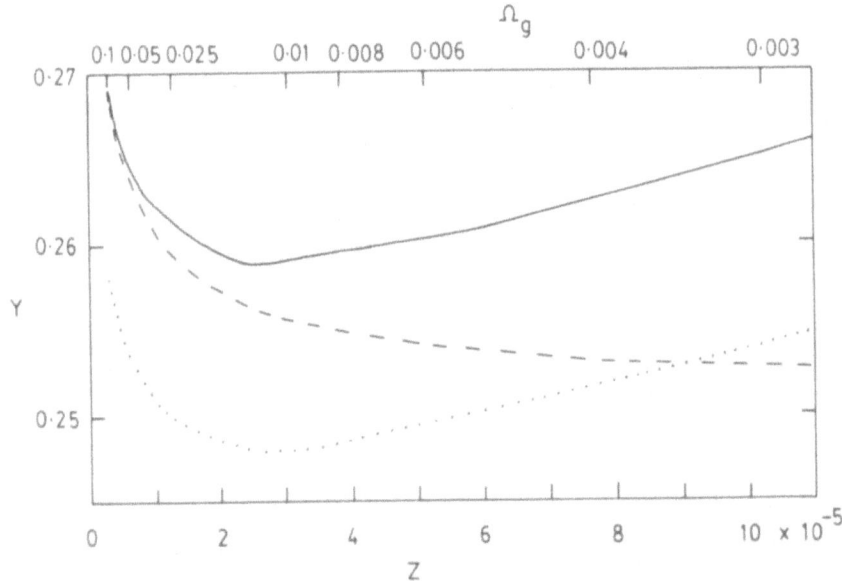

Figure 6. The total helium abundance, Y (solid
curve), and the primordial helium abundance, Y_p
(broken curve), plotted as a function of the heavy
element abundance, Z (lower scale). The contribution to Ω_o due to visible matter, Ω_g, is labelled
on the upper scale. The dotted curve shows the
2-neutrino case. Model parameters are given in the
text.

background spectrum for this case in more detail, taking into account the staggered way that the radiation from different stellar masses is generated, both to demonstrate clearly the capability of free-free absorption to thermalise the radiation at $\lambda > 3$ mm and to check that the distortion predicted at shorter wavelengths is consistent with observations.

7. PROMPT INITIAL ENRICHMENT

The final role of Population III considered here is to enrich the protogalactic gas with heavy elements and thus explain the lack of low metal abundance stars in our Galaxy (Truran and Cameron 1974, Hartquist and Cameron 1977). Here we want to generate a metal abundance of about $Z = 10^{-4}$, without significantly adding to the primordial helium. The restriction on β is only that it should be much less than unity, so that no unwanted distortion is produced. We have already seen that models which give the Woody and Richards distortion or explain the whole microwave background are capable of giving the correct initial enrichment, but here we consider only $\beta \ll 1$. Since galaxies are believed to condense out at redshifts of about 10, z_f is reduced to this value (though in this calculation this affects only the relationship between β and Ω_r, Ω_g, not the yields). We have looked for models which generate values of Z in the range 5×10^{-5} to 5×10^{-4}, the upper end of the range corresponding to the mean value for the oldest halo stars (Bond 1981). Many sets of parameters give satisfactory results a few are illustrated in Table 2 for $H_0 = 100$. In each case the minimum value of Y giving Z in the required range is chosen. With $\beta = 10^{-4}$, corresponding to $\Omega_{III} \sim 10^{-5}$, and $\beta = 10^{-3}$, models with a mass-function slope $\alpha = 2.5$, similar to the Salpeter value, and with $M_2 = 100\ M_\odot$, give

Table II. Parameters for PIE models.

M_2	μ_B	Ω_r	Ω_g	Z	Y	β	α
100	-	5.9×10^{-7}	0.005	4.8×10^{-4}	0.236	10^{-4}	2.5
500	-	4.0	0.005	4.4	0.236	10^{-4}	2.5
1000	1	4.6	0.003	3.8	0.224	10^{-4}	2.5
100	-	5.9×10^{-6}	0.051	4.7	0.262	10^{-3}	2.5
500	-	4.0	0.046	4.8	0.261	10^{-3}	2.5
1000	1	4.6	0.026	4.4	0.256	10^{-3}	2.5
1000	5	3.1×10^{-5}	0.006	4.2	0.241	0.01	1
1000	1	4.9	0.027	4.4	0.257	0.01	1
1000	5	3.1×10^{-4}	0.052	4.8	0.263	0.1	1
10^4	1	5.2	0.017	3.9	0.258	0.1	1

satisfactory yields. This corresponds to star formation
similar to that in the solar neighbourhood except that the
minimum mass has been taken much higher for Population III
($M_1 = 12\ M_\odot$). As the value of β is increased, the required
values of the parameters are shifted: either M_2 has to be
increased, or M_B is increased, or α is reduced, or all three.
To generate a significant distortion of the microwave background, PIE models involve star formation very different
than occurs today.

To conclude, there are several viable and interesting
roles for Population III and the best hope of testing these
ideas observationally is improved measurements of the
spectrum of the microwave background near the peak. Let us
finish with a speculation: could the more massive of the
Population III remnants sink into galactic nuclei and
provide the powerhouse for quasars and active galaxies ?

REFERENCES
Appenzeller I., 1970a, Astron.Astrophys. 5,355
Appenzeller I., 1970b, Astron.Astrophys. 9,216
Arnett W.D., 1978a, Annals of NY Acad. of Sci. 336,366
Arnett W.D., 1978b, Astrophys.J. 219, 1008
Arnett W.D., Bond J.R. and Carr B.J.,1982, NATO summer
        school on Supernovae, Cambridge, England(in press)
Audouze J. and Tinsley B.M., 1976, Ann.Rev.Astron.Astrophys.
        14, 43
Barkat Z., Rakavy G. and Sack N., 1967, Phys.Rev.Lett.18,379
Barrow J.D. and Turner M.S., 1981, Nature 291, 469
Bond H.E., 1981, Astrophys.J. 248, 606
Bondarenko L.N.,Kurguzov V.V., Prokof'ev Yu.A., Rogov E.V.
        and Spivak P.E., 1978, JETP Lett. 28, 303
Carr B.J., 1981, Mon.Noy.R.astr.Soc. 195, 669
Cowsik R., 1981, in Cosmolgy and Particles, ed. J.Audouze
        et al, p.157, Edition Frontieres.
Eggen O.J., Lynden-Bell D. and Sandage A., 1962, Astrophys.
        J. 136, 748
French H.B., 1980, Astrophys.J. 240, 41
Epstein R.L., 1977, Astrophys.J. 212, 595
Fricke K.J., 1973, Astrophys.J. 183, 941
Hartquist T. and Cameron A., 1977, Astr. Sp.Sci. 48, 145
Humphreys R.M. and Davidson K., 1979, Astrophys.J. 232, 409
Kahn F.D., 1974, Astron.Astrophys. 37, 149
Kunth D., 1981, in Cosmology and Particles, ed. J.Audouze
        et al, p.241, Edition Frontieres
Lequeux J., Peimbert M., Rayo J.F., Serrano A., Torres-
        Peimbert S., 1979, Astron.Astrophys. 80,155
Negroponte J., Rowan-Robinson M. and Silk J., 1981, Astrophys.
        J. 248, 38
Olive K.A., Schramm D.N., Steigman G., Turner M.S. and
        Yang J., 1981, Astrophys.J. 246, 557

Papaloizou   J., 1973, Mon.Not.R.astr.Soc. 162, 169
Peebles P.J.E., 1980, Large-scale structure of the universe,
          Princeton Univ. Press
Peebles P.J.E. and Dicke R.H., 1968, Astrophys.J. 154, 891
Puget J.L. and Heyvaerts J., 1980, Astron.Astrophys.83,L10
Rees M.J., 1978, Nature 275, 35
Rowan-Robinson M., 1982, Mon.Not.R.astr.Soc.(in press)
Rowan-Robinson M., Negroponte J., and Silk J., 1979,
          Nature 281, 635
Schramm D.N. and Steigman G., 1981, Astrophys.J. 243,1
Schwarzschild M. and Harm R., 1959, Astrophys.J. 129, 637
Silk J., 1967, Nature 215, 1155
Silk J., 1977, Astrophys.J. 211, 638
Stothers R., 1972, in Stellar Evolution, ed. Hong-Yee Chiu
          and Muriel A., MIT Press
Stothers R. and Simon N.R., 1970, Astrophys.J. 160, 1019
Talbot R.J., 1971a, Astrophys.J. 163, 17
Talbot R.J., 1971b, Astrophys.J. 165, 121
Talbot R.J. and Arnett W.D., 1971, Nature 229, 150
Tarbet P. and Rowan-Robinson M., 1982, Nature(in press)
Truran J.W. and Cameron A.G.W., 1974, Astrophys.J. 190, 605
van Riper K.A., 1979, Astrophys.J. 232, 558
Wagoner R.V., 1973, Astrophys.J. 179, 343
Weaver T.A. and Woosley S.E., 1978, Annals of NY Acad. of Sci.
          336, 335
Weaver T.A., Zimmerman G.B. and Woosley S.E., 1978, Astrophys.
          J. 225, 1021
White S.D.M. and Rees M.J., 1978, Mon.Not.R.astr.Soc. 183,341
Wilkinson D.H., 1980, in Proceedings of the Erice Summer
          School on Nuclear Astrophysics
Woody D.P. and Richards P.L., 1979, Phys.Rev.Lett. 42, 925
Woody D.P. and Richards P.L., 1981, Astrophys.J. 248, 18
Woosley S.E. and Weaver T.A., 1982, NATO summer school on
          Supernovae, Cambridge, England (in press)
Yang J., Schramm D.N., Steigman G. and Rood R.T., 1979,
          Astrophys.J. 227, 697
Doroshkevich A.G., Zeldovich Ya.B. and Novikov I.D., 1967,
          Sov.Astron. - AJ 11, 233

OBSERVATIONAL ASPECTS OF THE MICROWAVE COSMIC BACKGROUND SPECTRUM

D.H. Martin
Physics Department, Queen Mary College, Mile End Road,
London E1 4NS, U.K.

The discovery of the isotropic microwave background, in 1964, was followed by a decade of careful measurements of the background flux throughout the centimetric and millimetric ranges of wavelength. The results of these measurements are not inconsistent with a Planckian spectrum but the absolute precision of the measurements is not as high as is frequently assumed. More recently attention has turned to searches for variations in the flux density with direction in the sky, while preparations are made in laboratories around the world for a second wave of measurements of the spectrum which are to have a much higher absolute precision. I point out in this article the limitations in our present knowledge of the microwave background, identify the observational difficulties in improving that knowledge and report on some of the plans for future measurements. The excellent recent critical review of background measurements by R.J. Weiss (1980) and the papers presented at a 1979 Copenhagen Symposium (Kalchar et.al. 1979) should be consulted for further detail.

Figure 1 summarises the results of measurements of the isotropic microwave background flux in terms of the equivalent black-body temperature, T_{CBR}, at each measurement frequency. The experimenters in each case recognised various contaminating signals from sources other than the background and the bars in the figure indicate their estimates of the consequent uncertainty in the measured T_{CBR}. The data can be seen to be broadly consistent with a Planckian spectrum, i.e. the same value of T_{CBR} at all frequencies, but there is a large spread in the magnitudes of the uncertainties, and the measurements with relatively low uncertainty (the most recent) test the constancy of T_{CBR} only over a small range of frequency. Moreover, most of the measurements are in the range of wavelength longward of 3 mm where a 3 K Planckian spectrum deviates little from a simple ν^2 spectrum (the Rayleigh-Jeans limit), as can be seen in figure 2. A spectrum of that simple form does not differentiate between, on the one hand, emission from an optically thick emitter at a single temperature and, on the other, superposed emissions from optically thin thermal emitters,

A. W. Wolfendale (ed.), Progress in Cosmology, 119–143.
Copyright © 1982 by D. Reidel Publishing Company.

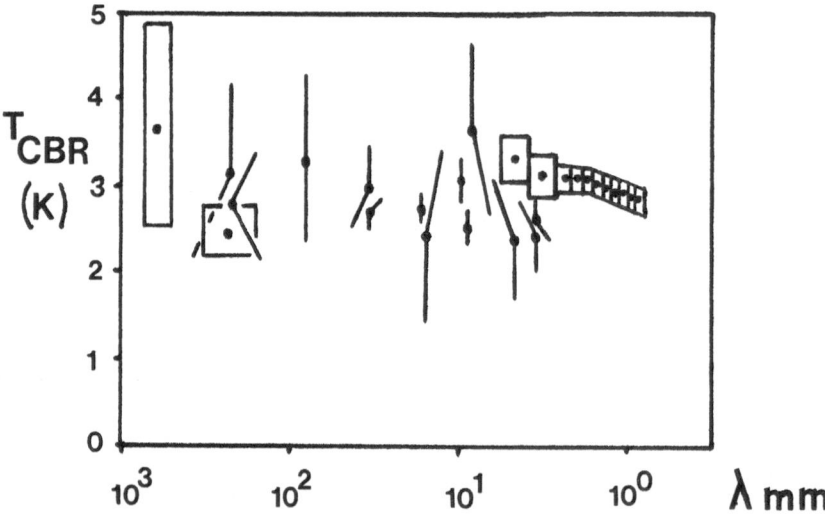

Figure 1   Measured values of T_{CBR} with ranges of experimental uncertainty
indicated by bars and boxes.   These data are taken from Table 1
of Weiss (1980) and fully referenced there.

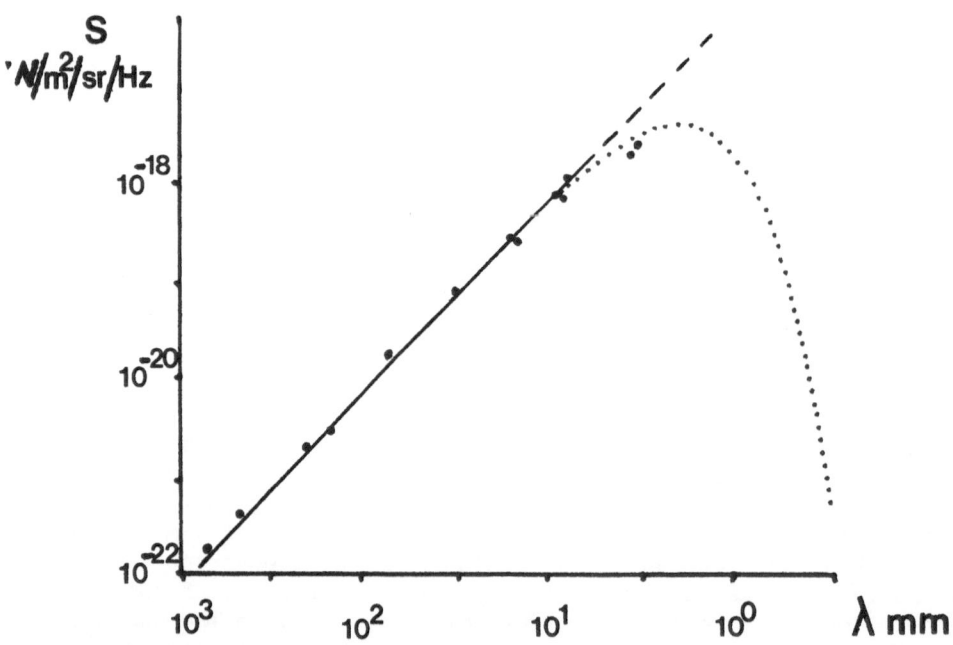

Figure 2   Measured background fluxes for wavelengths exceeding 3 mm.
The dotted line is the black-body spectrum for 2.73 K.

at various temperatures. The points for 3.3 mm wavelength deviate
significantly from the ν^2 best-fit to the data for lower frequencies
but measurements at wavelengths shortward of 3 mm are clearly of
crucial importance. Here severe observational problems arise from
atmospheric emission, and experiments have to be conducted on high-
altitude balloon-borne platforms. There are also technical difficulties,
at present, in using coherent detection methods at the shorter
wavelengths; the data in figure 1 shortward of 3 mm were obtained
using quasi-optical incoherent detection and broad-band signal analysis.
They clearly establish that the background flux falls well below the
extrapolated ν^2 fit to the longer wave data (see figure 14) but the
apparent T_{CBR} may be slightly higher than that at longer wavelengths
(see later).

In order to identify the difficulties that must be overcome if our
knowledge of the background spectrum is to be refined, it is instructive
to examine the sources of uncertainty in the experiments which gave the
data of figures 1 and 2. First it should be noted that it is not the
detectivity of receivers that sets the important limitations, as can be
seen from the fact that the receivers used in molecular-line astronomy
give line-strengths to $\sim 10^{-2}$ K in integration times of several minutes.
The essential difference, in this context, between molecular-line
astronomy and cosmic background radiometry is that, whereas molecular-
line signals are narrow-band and the sources are usually of small
angular diameter, the cosmic background flux is spatially extensive
(in fact, almost isotropic) and spectrally broad-band. As a consequence,
neither of the techniques used in molecular-line astronomy to selectively
modulate the signal from the source, so as to distinguish it from
contaminating thermal emission signals from objects inside and outside
the radiometer, can   be applied in measurements of the background;
these techniques are "sky-chopping" (pointing the telescope alternately
on and off the source) and frequency switching (setting the receiver
frequency alternately on and off the molecular-line frequency). A
primary requirement therefore, when developing the design of a radiometer
for background measurements, is to see that contaminating signals are
minimised and that those that cannot be eliminated can be assessed.

Figure 3 illustrates·in a schematic way the main features of any
radiometer system. A (Dicke) switch directs either the signal from the
antenna or that from a stable reference source to the signal processing
and detection system; the switch operates periodically, and the detector
synchronously, so that the recorded output is proportional to the
difference between the antenna signal and the reference signal. This
nullifies any contaminating signals which originate in the radiometer
after the switch - the switch is therefore best incorporated as early
in the radiometer system as practicable. To keep the total recorded
signal small, the reference source should be at a low temperature,
comparable to the signal proper (i.e. liquid helium). The switch must
be efficient and extremely well balanced thermally and optically,
otherwise it introduces its own, modulated, contaminating signal.

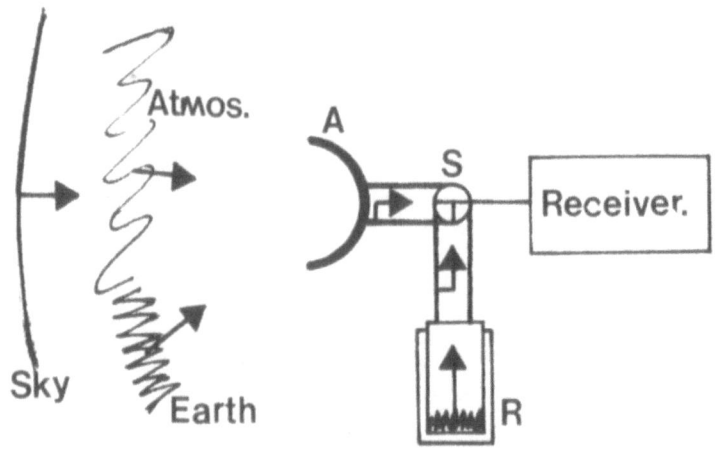

A: Antenna   S: Dicke Switch,   R: Cold Reference

Figure 3   Main features of a radiometer system.   The arrows indicate
          sources of detected power.

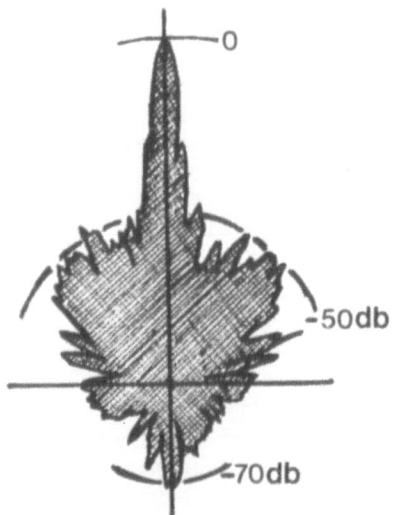

Figure 4   Polar diagram illustrating the general character of the
          antenna pattern of the antenna used by Penzias and Wilson
          (see Wilson, 1980 in Kalckar, 1980).

Four sources of contaminating signals prior to the switch can be identified in figure 3. First there are the warm objects in the periphery of the antenna's field of view, notably the earth. Second, there is the thermal emission of the atmosphere, which enters via the main beam of the antenna. Third is the thermal emission from the antenna, and from the feed system taking the signal to the switch. And fourth is the thermal emission from the feed system taking the signal from the reference source to the switch.

In radiometers for use at wavelengths greater than about 3 mm, the feed systems will usually be wave-guides and the thermal emission in them can be inferred form their measured attenuations. For shorter wavelengths, open quasi-optical feeds may be used and for these the diffraction into the beam of the thermal emissions from peripheral objects has to be assessed theoretically and/or measured in subsidiary test experiments. Cooling these parts of the radiometer reduces such contamination but it would be necessary to reach temperatures in the liquid helium range to gain significantly.

In order to minimise the contaminating signal from the earth and any other warm objects in the periphery of the radiometer's field of view, the radiometer system must be designed so that the large-angle sidelobes in its antenna pattern are extremely weak. The main-lobe of the antenna pattern must be sufficiently narrow for it to be possible to resolve changes in recorded signal on varying the elevation of the radiometer's optical axis, i.e. on looking through a varying path-length through the atmosphere, so as to allow an assessment to be made of the contaminating signal from the atmosphere - which cannot be eliminated because it enters by the main beam together with the back-ground signal. Estimates of the performance required in the antenna pattern can be made as follows. A black-body temperature $T_{CBR} = 3$ K, say, corresponds to an antenna temperature T_A which changes from 3 K for wavelengths in excess of about 3 mm, to about 1 K at 2 mm, 0.1 K at 1 mm, and to rapidly decreasing values into the submillimetre spectrum. Let us suppose it is required to keep the antenna temperature of the contaminating signal from the earth to below 0.01 K. Then we have

$$R_E \frac{\Omega_E}{\Omega_{MB}} \cdot \epsilon_E \cdot T_E < 0.01 \text{ K}$$

where Ω_E is solid-angle subtended by the earth (2π), Ω_{MB} is the solid angle of the primary lobe of the antenna pattern ($\sim 10^{-2}$ sterad, i.e. an angular resolution on the atmosphere of $\sim 3°$) and, ϵ_E, T_E are the emissivity and temperature of the earth (1; 300 K). R_E is a mean-value for the antenna pattern amplitude, relative to the on-axis value, over the directions spanned by the earth. Thus

$$R_E < 5 \times 10^{-8}$$

Together with similar considerations for other objects in the

peripheral field of view (e.g. a balloon) this leads to the following
guidelines for the antenna rejection factors: $>10^4$ for off-axis angles
greater than 30°, $>10^6$ for $>60^\circ$, and $>10^8$ for $>90^\circ$ (i.e. all back-angles).
Figure 4 shows the antenna pattern for the antenna used by Penzias and
Wilson in the measurements that led to the discovery of the cosmic
background, at 7.35 cm wavelength. This large horn-and-reflector
antenna, with a 20-feet aperture, had been designed for satellite
communications and it was the known quality of the antenna surfacing
and particularly the very low levels of the large-angle side-lobes,
that led Penzias and Wilson to persist in attempts to explain the small
excess of their measured antenna temperature above what they had been
expecting from recognised sources, local, remote and astronomical. At
millimetric, as opposed to centimetric, wavelengths an antenna that
will meet the requirements above can be much smaller, of course.
Relatively small horn antennae, with corrugated internal surfaces, can
be deisgned to have the required off-axis rejection factors. For
submillimetre wavelengths antennae which couple the receiver/detector
to many modes, rather than to one or two, are required and approaches
to this problem are referred to later.

The extent to which the various contaminant signals can enter into
measurements for wavelengths above 3 mm can be illustrated by reference
to two of these measurements - that of Penzias and Wilson (1965) at
7.35 cm, that of Wilkinson, Stokes and Partridge (1967) at 8.56 mm.

The radiometer for the Penzias and Wilson measurement is
illustrated schematically in figure 5. The small end of the large horn
antenna is shown connected to a wave-guide junction, P_1, where a linearly
polarised component of the signal beam was overlaid with an orthogonally
polarised signal from the reference source; the Dicke switch was a length
of twist wave-guide which could be rotated about its axis so as to
rotate the planes of polarisation thereby switching first the signal,
and then the reference, beam into one of the output ports of a second
wave-guide junction P_2 (while the other beam left through the second
output port to be dumped, unwanted, in a matched load D). The output
is detected synchronously with the switching, by a low-noise maser
amplifier and detector. The reference source was a carefully designed
matched load cooled by liquid helium and served, not only to give a
stable reference and to reduce the dynamic range of the measured signal,
but also to allow the signal power to be determined absolutely in a
null measurement. To do this, the reference source was connected to
the junction P_1 by a length of wave-guide which incorporated a variable
attenuator at about 300 K and the attenuation was varied to find a
null output. The budget for the identifiable sources at the null was
quoted by Wilson (1979) as follows (the atmosphere, antenna and
attenuator not only attenuate the signals passing through them but
also contribute thermal emission signals of their own, of magnitudes
determined by their attenuation coefficients and temperatures - the
latter is the greater contribution in each case because 300 K >> 3 K).

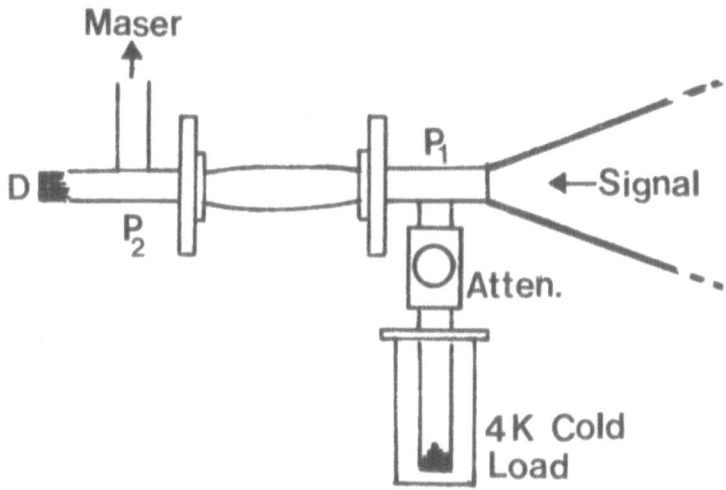

Figure 5   Schematic illustration of a radiometer of the type used by
Penzias and Wilson (see Wilson, 1980 in Kalckar, 1980).

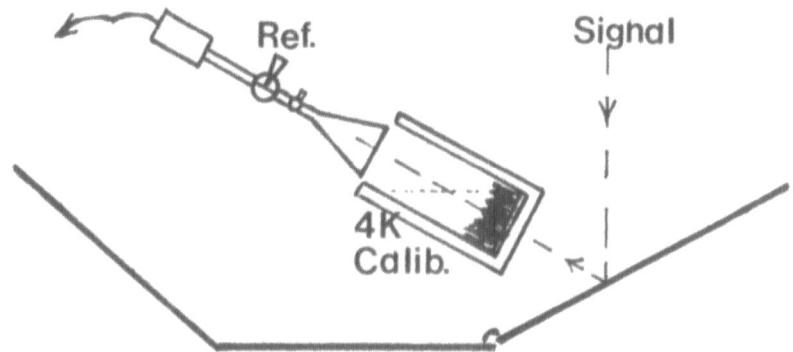

Figure 6   Schematic illustration of the radiometer used by Wilkinson,
Stokes and Partridge, 1967 (see Wilkinson, 1980 in Kalckar,
1980).

Signal channel: Atmosphere   2.3 ± 0.3 K
                     Antenna and feed   1.8 ± 0.3 K
                     Earth   0.1 ± 0.1 K
                                                   ‾‾‾‾‾‾‾‾‾‾‾
                                                   4.2 ± 0.7 K
                                                   ———————————

Reference channel: Cold reference source   4.22 K
                     Reference feed   0.70 ± 0.2 K
                     Attenuator   2.40 ± 0.1 K
                                                   ‾‾‾‾‾‾‾‾‾‾‾
                                                   7.32 ± 0.3 K
                                                   ———————————

Since the radiometer gave a null output, the difference in the temperatures above for the signal and reference channels was attributed to
the background, giving

$$T_{CBR} = 3.1 \pm 1 \text{ K} \quad \text{at } 7.35 \text{ cm}$$

The uncertainties estimated for the several contributions were simply
added together because the errors were not statistical; this possibly
exaggerates the experimental uncertainty. It is clear that the
precision of the experiment is limited by the systematic uncertainties
arising in estimating the contaminating signals; it was crucial that
the experimenters' thorough understanding of their overall system was
such that they were able reliably first to identify, and then to assess,
all sources of contaminating signals. It was also established clearly
in this experiment that the background was isotropic to well within
this uncertainty.

    In the measurements of Wilkinson, Stokes and Partridge at 8.56 mm,
a much smaller antenna system, illustrated schematically in figure 6,
was used since there is an order of magnitude difference in the wavelength. The scalar feed horn was used in conjunction with reflecting
screens placed so as to replace warm earth with cold sky in the side-
lobes. The Dicke switch had a skyward-pointing horn as its stable
reference channel. With an antenna as small as this it was possible
to place a dewar vessel, containing an absorber cooled with liquid
helium, over the feed horn in order to calibrate the radiometer and then
remove it for sky measurements; to facilitate this the radiometer pointed
downwards, fixedly, receiving the sky signals by reflection from a large
reflector which was pivoted so that the atmospheric path could be varied
to provide an estimate of the atmospheric emission. All measured antenna
temperatures were thus referred to the horn aperture (so wave-guide
losses were of reduced significance) and the calibration source was
sufficiently large that all but the weak side-lobes of the horn's antenna
pattern were filled by the cooled absorber, thereby keeping to very low
values the allowance that had to be made for emission from the warm parts
of the dewar. The budget for this experiment was:

    Reflector: 0.12 ± 0.04 K
    Warm parts of calibrator: 0.28 ± 0.11 K
    Atmosphere: 6 K

The atmospheric contribution at a wavelength of 8.56 mm is much greater that at 7.35 cm, even in the best conditions at the high altitude site at which the experiment was conducted, and it was the variability in this (e.g. 0.3 K change between runs separated by about one hour) that mainly set the final uncertainty in the measurement:

$$T_{CBR} = 2.56^{+0.17}_{-0.22} \text{ K at } 8.56 \text{ mm}$$

Similar accuracy in antenna temperature was obtained at 3.3 mm by Boynton, Stokes and Wilkinson (1968). However, ±0.2 K in antenna temperature translates to an uncertainty of ±0.5 K in thermodynamic temperature, because 3.3 mm lies outside the Rayleigh-Jeans range over which antenna and thermodynamic temperatures are equal.

$$T_{CBR} = 2.48^{+0.50}_{-0.54} \text{ K at } 3.30 \text{ mm}$$

At shorter wavelengths still there is a dramatic increase in the accuracy required in the measured antenna temperature in order to achieve a given accuracy in thermodynamic temperature. This implies increasing experimental difficulty because contaminating signals derive usually from objects at about 300 K for which the available power increases as ν^2 throughout the frequency range of interest here whereas the power spectrum for a source at ~3 K falls below the ν^2 variation at wavelengths less than ~3 mm, dramatically so for wavelengths less than 2 mm; this is illustrated by the variations in the antenna temperature, T_A, corresponding to thermodynamic temperatures of 300 K and 3 K, shown in figure 7. Suppression of contaminating signals must be more and more effective if the equivalent thermodynamic temperature of the background is to be determined at shorter and shorter wavelengths. There is another major impediment to making measurements at shorter wavelengths and that is the increasingly strong absorption and emission from the atmosphere. The data in figure 8 show clearly that, while precision measurements from high (4 km) dry sites may be possible for wavelengths in excess of 3 mm, belloon altitudes (40 km) are necessary for shorter wavelengths - or space-borne platforms, of course.

There are, furthermore, other basic questions of technique to be faced when moving to wavelengths below 3 mm. For a coherent detector the largest permissible value for $A\Omega$ (where A is the effective receiving area of the antenna and Ω is the solid angle of the main lobe of the antenna pattern on the sky) is $\sim\lambda^2$. For wavelengths shorter than 3 mm this may be an unwelcome restriction on throughput; incoherent detectors, such as bolometers, are not subject to this limitation and have been used efficiently in optical systems with $A\Omega \sim 25\lambda^2$ at λ 1 mm (e.g. a bolometer of area 5 x 5 mm, in an optical mount which receives over ~ 1 rad). Moreover, bolometric detectors can be used in quasi-optical systems which allow a range of signal frequencies to be covered simultaneously. These advantages of incoherent methods would be difficult to achieve at longer wavelengths because a radiometer that passes many beam-modes would necessarily be larger in size than a single-mode coherent system.

D. H. MARTIN

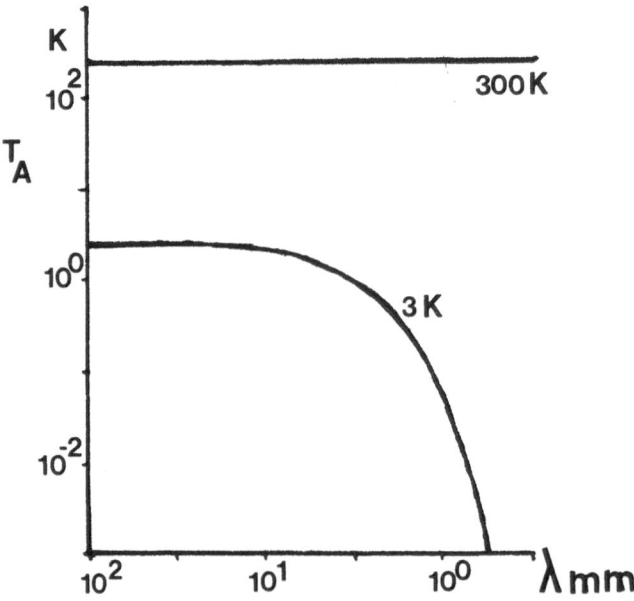

Figure 7   Antenna temperatures for black-body spectra at 3 and 300 K,
           showing the rapid decrease of background flux relative to
           that from objects at 300 K, at wavelengths less than 3 mm.

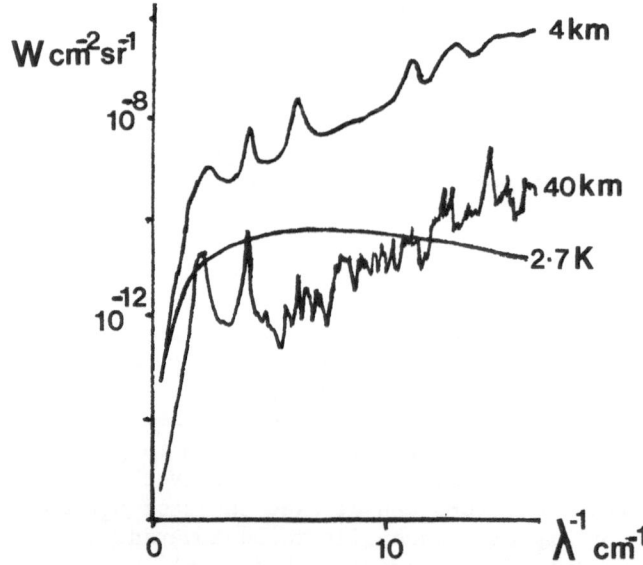

Figure 8   Flux of zenith atmospheric emission at 4 and 40 Km
           altitude compared with that from a black-body at
           2.7 K (after Weiss, 1980).

Below a wavelength of about 3 mm, however, the required dimensions are not a great impediment. Furthermore, current noise-temperatures of coherent receivers operating at wavelengths less than 2 mm are much poorer than those of longer-wave receivers and local oscillators currently available require high-stability, high-voltage power supplies which would not easily be operated on a high altitude balloon-borne platform.

For the reasons above, the radiometers used in the few measurements that have been made of T_{CBR} shortwards of 3 mm were cooled to liquid helium temperature, operated on remotely-controlled high-altitude balloon-borne platforms or, in the most recent case, a rocket, and they were based on quasi-optical antennae and Dicke switches, incoherent detectors and, in the earliest measurements, filters to isolate spectral bands (Muehlner and Weiss, 1973; Dall'Oglio et.al. 1976) and, in later experiments, frequency multiplexing, i.e. Fourier transform spectrometry (Robson et.al. 1974; Woody et.al. 1975; Woody and Richards, 1979; Gush 1981). I shall deal here only with critical aspects of the recent spectrometric measurements.

The first of the spectrometric measurements was made in 1974 by Robson, Vickers, Huizinga, Beckman and Clegg (1974). Their radiometer, in a liquid helium cryostat, is shown schematically in figure 9. The broad-band optical Dicke switch was a Martin-Puplett two-beam inter-ferometer (Martin and Puplett, 1970; Martin, 1982). This incorporates efficient and well-balanced wire-grid polarising components. As the path-difference in the interferometer changes smoothly, the detector receives power alternately from the background and from a reference source in the radiometer. The switching results from optical inter-ference and occurs each time the path-difference changes by $\lambda/2$. This means that there is no periodic motion in the system at the switching frequency, such as might give a spurious modulated signal. Also, the output is the Fourier transform of the power spectrum of the incident signal because each spectral component is switched at a frequency det-ermined by its wavelength; conversely, the signal's spectrum is given by Fourier transforming the interferogram. A moving reflector gives the changing path-difference and the maximum movement of the reflector sets limits to the spectral resolution; in the Robson et.al. experiment the resolving power was about 10, and the spectral range covered was from 3 mm to 0.2 mm, in wavelength. The wings of the radiometer's antenna pattern were determined by diffraction at the sequence of apertures through the optical train; a mirror imaged the detector in the outermost window aperture to reduce this. The raw data of the experiment are illustrated in figure 10, which is a mean of spectra obtained over a period of about 1 hour. The broad peak from 15 - 40 cm^{-1} (10 cm^{-1} ≡ 1 mm wavelength) is

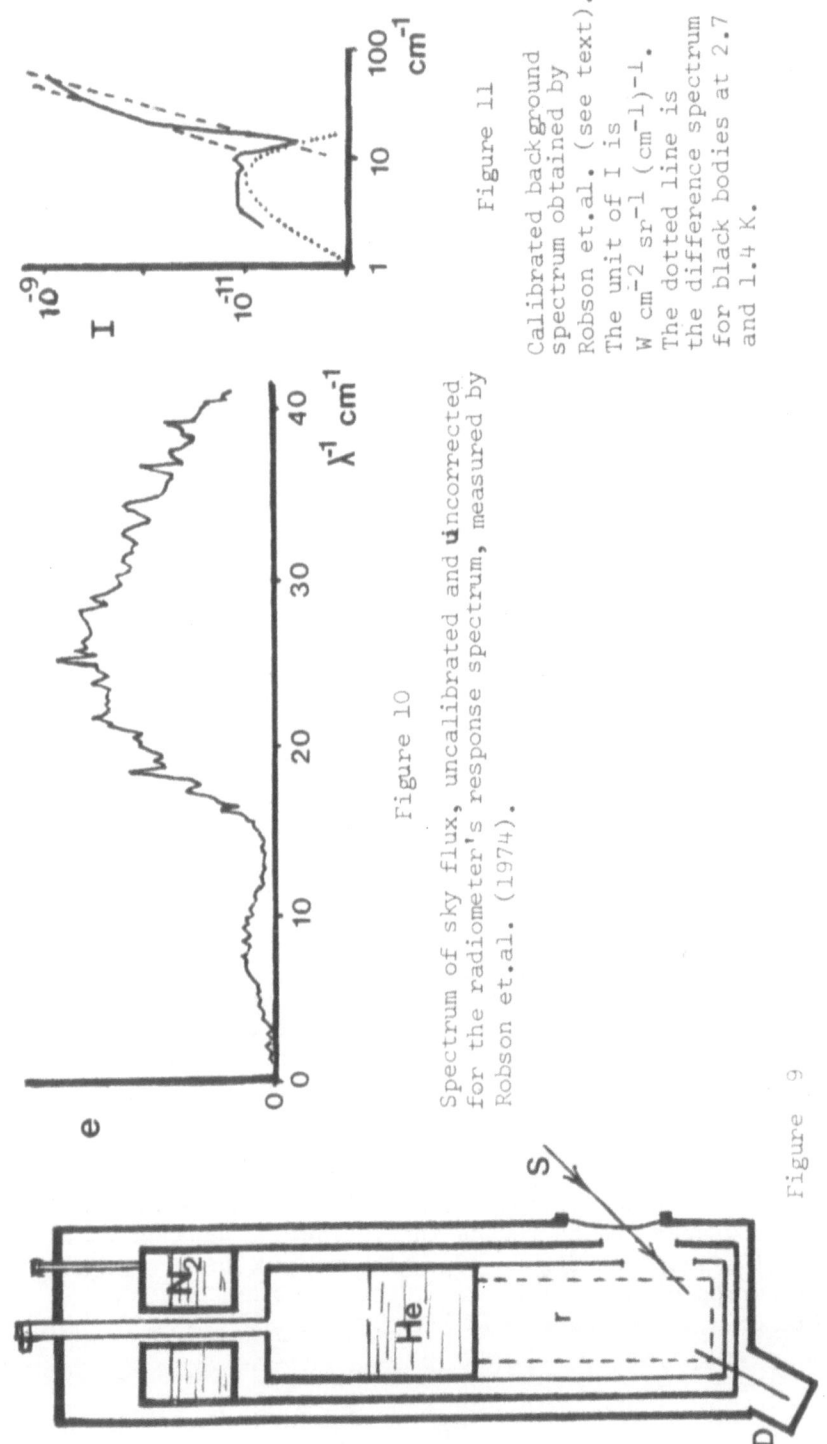

Figure 11

Calibrated background
spectrum obtained by
Robson et.al. (see text).
The unit of I is
$W\ cm^{-2}\ sr^{-1}\ (cm^{-1})^{-1}$.
The dotted line is
the difference spectrum
for black bodies at 2.7
and 1.4 K.

Figure 10

Spectrum of sky flux, uncalibrated and uncorrected
for the radiometer's response spectrum, measured by
Robson et.al. (1974).

Figure 9

General configuration of the cryogenic system used by Robson et.al.(1974) in a high-altitude
balloon experiment; the radiometer is mounted in the space r, the moving mirror in the
interferometric switch is driven by an externally mounted motor at D. The background signal
S enters through staged apertures.

attributable to thermal emission from the very thin polyethylene window
in the outer can of the cryostat and superimposed on it are sharp lines
attributable to emission from atmospheric oxygen, ozone and water-vapour.
Robson et.al. attribute the weak broad peak between 3 cm^{-1} and 12 cm^{-1}
to the background. It is technically difficult to calibrate absolutely
a cooled radiometer in the laboratory. Robson et.al. calibrated their
measurements by reference to the in-flight thermal emission from the
window. Laboratory measurements had been made of the attenuation
coefficient of polyethylene at about 25 cm^{-1}. The samples used were
thicker than the window material (in which the attenuation would be
extremely small) but the results were scaled to the window thickness.
From this, the absolute level of thermal emission from the window could
be inferred, knowing the ambient temperature at the flight altitude.
This provided an absolute calibration of the flight spectrum at about
25 cm^{-1}; laboratory measurements with an external liquid-nitrogen source
were used to determine the relative system sensitivities, from 3 cm^{-1}
to 25 cm^{-1}, and this served to complete the absolute calibration of the
flight data. The curve plotted in figure 11 is the result (the atmospheric lines superimposed on the window emission having been subtracted).
The broken lines show the range of uncertainty in the window emission
corresponding to variations in measured attenuations for different
(thick) samples of polyethylene. The dotted line shows a 2.7 K black-
body curve (relative to the 1.4 K reference source). The data from this
experiment for the first time showed a background spectrum which not
only clearly departed from a ν^2-dependence, but registered a maximum,
and the absolute level was consistent with

$$T_{CBR} = 2.94 \pm 0.06 \text{ K} \quad \text{at} \quad 1.25 \text{ mm}.$$

There can now be more confidence about the levels of stratospheric ozone
and water-vapour than was the case in 1975 and the self-consistency of
the Robson et.al. data has been criticised on the grounds that the
atmospheric emission lines in the spectra are weaker than would be
expected if their absolute calibration were correct, perhaps by as much
as a factor of five. However, even if one uses the atmospheric lines
themselves for the calibration (rather than the window emission, which
then becomes anomalously large) the measured flux densities in the range
3 - 12 cm^{-1} would set upper limits on the background that clearly
establish the rapid decrease of the background flux below the
ν^2-dependence for wavelengths shortwards of 2 mm. This is illustrated
by the data labelled QMC in figure 12 which were obtained from the
Robson et.al. fluxes in figure 11 by scaling upwards by x5. The excess,
above a 3 K spectrum, which the measurements would then indicate between
3 and 12 cm^{-1} could perhaps be the result of an incomplete control of
diffraction inputs which, as stressed above, must be extremely good if
a submillimetre background at $\sim$3 K is not be masked. Detailed antenna
analysis for a multi-mode system, leading to the design of an antenna
with the necessary very weak side-lobes, and followed by the demonstration
of its satisfactory performance, was a major step forward taken in the
next spectrometric measurement, later in 1974, by Woody, Mather, Nishioka
and Richards (1975) and, with further improvements, in 1977 by Woody and
Richards (1979).

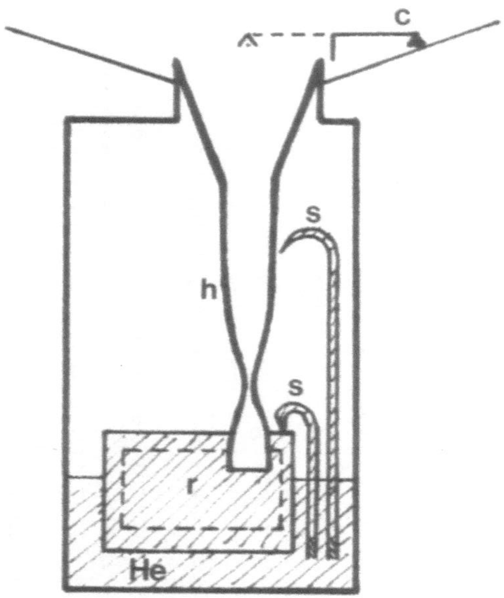

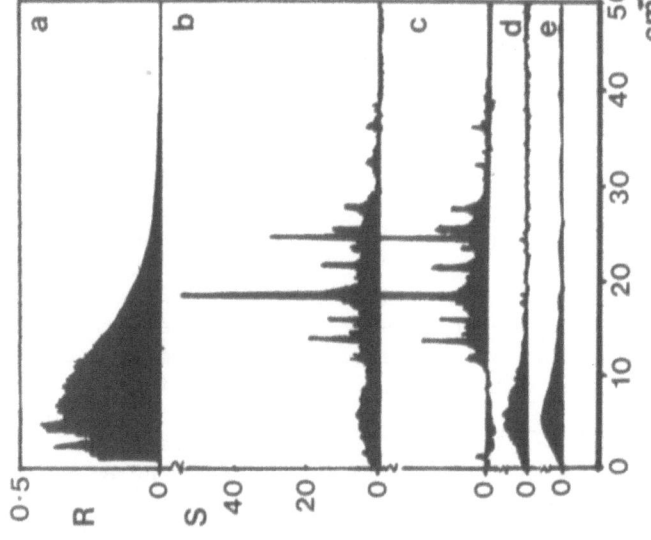

Figure 13 To illustrate the general form of the data obtained by Woody and Richards (1979) (see text). R is the instrument responsivity (units: 10^6 V cm^2 sr W^{-1}) and S is the detector response (units: 10^{-6} V $(cm^{-1})^{-1}$). The note on the final page regarding the figures prepared for this article applies particularly to this figure, which does not show the extra-ordinary precision and detail of the data obtained in the experiment.

Figure 12 Schematic illustration of the cryogenic configuration for the radiometer used by Woody and Richards, 1979, 1981 (not to scale). The shaped horn antenna, h, and the enclosure for the radiometer, r, are kept cool by liquid helium pumped by the superfluid fountain-effect (S, S). A small source at ambient temperature, C, could be swung into the beam as shown. The primary calibration was based on laboratory measurements in which a carefully designed black surface was inserted to fill cross-section of the horn (at about the point h).

The radiometer used in the 1977 measurements is illustrated schematically in figure 12. The spectrometric Dicke switch was a Martin-
Puplett two-beam interferometer, cooled to liquid helium temperature
and incorporating the cold reference source, as in the Robson et.al.
experiment. The signal beam entered, not through a window in the
vacuum wall of the cryostat, but through a windowless aperture in the
top of the helium reservoir; i.e. the interferometer was immersed in
the liquid helium and cold He gas streamed upwards and through the
windowless aperture, maintaining the temperature gradient between the
liquid helium and the outside of the cryostat. A long reflecting multi-
mode horn gave an antenna pattern of the required form; its shape served
to de-couple the signal beam from the walls at the upper end of the horn
where it is warm and therefore emissive. Most of the horn, where the
beam was essentially formed, was kept at very low temperature by a stream
of liquid helium delivered by a superfluid fountain-effect pump. Outside
the cryostat there was a reflecting screen to put cold sky, rather than
warm earth and gondola, into the antenna side-lobes. The efficacy of
this type of antenna was demonstrated by laboratory measurements, which
showed a rejection ratio $>10^8$ at angles greater than 180°, and $>10^{-2}$ at
angles greater than 5°. The upper and lower bounds (i.e. including the
estimated systematic error) for the data obtained in the 1974 flight are
plotted in figure 14. The data for the 1977 measurement are illustrated
in figure 13. First is shown the absolute responsivity of the radiometer
system over the full spectral range as measured in the laboratory by
methods described below. Of note here is the low-frequency cut-off and
the deep reproducible structure at the long-wave end, arising from inter-
ference effects in the throat of the long-horn. Next in the figure is a
plot of the measured emission spectrum of the background and atmosphere,
and then a calculated atmospheric spectrum based on assumed stratospheric
concentrations of oxygen, ozone and air, chosen so as to optimise
the fit to the measured data above 10 cm^{-1}. Finally plots at two resolutions are given for the differences between the measured and calculated
spectra. These reveal a clear broad peak in the range 2 - 10 cm^{-1}
attributable to the background. Both the absolute magnitude and the
spectral form are fairly close to the black-body spectrum for 2.96 K;
this is shown in figure 15. The data from this excellent experiment have
been analysed with great care because, as the figure reveals, they do
suggest that the spectrum is not perfectly Planckian, to an extent that
would, if correct, be bound to have major cosmological implications.

Two aspects of the recorded spectrum call for attention. First the
temperature that gives a best-fit black-body curve is, at 2.96 K,
decidedly higher than that which gives a best-fit to the longer-wave
data, say 2.73 K; and second there is a clear excess above the 2.96 K
curve around the peak together with a shortfall beyond the peak. Possible cosmological explanations for such a spectrum are being explored
theoretically (see Rowan-Robinson, 1982). There is, however, a
tantalising aspect of this pattern of deviation from black-body form.
If in a measurement of T_{CBR} over a range of frequency, the calibration
for power-flux should be in error by a scaling factor independent of
frequency, that error would transform to a frequency-dependent error
in the inferred T_{CBR}, e.g. a black-body spectrum would be distorted from

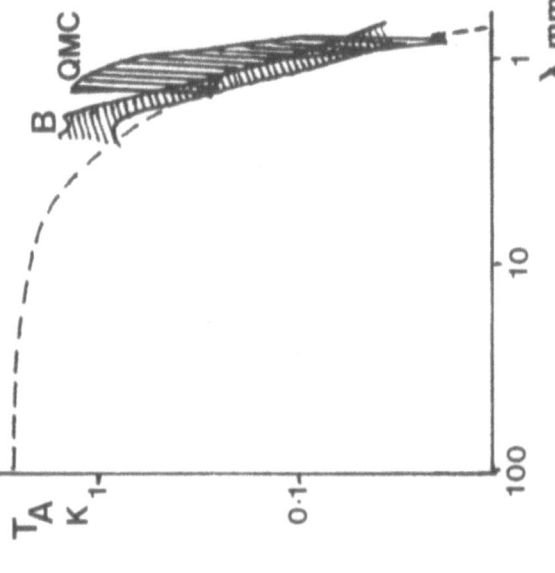

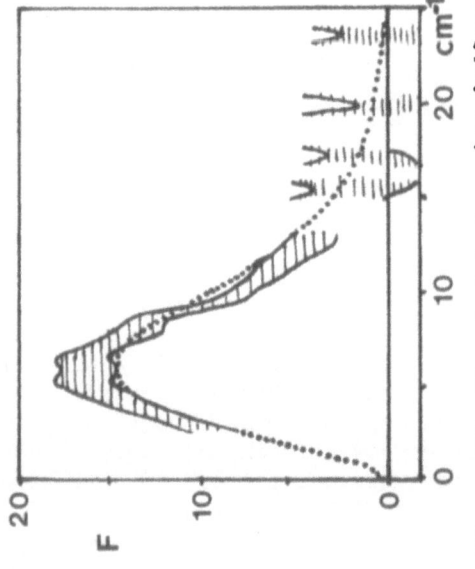

Figure 14 The hatched regions, QMC and B, show respectively the ranges of experimental uncertainty in the measurements of the background flux (expressed here as an antenna temperature) by Robson et.al. 1974, and by Woody et.al. 1975. For B the experimenters estimate of the 2σ limits are drawn; for QMC the lower bound is the 1σ limit estimated by the experimenters but the upper bound here is that given by a 5 × scaling of the flux quoted by the experimenters (the reasons for doing this are given in the text). The broken line is for a black-body spectrum at 2.8 K.

Figure 15 The hatched region indicates the experimenters' estimates of the range of overall experimental uncertainty (1α) in the measurements of the background flux by Woody and Richards (1979, 1981) (units of F: 10^{-12} W cm^{-2} sr^{-1} $(cm^{-1})^{-1}$). Gaps appear above 13 cm^{-1} where atmospheric emission is so strong, even at 42 km altitude, that no significant determination of F is possible. The dotted line shows the black-body spectrum for 2.96 K.

black-body form. The curves in figure 16 illustrate this point. The lowest curve is a black-body spectrum for 2.73 K whilst the curve with the highest peak records the same spectrum but scaled upwards by 38% at all frequenciess. The remaining curve is the black-body spectrum for 2.96 K.

The scaling factor (1.38) is such that the area under the scaled-up 2.73 K spectrum equal to that under the 2.96 K curve. Comparing this figure with the data in figure 15 indicates that, had there been a 1.38 scaling error in the calibration of the flight data, correction of the error would give a near black-body spectrum with T_{CBR} = 2.73 K, in accord with the long-wave measurements of T_{CBR}. Woody and Richards appreciated the need for good calibration, of course, and carried through a thorough examination of their calibration procedures. The primary calibration was based on post-flight measurements in the laboratory using a variable-temperature, grooved, black emitter inserted part-way down the horn-antenna (figure 12). The temperature of the emitter could be changed, from 4 to 20 K. The results of many laboratory measurements have been thoroughly analysed (Woody and Richards, 1981) and show persuasive self-consistency, and indicate overall rms calibration errors of $^{+4}_{-7}$%. Meas-urements made before and after the flight of the 300 K spectrum entering from the laboratory (a black-body) showed changes of less than 3% - the detector response was non-linear for signals as large as this but that does not spoil the conclusion that the system's configuration changed little before, during, and after flight. However there had to be some in-flight monitoring because the responsivity of a germanium bolometer is very sensitive to its operating temperature and bias current. A small source at external ambient temperature was included in the flight system and was moved into and out of the aperture of the radiometer under command from the ground; this gave a useful check on performance but had a rather large error (± 20%) as a relative calibration. The main check on flight performance was that proved by the direct monitoring of the bolometer heat-sink temperature, of the bias current through the bolometer and of the bias voltage. These were steady through the in-flight meas-urements. The post-flight laboratory calibration measurements were made under closely similar bolometric conditions. In assessing the overall frequency-independent calibration error at $^{+4}_{-7}$%, Woody and Richards included ± 3% for bolometer responsivity, presumably based on these monitoring data. The hatched region in figure 15 thus registers the results of careful assessments of all identified sources of error. Nevertheless, the experimenters conclude that "there are serious limit-ations to the statistical analysis where systematic errors are likely. It is clear that another generation of measurements with better accuracy is required before any deviation from a Planck spectrum can be firmly established". (See also Pickett et.al.1981 concerning the atmospheric lines)

A 1978 rocket-borne experiment has recently been reported by Gush (1981). A measurement from a rocket platform at 150 - 370 km altitude is free of the complication of atmospheric emission and (since there is no balloon above) the radiometer can be pointed so as to place the earth, sun and moon in the back-lobes of the radiometer's antenna pattern (i.e. at $\theta > 90°$). These advantages are gained at the expense of a severe reduction in the total measurement time, to no more than a few minutes.

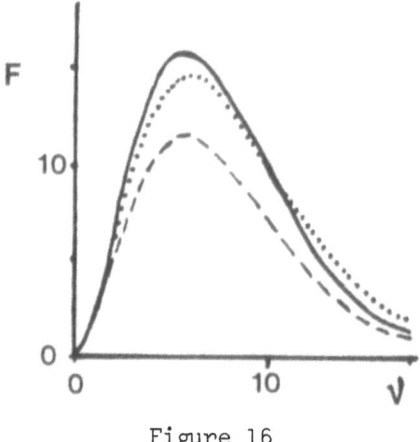

Figure 16

The dotted and broken line show black-body fluxes (units as in figure 15)
for 2.96 and 2.70 K respectively. The full line is the 2.70 K spectrum
scaled up by a factor 1.38.

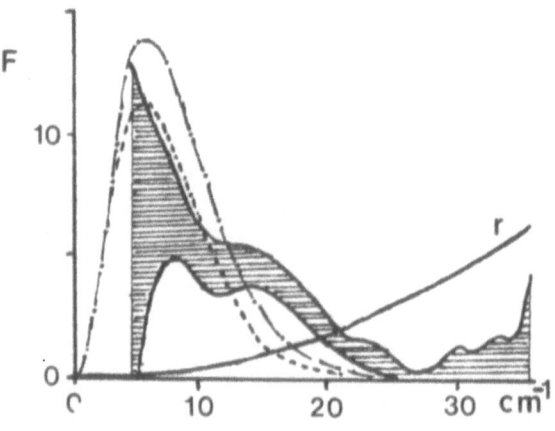

Figure 17

The hatched region indicates the experimenter's estimates of the 90%
confidence level for the measurement of the background flux by Gush
(1981). (units of F: 10^{-12} W cm^{-2} sr^{-1} $(cm^{-1})^{-1}$). The data have
been corrected for the unwanted contribution from the rocket motor
by subtracting an estimate of the upper limit for this (curve r).
The dotted and chain lines are black-body spectra for 2.7 and 2.96 K
respectively.

The Gush radiometer was cooled by liquid helium and incorporated a two-beam interferometer as a broad-band Dicke switch with a ^{3}He-cooled Ge bolometric detector. The antenna pattern was controlled by limiting apertures. The interpretation of the data obtained in flight was compromised by an unintended incursion of the de-coupled rocket motor into the periphery of the field of view. Thermal emission from the motor not only contributed a contaminating signal to the recorded spectrum but it also gave the interferogram a varying baseline because the radiometer was spinning about an axis at 6° to its optical axis. The phase of the large 1 Hz base-line ripple in every 7 s scan of the interferometer was not correlated with that of the interferometer scan cycle and it was possible to model and subtract the ripple (though there has to be some uncertainty about its residual effects). Figure 17 shows the spectrum obtained from the data acquired in the 100 s of measurement time. A progressively rising signal above 20 cm^{-1} was thought to be attributable to the ν^2 contaminating signal from the rocket motor. The available information about the location of the motor, its size and its temperature was used to set an upper limit for this contribution and it was found that subtraction of this from the recorded spectrum, does result in an intensity that falls to zero above 25 cm^{-1}, as shown in figure 17. The intensity of residual spectrum can be seen to have the right order of magnitude for a ∿3 K background but it departs markedly from black-body form. The radiometer was calibrated in the laboratory using an external conical cavity of sand-blasted stainless steel, cooled by liquid helium; the temperature of the cavity was set successively at various temperatures in the range below 4.2 K and the corresponding changes in recorded signal were used to determine the responsivity to signals entering the port of the radiometer. The internally generated and detected signal at each frequency was separated, to serve as a baseline, by extrapolation to zero cavity temperature; the cavity was not, however, highly black, its inferred emissivity was 0.61, and it therefore reflected to a significant degree and care has to be exercised in interpreting the calibration measurements. This experiment demonstrates that the problems arising from the short observing time of a rocket-borne experiment and from the need to control the liquid refrigerants in the varying-g environment have been mastered and the experimenter plans further flights "with obvious improvements in the deployment of the apparatus at altitude" and with polarising interferometer components.

It can be seen from what I have written so far that the errors in measured values of T_{CBR} arose mainly from the uncertainties in the magnitudes of the contaminating signals coming from objects inside and outside the radiometer, and/or from difficulties in making absolute calibrations at the time of measurement. Such limitations are much reduced, however, if the quantity of interest is the difference in the values of T_{CBR} in two directions in the sky. The absolute precision (in °K) can then be much higher because the contaminating signals may be unchanged on changing the direction of observation and because by making differential measurements, the uncertainty in the calibration scaling factor will apply to the difference itself. For this reason, the same kinds of radiometer that had been used in the late 1960s and early 1970s to measure T_{CBR} at centimetric and millimetric wavelengths with an uncertainty of several tenths of a degree, were used in the late 1970s and early 1980s to measure, or to set bounds on, the directional variations of T_{CBR} to precisions of several millidegrees. There is not sufficient space here to deal with such measurements, important though they be (Silk, 1981). The development of radiometers with improved absolute precision held fire while the background anisotropy was explored. In the last few years, however, plans have been laid in several laboratories around the world, for new determinations of T_{CBR} in the centimetric range with an absolute precision of a few centidegrees Richards et al   are soon to make new measurements in the near-millimetre range using a ^{3}He-cooled bolometer with band-isolating transmission filters and Gush hopes to re-fly his rocket-borne experiment in 1984. There is also to be an Explorer satellite (COBE) to make measurements of T_{CBR} in the near- and submillimetre ranges with a precision of a few millidegrees,and of the anisotropy at centimetric wavelengths (Weiss, Mather and Kelsall, 1980); subject fo final decisions on funding, COBE should be launched in 1987 or soon after. Another space-borne experiment to measure the spectrum at wavelengths shortwards of the peak (CIRBS) has been proposed for launch in 1986 and is under consideration by ESA.

My group at QMC is building a balloon-born radiometer to measure T_{CBR} near the peak at 1.9 mm. We expect to fly this in 1982. I finish this account of background measurements with a description of our system to illustrate the kinds of advance in technique that will bring, during the next few years, measured values of T_{CBR} in the millimetric and sub-millimetric ranges having much smaller experimental uncertainties.

The immediate need is for an experiment that will give a greater precision,or less uncertainty, in the absolute value of T_{CBR} shortwards of 3 mm, even if the spectral range covered has to be restricted in order to achieve this. A broad-band antenna that admits many modes and has very weak far-field side-lobes necessarily has a large aperture diameter, say 50 λ (Mather, 1981). Its near-in field is thus widely spread and that makes it difficult to ensure that there is no significant contamination of the recorded signal by thermal emission from warm parts of the experiment package, especially from the apertures at the mouth or window of the cryostat. If single-mode operation is acceptable, a small antenna with a narrow pencil-beam near-in is possible, and that should lead to a smaller systematic error for the radiometer. The smaller throughput, however, would result in a larger detector-noise error in a given measurement time, and the advantage of having a smaller systematic error would be realised only if the detector noise-equivalent-power (NEP) were sufficiently low. The NEP of a good ^{3}He-cooled bolometric detector is such that, for the throughput of a single mode ($A\Omega \simeq \lambda^2/2$, see the discussion of antenna patterns above), the contribution of detector-noise to the error of measurement of T_{CBR} would be a few millikelvin for an integration time of 15 minutes. Single-mode operation is, on this ground, acceptable. It has to be noted, however, that a given single-mode antenna will have the required narrow-pencil beam and good transmission efficiency only over a restricted frequency range, usually less than an octave. The radiometer described below is designed for the wavelength range 1.7 to

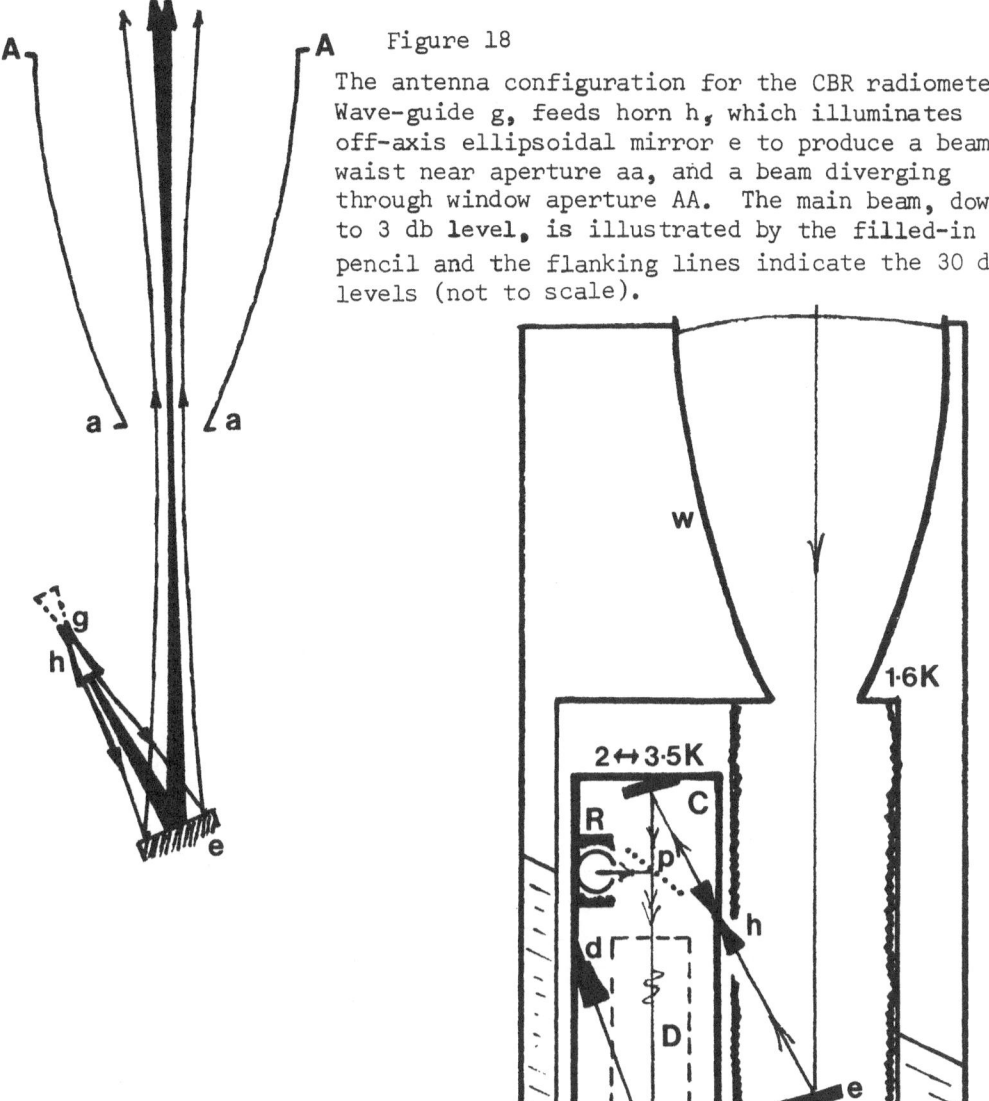

Figure 18

The antenna configuration for the CBR radiometer.
Wave-guide g, feeds horn h, which illuminates
off-axis ellipsoidal mirror e to produce a beam-
waist near aperture aa, and a beam diverging
through window aperture AA. The main beam, down
to 3 db **level**, is illustrated by the filled-in
pencil and the flanking lines indicate the 30 db
levels (not to scale).

Figure 19

Main features of the CBR radiometer. W is a reflecting
cone, the bottom end of which is at 1.6 K. The radiometer is in the cavity C, whose temperature (together
with that of the components inside and the horns h and d)
can be varied between 2 and 3.5 K. A black cavity is at
R and a wire-grid polariser at p; the interferometric
switch (not shown) is at location D and the detector horn
at d.

2.3 mm at a spectral resolving power of 25. This range is close to the peak of a ∿ 3 K black-body spectrum and, at balloon altitudes, there should be no complication from atmospheric emission (see figure 8).

The basic antenna system of the radiometer is illustrated in figure A scalar-horn fed by a very short length of corrugated cylindrical wave-guide is combined with an off-axis ellipsoidal mirror. Over the wave-length range involved, the wave-guide passes only one type of mode and the dimensions of the horn (54 mm long, 5.5 mm aperture radius) are such that this mode is transformed by the horn into a freely propagating axially symmetric beam with an amplitude distribution in the transverse plane close to a Gaussian function. Such a beam maintains the Gaussian form as it propagates, but with changing width. Taking the time-reversed, or transmit, view for convenience, the beam from the horn reflects from the ellipsoidal mirror and passes through the two apertures in the upper half of the cryostat (the antenna, and the rest of the radiometer, are at liquid helium temperature). The beam converges to a minimum beam-width near the smaller, lower, aperture and then diverges again, passing through the larger, upper, aperture to give the external Gaussian main beam having a full angular width ($^1/e$, amplitude) of 6° in the far field. The external main-beam is thus Gaussian but at large angles the antenna pattern will be determined by multiple diffraction of the beam at the three apertures - i.e. at the edges of the ellipsoidal mirror and those of the upper and lower apertures in the upper half of the cryostat. The same considerations govern the contamination of the recorded signal by thermal emission from objects within the radiometer. In this regard it is important that the beam is heavily tapered at the edges of each aperture and that no point on the upper aperture can be joined to any point on the edge of the ellipsoidal mirror by a straight line passing through, i.e. within, the lower aperture. Also, the large shaped cone is highly reflecting and directs any ray that comes from below the lower aperture into a direction within 16° of the optical axis, i.e. into the background, not the earth; furthermore this cone, though warm at its upper, wider, end, is cold (2 K) at its lower, narrower, end.

The antenna horn is, in fact, one of a pair of horns, back-to-back, joined by the short length of wave-guide. Wave-guides for 2 mm wave-length are too small to contain signal processing components. The beam is, therefore, (I now take the time-forward view) coupled out of the guide by the second horn (broken lines in figure 15) to form a freely propagating Gaussian beam which is to pass through a quasi-optical calibration unit on its way to the detector, as illustrated in figure 19. The short length of wave-guide thus serves essentially as a mode filter and so defines the received beam.

The calibration unit includes a two-beam polarising Martin-Puplett interferometer which serves as a Dicke switch, directing to the detector, alternately, the signal beam from the back-to-back horn-pair and that from an optically black cavity incorporated in the calibration unit (details of the interferometer are omitted from figure 19 for the sake of clarity).

The two signals are combined with orthogonal polarisations at a
wire-grid and, when a mirror in the Dicke switch is moved at constant
speed, each signal is sinusoidally modulated interferometrically, but
in anti-phase; if the two signals are equal there will be no resultant
modulation at the detector and a null will be indicated. It is simpler
first, to discuss the calibration process assuming that the detector is
in the small receiving horn shown in figure 19 and at the same temperature.
The temperature of the black cavity can be varied and that provides the
means for a null calibration of the background signal. In fact the
temperature-active unit includes, not only the black cavity, but also
the wire grid, the back-to-back horn, the interferometric switch and
the detector horn, all of which are to be at closely the same temperature. The absolute calibration does not then depend on optical
equivalence of the two signal channels - for example, the transmitting
and reflecting polarisation efficiencies of the grid need not be equal,
the horn antennas need not be perfectly transmitting, and thermal
emission from the horn antennas and from the rest of the optical system
has no consequence. Such deviations from ideality reduce only the
sensitivity of the null detection, not its absolute character. This can
be most directly understood as follows. From within the calibration unit,
it would not be possible to tell that the back-to-back horn pair was
not looking into a black cavity at T_{CBR} provided, of course, the control
of the antenna pattern is as good as intended (if it isn't the radiometer
is in any case inadequate). If, therefore, the temperature of the
calibration unit is brought into equality with T_{CBR}, the unit would be
filled with black-body radiation at T_{CBR} whatever the transmissive,
reflective and absorptive properties of the horns, or of any other optical
component in the unit. In that circumstance, movement of the mirror in
the interferometer could not produce any modulation of the power being
received at the detector horn; the detector would register a null. In
practice, of course, the temperature of the calibration unit would be
set at a series of values above and below T_{CBR}, and null conditions
determined by interpolation. Since the frequency of interferometric
modulation in the switch is determined by the wavelength, Fourier-
transformation of the detector output as a function of mirror position
will give a spectrum and the null temperature can be determined, by
interpolation, for each spectral frequency individually.

The calibration process has to be examined more carefully when account
is taken of the fact that the detector is not at the temperature of the
calibration unit (it was assumed above to be so). The thermal emission
signal from the detector passes through the calibration unit and is
divided between the black cavity and the beam leaving the radiometer,
and if this happens without partial reflection back into the detector
horn, there is no problem. Care is taken to design and manufacture the
system so that is, as nearly as possible, the case. The part of the
system requiring special attention in this regard is the wave-guide
joining the back-to-back horns because any reflecting discontinuity
there will lead to a reflected beam having the form that is optimally
coupled optically to the detector horn. A check can be made of the
adequacy of the system in this respect by covering the outer end of the
horn-pair with absorber at the temperature of the calibration unit and

measuring the residual interferometric modulation. Ideally there would
be none but if there were a small residual modulation, that could be
used as an off-set zero. In fact, there is a number of optical remedies
for this reflection effect but there is not space here to discuss them.

In the actual radiometer the detector horn, too, is a back-to-back
pair and the **emerging** Gaussian beam is condensed by an off-axis ellipsoidal mirror onto a bolometer placed, in its ^{3}He-cooler, at a beam-
waist, where the signal phase-front is plane and can be optimally matched
to the absorbing film of the detector. The same radiometric system could
be used with a heterodyne receiver (though this would bring additional
technical problems to be handled) and that gives some possibility that
precision measurements could be made at balloon altitudes at submillimetre
frequencies between the atmospheric emission lines.

I hope, of course, that the radiometer described in outline above
will soon be operative and will successfully measure T_{CBR} near the peak
in the spectrum with a greatly improved precision. But it is, in any
case, clear that the next few years will see renewed attempts, around
the world, to determine the background spectrum absolutely with centi-
kelvin precision, over the centimetric and millimetric ranges. Beyond
that (especially, but not exclusively, with space-borne systems) the
precision of measurement may be taken to millikelvin levels and into
the submillimetre range. The full cosmological significance of the
microwave background will then be open to real test and exploration.

I am grateful to many fellow experimenters for willingly giving
information about their experiments and plans for future measurements.
In particular R.B. Partridge and G.F. Smoot have told me of their planned
experiments, in collaboration with N. Mandolesi and G. Sironi, to measure
T_{CBR} at 3, 6 and 12 cm with $\sim$2% accuracy, and P.L. Richards and H.P. Gush
of their intentions to re-fly their submillimetre experiments in the near
future.

N.B.   The figures were prepared so as to make simple, bold, transparencies
for projection, and they may not be quantitatively reliable. The original
sources should certainly be referred to for quantitative information,
especially for experimental data.

REFERENCES

Boynton, P.E., Stokes, R.A., Wilkinson, D.T. 1968, Phys.Rev.Lett. 21, 462.

Dall'oglio, G., Fonti, S., Melchiorri, B., Melchiorri, F., Natale, V., Lombardini, P., Trivero, P., Sivertson, S. 1976, Phys.Rev.D, 13, 1187.

Gush, H.P. 1981, Phys.Rev.Lett. 47, 745.

Kalckar, J., Ulfbeck, O., Nilsson, N.R. 1980 (Ed.s) The Universe at Large Redshifts, Physica Scripta 21, No. 5.

Martin, D.H. 1982, Ch. 2 of Infrared and Millimetre Waves, Vol. 6, Ed. K.J. Button, Academic Press.

Martin, D.H. and Puplett, E. 1970, Infrared Physics 10, 105 (and Martin, D.H. 1972, in "Infrared Detection Techniques for Space Research", Eds. V. Manno and J. Ring, p.267, Reidel).

Mather, J.C. 1981, IEEE Trans. AP29, 967.

Muehlner, D.J. and Weiss, R. 1973, Phys.Rev.Lett. 30, 757.

Penzias, A.A. and Wilson, R.W. 1965, Astrophys.J. 142, 420.

Pickett, H.M., Cohen, A., and Brinza, D.E. 1981, Astrophys.J. 248, L49.

Robson, E.I., Vickers, D.G., Huizinga, J.S., Beckman, J.E., Clegg, P.E. 1974, Nature 251, 591 (and P.E. Clegg 1978, in "Infrared Astronomy", Eds. G. Setti and G.G. Fazio, Reidel).

Rowan-Robinson, M. and Tarbet, P. 1982, This Volume.

Silk, J., Proceedings of Conference on Astrophysics and Elementary Particles & Cosmology, 1981, Eds. J. Audouze and others.

Weiss, R. 1980, Ann.Rev.Astron.Astrophys. 18, 489.

Weiss, R., Mather, J. and Kelsall, T. 1980, Physica Scripta 21, 670.

Wilkinson, D.T., Stokes, R.A. and Partridge, R.B. 1967, Phys.Rev.Lett. 19, 1199.

Woody, D.P., Mather, J.C., Nishioka, N.S. and Richards, P.L. 1975, Phys.Rev.Lett. 34, 1036.

Woody, D.P. and Richards, P.L. 1979, Phys.Rev.Lett. 42, 925.

Woody, D.P. and Richards, P.L. 1981, Astrophys.J. 248, 18.

THE COSMIC MICROWAVE BACKGROUND RADIATION AS A PROBE
OF THE LARGE-SCALE STRUCTURE OF THE UNIVERSE

Joseph Silk
Department of Astronomy
University of California
Berkeley, CA 94720, U.S.A.

SUMMARY: Over scales in excess of a few percent of our present horizon, the search for anisotropy of the cosmic microwave background radiation promises to provide unique information on the existence of inhomogeneities in the matter distribution. Anisotropy at some level seems inevitable in any plausible scenario for the origin and evolution of structure in the early Universe. My goal here is to review the implications of galaxy formation theory for the presence of large-scale fluctuations. The dipole and recently discovered quadrupole anisotropy of the microwave background radiation severely constrain such theories. Interesting modifications of the predicted radiation anisotropy arise from the effects of spatial curvature and the possible presence of massive neutrinos.

I. INTRODUCTION

The cosmic microwave background radiation provides us with a unique glimpse of the very early Universe. The Universe could not have been opaque at a redshift less than 10, and most cosmologists place the last scattering surface of the radiation at the matter-radiation decoupling epoch, at a redshift $z \approx 1000$. This means that detection of angular anisotropies in the radiation could provide a probe of the matter distribution at these early epochs.

In fact, there is also the intriguing possibility of inferring the state of the Universe at a far earlier epoch. The standard hot Big Bang model is generally considered to provide a good description of the Universe back to the epoch of nucleosynthesis, at $z \sim 10^9$. Now one of the fundamental properties of the Friedmann cosmology is its horizon structure. The region that any given observer can communicate with and affect causally increases in scale as proper time t. Since the scale factor $a(t)$ varies as $t^{2/3}$ during the present matter-dominated era and as $t^{1/2}$ in the early radiation-dominated era (at $z \geq 1.1 \times 10^4 \ \Omega^{-1}$), one infers that progressively smaller proper volumes (which vary as a^3) are

A. W. Wolfendale (ed.), Progress in Cosmology, 145–159.

causally connected within a single horizon as one approaches earlier
epochs.

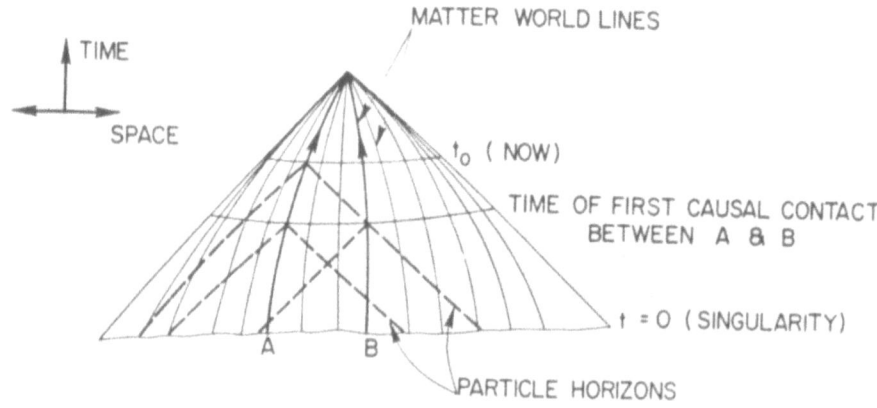

Figure 1: Conformal space-time diagram for Friedmann universe. This
demonstrates (a) that the world-line of any observer such as A first enters the (particle) horizon of another observer B after a finite time
has lapsed, and (b) that our particle horizon now includes many horizon
scales at an early epoch.

     Consider now an observer who measures microwave background radiation
from different directions. Photons from these directions arrive from
regions that have not causally interacted within the age of the Universe
(Figure 1). Thus, by studying the large-scale anisotropy of the background radiation, one will inevitably probe the intrinsic large-scale
structure of the Universe. Radiation anisotropies, if detected, could
not have been produced (or erased) by causal processes above some angular scale. This angular scale corresponds to the horizon size on the
surface of last scattering, or about $(\Omega/z)^{1/2} \sim 2\Omega^{1/2}$ degrees. The
search for larger angular scale structure in the radiation therefore
probes the structure of the Universe back to the limit of the standard
model, prior to nucleosynthesis of the light elements at an epoch as
early as < 1 second after the singularity.

     With sufficient faith, one can even trace the evolution of the Uni-
verse back to a far earlier epoch. The success of grand unified theories of the elementary particles in accounting for the baryon number of
the Universe suggests that the Friedmann models may provide a valid description of the Universe back to the grand unified epoch, at $t \sim 10^{-36}$
second. The cosmic microwave background radiation accordingly provides
an important link between observation and speculation about the very
early Universe.

Anisotropy at some level seems inevitable in any plausible scenario for the origin and evolution of the observed large-scale structures in the Universe. One of the main goals here is to review the implications of galaxy formation theory for the large-scale anisotropy of the cosmic microwave background radiation. The observational evidence for large-scale anisotropy is summarized in §II, and the theoretical implications are described for both the dipole and recently discovered quadrupole anisotropy (§III). Interesting modifications of the predicted radiation anisotropy arise from the effects of spatial curvature (§IV) and the effects of massive neutrinos (§V).

II. OBSERVATIONS OF LARGE-SCALE ANISOTROPY

The first detection of anisotropy was that of the dipole component, expected on the sole assumption that the blackbody radiation was of extragalactic origin (Smoot et al. 1977). Independent experiments have confirmed the effect, both in amplitude and in direction of the intensity maximum, despite limited sky coverage (Cheng et al. 1979; Smoot and Lubin 1979; Fabbri et al. 1980). An unavoidable problem which arises because of the partial sky coverage is the presence of higher order moments that will bias the various experiments somewhat differently.

Taking a simple average of the Berkeley, Princeton, and Florence experiments leads to the following parameters for the dipole anisotropy: amplitude: 3.3 ± 0.6 mK and direction of peak intensity RA $1^h.5$ ($\pm 0^m.4$), declination $+0°.2$ ($\pm 7°$). If this were entirely due to the motion of the earth with velocity, v, relative to the blackbody radiation, one predicts a dipole anisotropy for the temperature angular distribution of the form

$$T(\theta) = T_0(1 + v \cos\theta/c), \text{ with } v = 340(\pm 60)(2.9K/T_0) \text{ km s}^{-1}.$$

Correction for galactic rotation (v = 275 km s^{-1} in the direction $\alpha = 21^h.2$, $\delta = +48°$) yields the motion of the Local Group relative to the cosmic background radiation: v = 540 (± 60) km s^{-1} towards $\alpha = 10^h.7$ ($\pm 0^h.4$), $\delta = -22°$ ($\pm 7°$). On the other hand, measurements of galaxy redshifts in the Local Supercluster have yielded information on the Virgocentric flow. The Local Group is falling towards the Virgo cluster ($\alpha = 12^h.5$, $\delta = +12°.4$) with a velocity estimated at between 180 (± 30) km s^{-1} (Yahil et al. 1980) and 440 (± 100) km s^{-1} (Tonry and Davis 1981): the difference is in part due to selection of the galaxy sample and to how the centre of mass of the Virgo cluster is defined. According to the former analysis, the motion of the Local Group relative to the background radiation must be largely determined by the matter distribution outside the Local Supercluster.

However, on the basis of the interpretations of the available redshift data, we cannot exclude the entire effect as being due to the Virgocentric flow. While the apex of the dipole motion is apparently some 45° away from Virgo, the effects of the partial sky coverage indicate that the errors in direction have most likely been underestimated, as is also implied by comparing the different results on the dipole anisotropy. Moreover, non-radial motions are very likely to be produced during the recollapse of a complex region like the Virgo Supercluster. The simple model of White and Silk (1979) indicates that a modest degree of shear could readily reconcile the radiation anisotropy with the dynamics of the Local Supercluster.

The motivation for believing that the dipole anisotropy should be explained by a simple Virgocentric flow model becomes questionable in view of the persistence of the Rubin-Ford effect. Rubin et al. (1976) found a motion for our galaxy of 600 (±125) km s^{-1} relative to a sample of Sc galaxies at a typical redshift cz = 5000 km s^{-1} and in a direction $\alpha = 2^h$, $\delta = 53°$. Efforts to explain this effect away have not succeeded. Large-scale inhomogeneity, beyond the Virgo Supercluster, represents a possible explanation of this effect; such structure would inevitably contribute to the gravitational acceleration of the local rest frame and therefore also affect the dipole anisotropy.

The discovery of a quadrupole moment in the cosmic microwave background radiation provides perhaps the strongest evidence for large-scale inhomogeneity or anisotropy in the universe. Two independent determinations of the quadrupole moment have recently been reported. Fabbri et al. (1980) utilized a balloon-borne far infrared bolometer to search for large angular scale anisotropy in the background radiation in the 0.05-3mm wavelength region. In addition to confirming the dipole anisotropy, a second harmonic term, identified with a quadrupole anisotropy of amplitude ∼ 1/3 the dipole moment, was detected. The peak amplitude, with very limited sky coverage, was 0.9 (±0.3) mK, and consistent with alignment along the dipole axis.

Balloon-borne radiometers were used by Boughn et al. (1981) at wavelengths in the range 0.65-1.6 cm to establish the thermal nature of the dipole anisotropy and to improve the accuracy of previous measurements of the dipole anisotropy. A quadrupole anisotropy was detected at the 4σ level. In general, the quadrupole moment contains 5 spherical harmonic coefficients. The Princeton experiment yielded a 4σ effect in Q_5, and a 2σ effect in Q_4, with upper limits on Q_2 and Q_3; the Berkeley experiment (Smoot and Lubin 1979) set an upper limit on Q_1. One can combine these coefficients to estimate the amplitude of the entire quadrupole matrix:

$$Q^2 = \frac{3}{2} Q_1^2 + 2\left(Q_2^2 + Q_3^2 + Q_4 + Q_5\right),$$

where Q = 1.1 (±0.3) mK, and is again about 30 percent of the measured dipole anisotropy.

This reasonably good evidence for quadrupole anisotropy can be explained with a minimum of assumptions in the context of the gravitational instability theory for the origin of large scale structure in the

Universe (§III). However, it may be premature to conclude that a definitive link between theory and observation has finally been forged. One
worrisome point is that the galactic non-thermal radio emission pos-
sesses a large-scale spatial distribution that contains a quadrupole-
like component; indeed, prominent radio loops occur at the warm lobes
where the quadrupole component Q_5 peaks. The Princeton experimenters
estimate that there should be negligible contamination by non-thermal
emission in the high frequency range of their measurement. A definitive
detection will require complete sky coverage as well as improved sensitivity.

III. THEORETICAL IMPLICATIONS

The dipole anisotropy, after correction for the motion of the sun
around our galaxy and the motion of our galaxy towards M31, is interpreted as due to the motion of the Local Group relative to the cosmic
microwave background radiation. Such a motion yields an angular anisotropy

$$\delta T/T = (v/c) \cos\theta.$$

The preceding discussion implies that it is unlikely that this can be
entirely associated with the infall of the Local Group relative to the
centre of the Virgo Supercluster of galaxies. First, the apex of the
motion is shifted some 45° from the centre of the Virgo cluster.
Second, the measured values of the infall velocity from redshift surveys
fall considerably below the velocity inferred from the dipole anisotropy. Third, the Rubin-Ford effect indicates the presence of a large-
scale shearing motion outside the Virgo Supercluster region yet in a
direction ~ 90° from the apex of the dipole anisotropy. It seems that,
while some, if not most of the dipole anisotropy could be attributed to
our infall relative to the Virgo cluster, there is likely to be a substantial component that cannot be accounted for by matter within 20 Mpc,
and by matter, moreover, where the density contrast $\delta\rho/\rho$ is relatively
small, and certainly in the linear regime.

Density fluctuations on scales much larger than the Virgo Supercluster can provide a substantial contribution to our peculiar motion
relative to the cosmic microwave background radiation. The gravita-
tional potential fluctuations due to such irregularities are $\delta\psi \sim G\delta\rho\ell^2$
on scale ℓ. Now if $\delta\rho/\rho$ decreases less rapidly than ℓ^{-2}, one can clearly have a large contribution from very distant regions. In terms of the
comoving mass-scale $M = \pi/6 \ \rho \ \ell^3$ associated with a fluctuation, we see
that $\delta\rho/\rho \propto M^{-2/3}$ is the critical power-law spectrum: a steeper spectrum would not induce a significant component of peculiar motion.

It is helpful at this stage to generalize the discussion of density
fluctuations to an arbitrary power-law spectrum. A power-law is chosen
in order to avoid introducing any preferred scale, although it will be
seen later that in a cosmological model with spatial curvature, for

example, there is indeed a natural scale associated with the curvature radius. Restricting the present discussion to a spatially flat cosmological model, one can write the linear density contrast in terms of a Fourier spectrum

$$\delta\rho/\rho = \int \delta_k \, e^{i\underset{\sim}{k}\cdot\underset{\sim}{x}} \, d^3k,$$

and express the Fourier power spectrum coefficient as

$$|\delta_k|^2 \propto k^n.$$

Transforming back to the comoving mass as an independent parameter, one obtains the power-law spectrum representation of the density fluctuations:

$$\delta\rho/\rho \propto M^{-1/2-n/6}.$$

Now the case of constant gravitational potential (or equivalently, constant metric or curvature) fluctuations corresponds to setting $n = 1$, and we recover $\delta\rho/\rho \propto M^{-2/3}$. On the other hand, the choice $n = 0$ corresponds to white noise. This is considered by some cosmologists to be a natural value that might arise from an unspecified mechanism for spontaneously generating density fluctuations. Any such process would have to be causal, and it is of interest to extrapolate back to very early epochs, when the horizon scale contained less mass than that of any structure of interest in the present Universe. For example, the mass of a galaxy was first encompassed by the horizon at $z \approx 10^8$. Now density fluctuations grow in amplitude on scales greater than the Jeans length, and in the radiation-dominated era this is effectively the horizon scale. The growth rate in this regime is $\delta\rho/\rho \propto t$ for the fastest growing mode, where t denotes proper time. Strictly speaking, this result holds for curvature fluctuations in a specified coordinate gauge, such as the comoving system. Now the scale-factor $a \propto t^{1/2}$, and so at a given comoving wavenumber k/a, the spectrum of density fluctuations must flatten within the horizon, where growth is suppressed due to radiation pressure. One can see that the spectrum in fact flattens by

$$\delta\rho/\rho|_{in} \propto \delta\rho/\rho|_{out} \cdot t \propto \delta\rho/\rho|_{out} \cdot (a/k)^2 \propto \delta\rho/\rho|_{out} \cdot M^{2/3}.$$

Consequently, a causally generated white noise spectrum is associated with a fluctuation spectrum $\delta\rho/\rho \propto M^{-7/6}$ on scales larger than the horizon. One can show in fact that such a spectrum, corresponding to $n = 4$, represents the minimal fluctuation level expected to arise from any rearrangement of the matter distribution into non-linear clumps (Peebles 1980). Finally, a minimum value of n arises from requiring that the r.m.s. density fluctuations do not diverge on large scales: this convergence criterion yields $n > -3$.

The only "natural" spectrum to emerge from this discussion was the one that initiated it: $n = 1$. The gravitational potential (metric) fluctuations diverge strongly if n ≠ 1: on large scales if n < 1 and on small scales if n > 1, and in either case, the constant index power-law assumption necessarily becomes invalid. What one can hope to eventually do is to directly measure n in different regimes, bearing in mind that even if the seed fluctuations are arranged in discrete clumps, one expects that a power-law tail to the fluctuation spectrum with n = 4 will have been generated.

The most direct attack on the density fluctuation spectrum is by measuring the autocorrelation function for a deep sample of galaxies with measured redshifts. This should directly probe the linear fluctuation regime. Considerable amounts of large telescope time are required, and only very recently have sufficient data been acquired. Preliminary indications are that there may indeed be structure in the linear regime, on scales beyond the Virgo Supercluster. In principle, information may be available on scales out to ~ 100 Mpc, but it seems likely that the quoted errors will be large.

Another, and currently promising, approach is to study the large-scale anisotropy of the cosmic microwave background radiation. Confirmation of the quadrupole anisotropy provides strong evidence that large-scale matter fluctuations are present: these provide the simplest interpretation of the observations. The only cosmological alternative is to invoke an anisotropic cosmological model (cf. Matzner 1980). Non-conventional cosmologies will not be considered here, and we shall explore the implications of the large-scale radiation anisotropy for the matter fluctuation distribution.

Over angular scales greater than a few degrees, the evolution of the radiation fluctuations takes a particularly simple form (Sachs and Wolfe 1967). This is because the scattering terms that generate secondary fluctuations due to Doppler motions are only important on scales less than the horizon size on the surface of last scattering. One is left with the gravitational potential fluctuations. The fact that linear curvature fluctuations are growing results in a net acceleration that perturbs the radiation field, leading to temperature fluctuations

$$\delta T/T \sim G \, \delta\rho \, \ell^2 \sim (\delta\rho/\rho)(\ell/ct)^2. \tag{1}$$

This contains contributions both from the epoch of last scattering of the radiation and from the present epoch of reception of the radiation: intervening fluctuations make a negligible contribution. In particular, it is clear that temperature fluctuations can be induced by potential fluctuations on scales much larger than the present horizon.

Now the angular distribution of these large-scale temperature fluctuations can be decomposed into spherical harmonics. To see how this goes, it is useful to approach the problem in a relatively formal way. The generating equation for fluctuations in the radiation temperature $\Delta \equiv \delta T/T$ is just the perturbed Boltzmann equation, which takes the form

$$\frac{\partial \Delta}{\partial t} + \frac{\gamma^i}{a} \frac{\partial \Delta}{\partial x_i} = \frac{1}{2} \frac{\partial h_{ij}}{\partial t} \gamma^i \gamma^j. \tag{2}$$

Here h_{ij} is the perturbed metric tensor and γ^i is a unit vector describing the radiation direction. The linearized Einstein equations enable h_{ij} to be expressed in terms of a generalized gravitational potential, which leads to the result (1) previously described. To investigate the detailed angular dependence of $\delta T/T$, we make a Fourier decomposition $\Delta = \int \Delta_k \, e^{i k \cdot x} \, d^3 k$, where $\Delta_k(k,\gamma,t)$. Define $\mu = k \cdot \gamma$, and let $K = k c t_0 / a$, where t_0 denotes the present epoch. The general solution to (2) takes the form (Wilson and Silk 1981)

$$\Delta(\gamma, t_0) = \frac{2\delta}{9K^2} \left[(1 - 3iK\mu\beta) \exp\{-3iK\mu(1-\beta)\} \right]_{\beta=z_s^{-1/2}}^{\beta=1}. \tag{3}$$

This contains contributions only from the surface of last scattering z_s and the present epoch of reception of the radiation. The exponential factor is due to the spatial dependence (proportional to $e^{ik \cdot r}$ in a spatially flat universe) of the fluctuations, integrated along a light ray. Two types of terms are contained within the square brackets. The term proportional to K^{-2} is essentially the gravitational potential fluctuation. This yields the usual Sachs-Wolfe (1967) effect. The second term, proportional to K^{-1}, is a dipole anisotropy that is proportional to the peculiar velocity. In general, the interpretation of this term is gauge-dependent (it can be included in the potential term, for example), but it results in a measurable radiation anisotropy: therefore, it is a physically significant term.

Expanding the expression (3) for Δ_k in powers of the normalized wave number K yields for $K \ll 1$,

$$\Delta = \delta \, \mu^2 + O(K\mu^3);$$

and for $K \gg 1$,

$$\Delta = -\delta \, \frac{2}{3} \, \frac{i\mu}{K} \left(1 - z_s^{-1/2} \, e^{-3iK\mu} \right).$$

Now one can take the direction of k to be fixed (this is equivalent to fixing the location of the observer), so that $\cos^{-1}\mu$ is the angle of photon arrival relative to some specified direction. Evidently fluctuations on scales larger than the present horizon contribute a quadrupole and higher order anisotropy, whereas the fluctuations with scales much less than the horizon contain a dipole component that is dominant over higher order anisotropies.

Hitherto, the emphasis has been on curvature fluctuations. Isothermal fluctuations constitute an independent mode of density fluctuations. Secondary metric fluctuations are induced by isothermal fluctuations on the horizon scale, and these also result in radiation temperature anisotropies. For a similar value of $\delta\rho/\rho$, the secondary metric fluctuations are reduced by a factor $\rho_m/(\rho_m+\rho_r)$ relative to the primary curvature fluctuations on the same scale.

Normalization of $\delta\rho/\rho$ is best achieved by computing the galaxy correlation function and comparing with observations in the linear regime. The result is uncertain by a factor of 2 or 3, in part because the predicted correlation function tends to be oscillatory for large values of n. The normalization can be quite different for isothermal and adiabatic fluctuations. The former extend down to much smaller mass scales

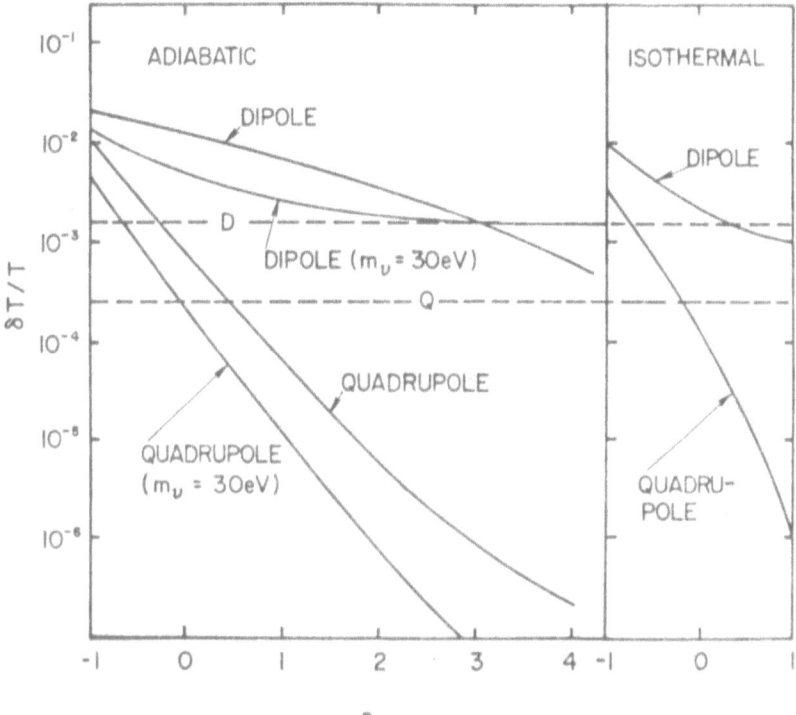

Figure 2: Predictions of dipole and quadrupole temperature anisotropy as a function of the power-law index of the fluctuation spectrum, compared with observed values. The dashed lines denote the observed amplitudes (D and Q are the rms values of the dipole and quadrupole anisotropies). Predictions by Silk and Wilson (1981) are for the standard Ω = 1 model (m_ν = 0) for adiabatic and isothermal fluctuations and for a neutrino-dominated cosmological model (Ω_ν = 0.98, Ω_b = 0.02).

(namely to scales above the Jeans mass after decoupling as opposed to the damping mass). This adds extra power to the correlation function and reduces the amplitude of isothermal relative to adiabatic fluctuations on large scales. N-body simulations of galaxy clustering yield a reasonable model for the non-linear correlation function if $0 > n > -1$: this is applicable to heirarchical clustering of isothermal density fluctuations. Comparison of the predicted correlation function with the

data indicates that the adiabatic fluctuation models are consistent only
if n = 3-4, in the standard model. For n=1, there is far too much power
on large scales.

A detailed comparison of the predicted dipole and quadrupole anisotropies with the observed values leads to the following conclusions
(Figure 2). The isothermal fluctuation model with n $\approx$ 0 can simultaneously account for the quadrupole and most of the dipole anisotropy.
It also is consistent with small angular scale anisotropy limits, the
most critical of which is one (reported as a detection) at 6° (Fabbri et
al. 1981). Now with n = 0, we see that $\delta T/T \sim (\delta\rho/\rho)(\ell/ct)^2 \propto \ell^{1/2} \propto \theta^{1/2}$. The observed values are $\delta T/T = 1.3\times10^{-4}$ (90°) and $\delta T/T = 3\times10^{-5}$
(6°), in good agreement with this model.

The adiabatic fluctuation model fails miserably, however. Fitting
the quadrupole anisotropy results in excessive dipole anisotropy; the
dipole constraint means that the quadrupole contribution is negligible.
This conclusion assumes a power-law fluctuation spectrum: only by arranging sizable fluctuations outside the present horizon could one boost
the quadrupole anisotropy significantly. This seems a totally artificial procedure. In addition, the small-scale anisotropy limits rule out
adiabatic fluctuations, at least in the absence of a reionization of the
Universe at high redshift, and the galaxy correlation data forces the
density fluctuation spectrum, if adiabatic, to be relatively steep (n >
3), consequently requiring a cut-off or flattening in spectral index at
comoving scales of $\sim$ 1 Mpc to avoid non-linearities in the gravitational
potential fluctuations. All of this leads one to assert that adiabatic
fluctuations are untenable in the standard model as the seeds for galaxy
formation, whereas an isothermal fluctuation spectrum satisfies all of
the observational constraints.

IV. CURVED COSMOLOGICAL MODELS

The preceding discussion assumed an Einstein-de Sitter cosmological
model. Cosmological curvature introduces a qualitatively new effect on
the radiation anisotropy. Since cosmologists are now converging on the
view that the Universe is nearly closed with 0.1 < Ω < 0.5, our conclusions about the type of initial density fluctuations will not be significantly modified. However, the new effect is of sufficient interest to
merit a brief review.

Throughout this discussion, the frequency-dependence of the radiation fluctuations has been suppressed: one assumes that it is Planckian. If electron scattering predominates, this should be an adequate
approximation. Then the radiation fluctuations at decoupling, if the
radiation is subsequently free-streaming, are related to the observed
fluctuations by Liouville's theorem. The radiation path is along null
geodesics, the coordinate distance between the points $\underset{\sim}{x}(t_s)$ and $\underset{\sim}{x}(t_o)$ on
a geodesic being given by $\tau = \int_{t_s}^{t_o} dt/a(t)$. Now $\Delta(t_s)$ is a function only
of x, and can in general be expressed as a superposition of eigenfunc-

tions of the Laplacian operator in the 3-space of curvature K (corresponding to the curved Friedmann model). The eigenfunctions are the
solutions of

$$\Box Q = -(k^2 - K)Q, \tag{4}$$

where k is the comoving wave number of the perturbation, with wavelength
$2\pi a/k$. For an open universe $K < 0$, and the allowed values of k are k ε

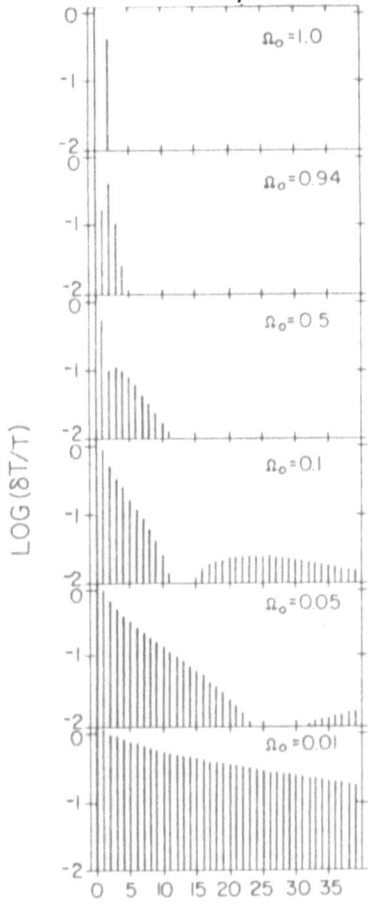

Figure 3: The contribution to the rms radiation fluctuations by
the lowest multipole moments for an infinite wavelength (k = 0)
perturbation of different Friedmann models, from Wilson (1981)

$(0, \infty)$. We immediately infer an important difference between the $k < 0$
and $k = 0$ models: the potential fluctuations, which must contain terms
like $(k^2 - K)^{-1}$, do not diverge as $k \to 0$. The convergence properties of
density and potential fluctuations are similar, and independent of the
fluctuation spectrum.

    Another property of (4) is that the eigenfunctions die away rapidly
for $r \gg (|K|)^{-1/2}$. This means that if the relic fluctuations can be

represented by a smooth superposition of eigenfunctions, they should fall off rapidly on angular scales larger than that corresponding to the curvature radius, Ω^{-1}.

A final effect that adds a further complication is that the density fluctuations themselves induce additional temperature fluctuations via gravitational potential fluctuations. Although there may be little power in the relic fluctuations, these secondary fluctuations can yield significant power on very large angular scales, including contributions to the dipole and quadrupole anisotropies (Figure 3).

One concludes from this that curvature, especially negative curvature, can profoundly affect the behaviour of radiation fluctuations. It is essential to take the curvature into account in interpreting the angular anisotropy of the cosmic microwave background radiation on intermediate angular scales, even if Ω is as large as 0.8 or 0.9.

V. MASSIVE NEUTRINOS

The tentative experimental evidence that the electron neutrinos have a mass of ~ 30 eV promises, if confirmed, to have a revolutionary effect on the evolution of density fluctuations in the early Universe and on the predictions of cosmic background radiation anisotropy. If the average neutrino mass (in three flavours) exceeds ~ 30 ($H_0/100$ km s^{-1} Mpc^{-1})^2 eV, the Universe is closed; the neutrinos need only possess a mean mass in excess of about ~ 0.2 eV to dominate over conventional forms of baryonic matter.

The standard theory of fluctuation growth is unaffected while the neutrinos are relativistic, but once they become non-relativistic at a redshift $z_{nr} = 1.8 \times 10^5$ ($m_\nu/30eV$), where m_ν denotes the mass in any one flavour, neutrinos play an important role. For the decreasing velocity dispersion of the neutrinos implies that one can define an effective Jeans mass which peaks at $M_{\nu m} \equiv 4 \times 10^{15}$ ($30eV/m_\nu$)^2 M$\odot$ at $z_m = 43000$ ($m_\nu/30eV$). It is found that all primordial neutrino fluctuations are erased on scales smaller than $M_{\nu m}$ due to phase-mixing (Bond et al. 1980). Larger primordial fluctuations survive, and density fluctuations grow after entering the horizon at $z < z_m$. On the other hand, baryonic fluctuations in the hot big bang model do not grow on sub-horizon scales until after decoupling at $z \approx 1000$, because of the large radiation pressure. Adiabatic neutrino fluctuations therefore gain an additional growth phase. Gravitational coupling to the baryonic component drives $\delta\rho/\rho$ to $\delta\rho_\nu/\rho_\nu$ after decoupling of matter and radiation when the matter can move freely. Consequently, the radiation temperature fluctuations are reduced by a factor ~ z_m/z_s for the adiabatic mode (Doroshkevich et al. 1980).

This reduction applies on small and intermediate angular scales. The effect is relatively smaller for the large angular scale anisotropy, because it is the gravitational potential fluctuations that drive the anisotropy, and these are also present in the neutrino component at de-coupling (Silk and Wilson 1981).

Isothermal fluctuations can also play an interesting role in a neutrino-dominated universe. As in the standard ($m_\nu = 0$) model, no growth occurs prior to decoupling. After decoupling, the baryonic fluc-tuations are immersed in a hot neutrino fluid. No growth of a fluctua-tion can occur until the neutrino Jeans mass has dropped below the appropriate comoving mass-scale. At present, the neutrino Jeans mass is 10^9 ($30eV/m_\nu)^{7/2}$ MΘ, so that this effect need not prevent galaxy forma-tion. However, the growth rate is suppressed on scales below $M_{\nu m}$ by a factor Ω_b/Ω_ν even for fluctuations that can grow. The net effect of all this is that the isothermal mode requires considerably larger initial amplitudes than in the standard ($m_\nu = 0$) model in order to account for galaxy formation. Consequently, the predicted amount of radiation ani-sotropy tends to be increased. However, isothermal fluctuations, be-cause of their weak coupling to the radiation anisotropy via scattering off moving inhomogeneities, are still consistent with observational con-straints.

A comparison of the predictions of the adiabatic fluctuation theory with the observed large-scale anisotropy leads to the following conclu-sions. If the neutrino mass is in the range 10–30eV, an adiabatic fluc-tuation spectrum with $0 < n < 1/2$ can account for the quadrupole aniso-tropy and avoid excessive dipole anisotropy. The constant curvature value n = 1 appears to give too small a quadrupole effect. Of interest also are the implications for the galaxy correlation function. Because most of the power is at $M_{\nu m}$ rather than the Jeans mass at decoupling as would be the case if $m_\nu \lesssim 1eV$, the correlation function is suppressed at large scales. Consequently, all of the observations are perfectly con-sistent with a white noise primeval adiabatic spectrum characterized by $n \approx 0$, in contrast to the standard model which requires n = 3–4. Be-cause of the strong damping on scales less then $M_{\nu m}$, this is tantamount to requiring a random distribution of density fluctuations of character-istic scale $M_{\nu m}$.

VI. CONCLUSIONS

The cosmic microwave background radiation provides a link between the largest scale inhomogeneities in the Universe and galaxy formation theory. The underlying assumption is that there are traces of the ini-tial seed fluctuations from which galaxies formed that persist on much larger scales. Since little evolution can have occurred on large scales, one has the potential to view the initial conditions from which non-linear structure arose. Any initial seeds, imposed in the quantum gravity era, should leave some structure on horizon scales in any plau-sible model. This should be visible today as large-scale anisotropy in the cosmic background radiation. Both the curvature of the Universe and the possible presence of massive neutrinos have important effects on the predicted magnitude of the anisotropy. At present, one can only say

that the isothermal fluctuation choice of initial conditions is preferable if the neutrino rest mass is less than a few eV. Otherwise, gaseous fragmentation of adiabatic pancakes constitute the foundation for
developing a viable scenario for galaxy formation and clustering. Primordial isothermal fluctuations are also consistent with observational
constraints in a neutrino-dominated Universe if the initial fluctuation
spectrum is nearly flat (n $\approx$ 0). In this case, the delayed growth on
scales below the neutrino Jeans mass at decoupling ($\sim$ 3x10^{13} MΘ for m$_\nu$ =
30 eV) results in the first fluctuations to go non-linear being on this
mass-scale. Consequently, one may again end up with a pancake-type
theory for galaxy formation.

There is one notable alternative that merits separate mention.
Consider a causal process for generating large-scale fluctuations, such
as the cosmic amplifier model of Ostriker and Cowie (1981). This latter
idea involves exploding galaxies that sweep up large shells of intergalactic matter which fragment and form more galaxies. This type of
model only generates a minimal level of curvature fluctuations on very
large scales, corresponding to a power spectrum with n = 4. Very little
large-scale radiation anisotropy is predicted. Consequently, if the
quadrupole anisotropy is confirmed, it would be very difficult to sus-
tain such a model.

This research has been supported in part by NASA. I acknowledge
several helpful discussions with M.L. Wilson.

REFERENCES

Bond, J.R., Efstathiou, G., and Silk, J. 1980, Phys. Rev. Lett., 45,
    1980.
Boughn, S.P., Cheng, E.S., and Wilkinson, D.T. 1981, Ap. J. (Letters),
    243, L113.
Cheng, R.S., Saulson, P.R., Wilkinson, D.T., and Corey, B.E. 1979, Ap.
    J. (Letters), 232, L139.
Doroshkevich, A.G., Zel'dovich, Yu. B., Sunyaev, R.A., and Khlopov, M.
    Yu. 1980, Sov. Astron. Letters, 6, 457.
Fabbri, R., Guidi, I., Melchiorri, F. and Natale, V. 1980, Phys. Rev.
    Letters, 44, 1563.
Fabbri, R., Guidi, I., Melchiorri, F. and Natale, V. 1981, Proc. 2nd
    Marcel Grossman Meeting (in press).
Matzner, R. 1980, Ap. J., 241, 851.
Ostriker, J.P. and Cowier, L.L. 1981, Ap. J. (Letters), 243, L127.
Peebles, P.J.E. 1980, The Large-Scale Structure of the Universe
    (Princeton: Princeton University Press).
Rubin, V.C., Thonnard, N., Ford, W.K., and Roberts, M.S. 1976, A.J.,
    81, 719.
Smoot, G.F., Gorenstein, M.V., and Muller, R.A. 1977, Phys. Rev.
    Letters, 34, 898.
Smoot, G.F. and Lubin, P.M. 1979, Ap. J. (Letters), 234, L83.
Tonry, J. and Davis, M. 1981, Ap. J., 246, 680.

White, S.D.M. and Silk, J. 1979, Ap. J., 231, 1.
Wilson, M.L. and Silk, J. 1981, Ap. J., 243, 14.
Yahil, A., Sandage, A., and Tamman, G. 1980, Physica Scripta, 21, 635.

THE COSMIC RAY ORIGIN PROBLEM

Jerzy Wdowczyk
Institute of Nuclear Research, Lodz, Poland

SUMMARY

A survey is made of the measurements of the various components of the
cosmic radiation: protons, nuclei and electrons. The form of the
energy spectra and anisotropies are considered. The problem of
the location of cosmic ray sources is examined and it is concluded
that there is still no agreement; what does appear to be confirmed
is that the bulk of the particles below 10^{14}eV are generated in the
Galaxy and those above 10^{18}eV are extragalactic. The energy at
which the transition occurs is very uncertain. A variety of features
of cosmological interest are discussed.

1. THE ENERGY SPECTRA AND ANISOTROPIES OF COSMIC RAY NUCLEI

The cosmic radiation is one of a number of important phenomena in
interstellar space. The energy density of cosmic rays in the
neighbourhood of the solar system is comparable with the energy
density of the stellar radiation there. In spite of the fact that
the existence of cosmic rays has been known for almost seventy
years the problem of their origin is still fully open. At the
present moment the localisation of their sources, and the mechanisms
of their acceleration and propagation are all unknown.

The present article starts with a review of the experimental
properties of cosmic rays. The main characteristic is their energy
spectrum; this spectrum is known in a very wide energy range from
energies as low as a few tens of megaelectronovolts to as high
as 10^{11} gigaelectronovolts, (i.e. 10^6 - 10^{20}eV). The direct
measurements of cosmic ray spectra extend up to several thousand
GeV. The spectra are shown in figures 1 and 2. The data on the
proton and helium spectra are taken from the paper by Ryan et al.
(1972)and from the recent American-Japanese-Polish collaboration
(Gregory et al. 1981). The data on heavier nuclei are taken from
Simon et al. (1979). The overall spectrum expressed in terms of

161

A. W. Wolfendale (ed.), Progress in Cosmology, 161–175.
Copyright © 1982 by D. Reidel Publishing Company.

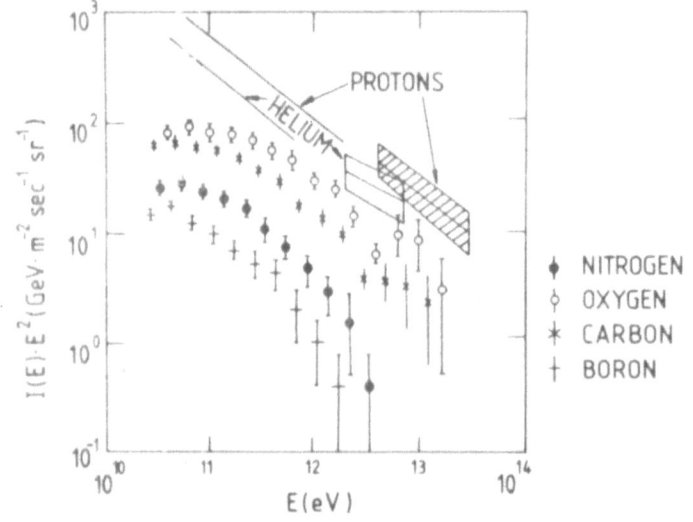

Figure 1

Energy spectra of
light and medium
nuclei in the
cosmic radiation.

energy per nucleus is obtained by summation of the spectra given in
figures 1 and 2 and is compared in figure 3 with the measurement of
that spectrum by Grigorov et al. (1971). In the latter measurement
the energy spectrum was obtained irrespective of the charge of the
cosmic ray particles using a calorimeter carried on a space satellite.

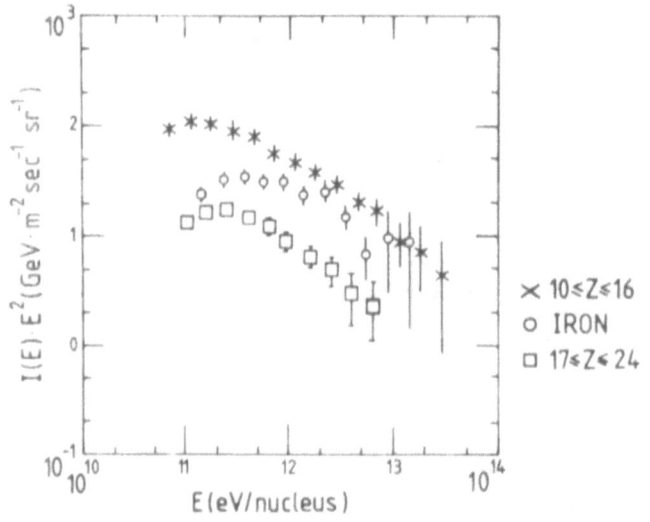

Figure 2

Energy spectra of
heavy nuclei in the
cosmic radiation.

At energies exceeding few hundred TeV practically only relatively
direct information about energy spectra can be deduced from
investigations of extensive air showers (EAS). The most direct
method, which is practically independent of the assumption about the
properties of high energy interactions and mass of the primary
particles, is the method based on the so-called 'equal intensity cuts'.
The essence of the method lies in the determination at various
depths of the sizes (total number of particles) of the showers with

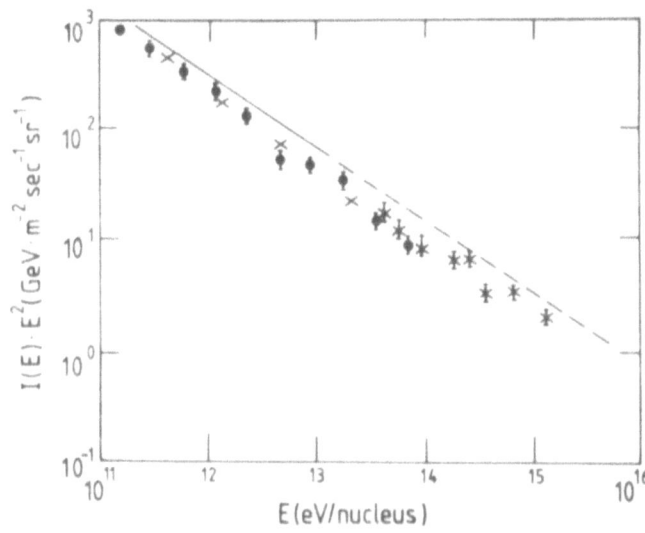

Figure 3

Spectrum of all cosmic
ray nuclei obtained
from the data
summarised in figs. 1
and 2 compared with
earlier measurements
of Grigorov et al.
The dotted line is an
extrapolation.

the same intensity. The requirement of the same intensity assures
that at all depths showers with practically the same primary energy
are registered. The method is illustrated in figure 4, where data
from various experiments are put together under the requirement that
the observed intensity of the showers amounts to 10^{-8} m^{-2} s^{-1} sr^{-1}.

The primary spectrum obtained using the data on the equal intensity
cuts recently summarised by Kikamoto et al., 1981, Hara et al., 1981
and Andrashitov et al., 1981, is shown in figure 5 together with
the spectrum from figure 3. At energies below 10^{16}eV(10^4TeV) the

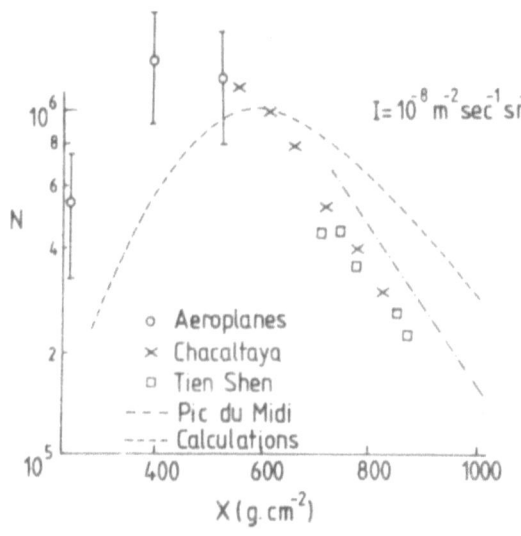

Figure 4

Equal intensity cut for
extensive air showers with
frequency $10^{-8}m^{-2}sec^{-1}sr^{-1}$.

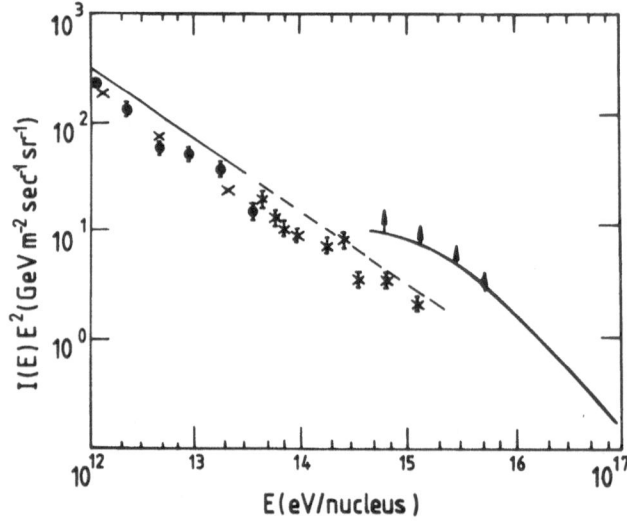

Figure 5

Spectrum deduced from
EAS compared with
that extrapolated from
lower energies.

intensities should be taken as a lower limit since the shower maxima
are reached above the highest observation level (Mt. Chacaltaya),
whereas in the evaluation it was assumed that the maxima are reached
at the Chacaltaya level. It is seen that there is a discrepancy
between the EAS and satellite data. In the author's opinion the EAS
data in that energy region are more reliable as the calorimetric
experiment at energies above 10^{14}eV is obviously dependent on an
assumed high energy interaction model.

If the extensive air shower data are taken together with low
energy direct measurements some flattening of the spectrum at
energies around 10^{14}eV (100 TeV) is indicated. That point will be
taken up later.

At even higher energies the information on the energies of the EAS
primary particles must be obtained on the basis of model calculations
as at the present moment there does not exist an EAS device located
at mountain altitudes which is sufficiently large for the highest
energy showers. The best data on the energy spectrum come from the
Haverah Park experiment (Cunningham et al., 1980). The data are
plotted in figure 6 together with the data from equal intensity
cuts. In view of the large uncertainty in energy determination due
both to unknown nuclear physics and to primary mass composition, the
agreement is moderately good.

In the overlap region the agreement of slopes is impressive. Insofar
as the Haverah Park result is more sensitive to the assumed nuclear
physics its spectrum has been displaced (diagonally) upwards, as
indicated by the dotted line. The resulting spectrum is, in the
author's view, the best available at the present time.

The spectra shown in figures 1 - 6 can be summarised as follows.

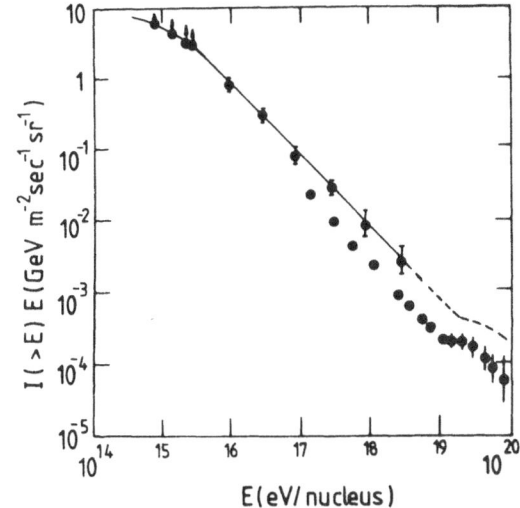

Figure 6

Sepctrum of the highest
energy cosmic rays (see
the text for the significance
of the dotted line).

Their general shape can be described by a power law with slowly
varying index over a very wide energy interval. The spectrum shows a
very pronounced 'kink' at an energy of $\simeq 3.10^{15}$eV. The kink seems
to be preceeded by a 'bump' in the range 10^{14}eV to 10^{16}eV. At the
highest energies (above 10^{19}) the spectrum seems to fall less
rapidly, the turn up being relatively well founded statistically.
Recently, authors from the Yakutsk experiment (Efimov, 1981) have
cast some doubt on the existence of the turn up but the Haverah
Park experiment seems to have confirmed it by, amongst other evidence,
registering a shower with energy clearly exceeding 10^{20}eV (Bower et
al., 1981).

The next parameter characterising the primary cosmic ray particles
is their mass composition. The only certain data relate to the
region of direct measurements, summarised in figures 1 and 2; at
EAS energies, so far only indirect information has been obtained.
On the basis of the data on fluctuations in EAS (measured mainly by
means of muon to electron ratio fluctuations) it has been concluded
that protons comprise a significant fraction of the primaries
initiating EAS with energies in the range $10^{15} - 10^{17}$eV (at least 40%;
see Dzikowski et al., 1979 and Kirov et al., 1980). The data at
higher energies are more scarce but the same tendency - a significant
fraction of protons - seems to be preserved.

The data on mass composition presented in figures 1 and 2 can be
used to deduce information about cosmic ray propagation in space.
Comparison of the relative intensities with the Universal elemental
abundances clearly shows that some of the elements are products
of the fragmentation of heavier elements in space. Among the most
clear secondaries we can place boron, which originates from the CNO
group, and the group $17 \lesssim Z \lesssim 24$ which comes from iron. In figure 7
there are plotted the secondary to primary ratios for the above

cases as a function of energy per nucleon. The last parameter is used here since the propagation of cosmic rays depends on value of that parameter and fragmentation of heavier nuclei produces lighter ones with the same energy per nucleon.

It is seen that the ratios clearly decrease with increasing energy. This proves that the path length traversed in space also decreases for particles with higher energies. Detailed calculation of the fragmentation probabilities allows an evaluation of the absolute values of the path length and also its variation with energy. Results of the calculation are summarised in figure 8. It is seen that the

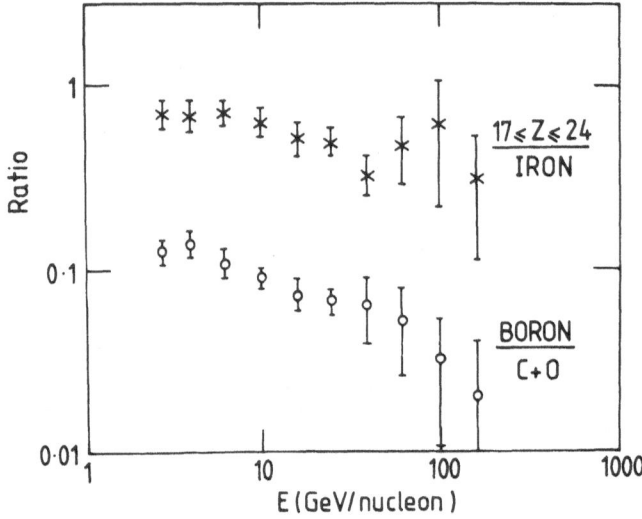

Figure 7

The ratio of secondary to primary cosmic ray intensities in space.

path length (and most likely the confinement time) decrease as $E^{-\alpha}$ with α between 0.3 and 0.5. If the confinement time decreases to that degree the ambient spectrum should be steeper than that actually produced in the sources. The power law index in the ambient spectrum is $\simeq 2.7$ so the spectrum at the sources should have slope 2.2 - 2.4.

An interesting point is that if at a certain energy, say around 10^{14}eV, the confinement time becomes energy independent the ambient spectrum may become relatively flatter. This picture may be the reason for the apparent flattening of the spectrum near 10^{14}eV. That would be a simple explanation for the existence of the earlier suggested "bump" in the primary cosmic ray spectrum.

Another interesting parameter of the primary cosmic rays is the distribution of their arrival directions. Due to the existence of the magnetic field in the Galaxy the potential anisotropies for low energy particles are very small. The anisotropies may increase when the energies of the particles increase as the effect of the field become weaker. A summary of the data on the anisotropies is given

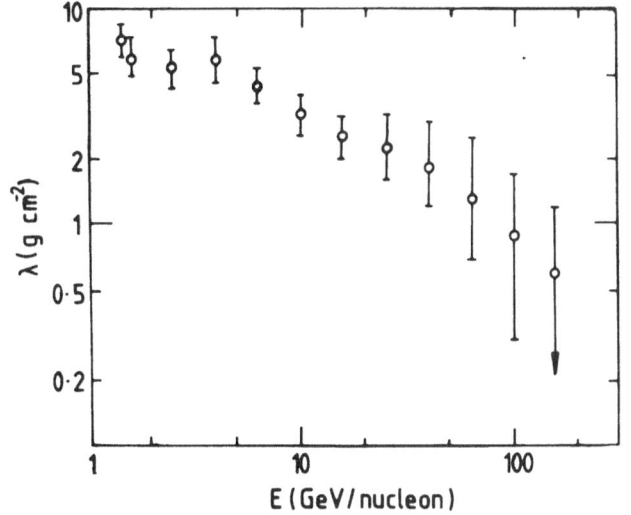

Figure 8

The energy dependence
of the mean path length
of cosmic rays

in figure 9; the anisotropies are seen to be increasing with primary
particle energy, as expected. In addition to the size of the
anisotropy, interesting information follows from an investigation
of the actual direction of the cosmic ray intensity enhancement. This
aspect is illustrated in figure 10 for the highest energy cosmic
rays. In that figure the average sine of the galactic latitude of
shower arrival direction is plotted as a function of primary
particle energy. It is seen that for energies below 3×10^{19}eV the
value agrees with expectation for semi-isotropic distribution. For
higher energies however, the showers tend to come predominantly
from the direction perpendicular to the Galactic plane. (See Giler
et al. 1980). This result, as will be discussed later, supports an

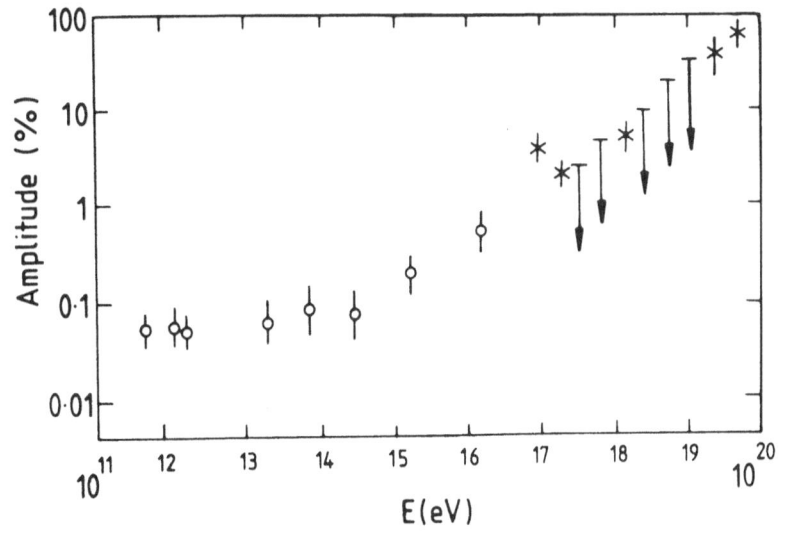

Figure 9

Anisotropies
of cosmic
rays.

extragalactic origin of the highest energy cosmic rays.

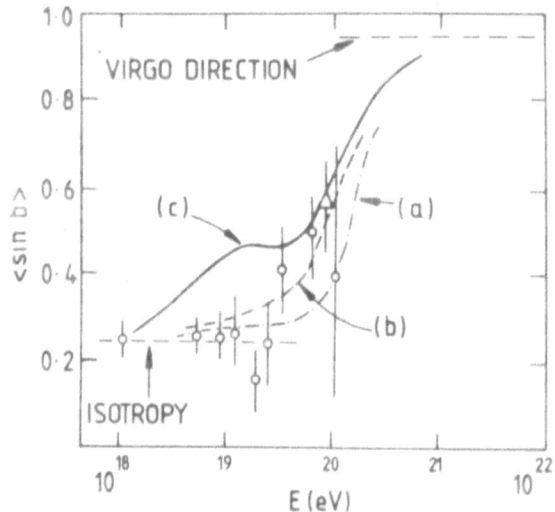

Figure 10

Mean values of the sine
of the Galactic
latitude. The curves
marked are the
predictions of the Virgo
origin model for various
assumptions (see Giler
et al., 1980a).

2.   THE ENERGY SPECTRUM OF PRIMARY ELECTRONS

The energy spectrum of electrons is indicated in figure 11. The
main feature of the spectrum is the very rapid change of slope at a
few GeV energy, this change being larger than expected on the basis
of the simple cosmic ray propagation model (the leaky box model).
There are two possible explanations for the large change of slope:
either it is simply a property of the production spectrum or it is

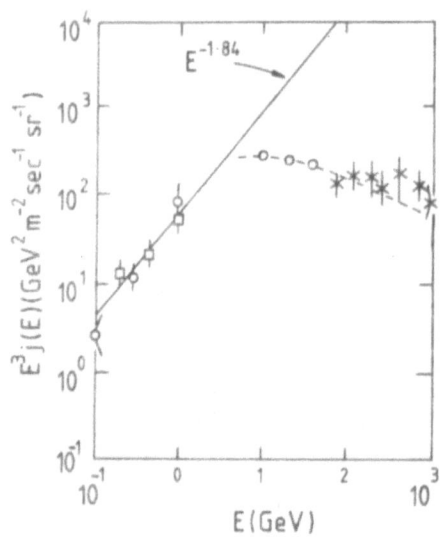

Figure 11

Energy spectrum of cosmic ray
electrons.

an effect of propagation. The second possibility has been fully
analysed by Giler et al. (1980b). The essence of the analysis lies
in the assumption that the path length distribution of the electrons
is lacking at short path lengths (there is, in fact, experimental
evidence favouring this situation). Such a condition would occur if
the electron sources were located at relatively large distances
from the solar system.

The most striking feature of the electron spectrum presented in
figure 11 is the fact that at low energies the slope of the spectrum
seems to be significantly lower than the slope of the proton
spectrum. After correcting for the energy losses of electrons, their
ambient spectrum seems to have a slope more like the production
spectrum of protons (which, as pointed out above, seems to be
significantly flatter than the ambient spectrum of protons).

3. THE PROBLEM OF SOURCE LOCATION

Generally, under the problem of 'cosmic ray origin', we understand
the location of the cosmic ray sources, their nature, including the
mechanisms of acceleration, and the mode of cosmic ray propagation.
The first fundamental question which has been discussed for many
years is: are the sources Galactic or extragalactic? It seems that
for the majority of cosmic rays the Galactic origin is much more
likely. The investigation of Galactic γ-rays indicate the
existence of a cosmic ray gradient towards t h e Galactic Anti-
centre which would not be present if cosmic rays were extragalactic
(for details see Wolfendale, 1980). Further support comes from the
fact that galactic models face difficulties such as the very high
energy content of cosmic rays and the prediction of too many
extragalactic gamma rays (at energies E >100 MeV). The last point
has been analysed extensively by Said et al. (1981) and will be
considered here in a little detail.

The recently moderately accurate estimation of the mass of hot gas
in clusters of galaxies allows an estimate of the expected universal
background flux of 100 MeV γ-rays coming from cosmic ray interactions.
Such particles interacting with matter in space produce π^0 mesons
which decay and produce γ-rays.

Said et al. have evaluated the flux of γ-rays assuming two models.
In the first model (E1) it was assumed that the whole Universe is
filled by cosmic rays with the local density: the second model (E2)
assumes that the locally observed cosmic rays come from Virgo
cluster. In the latter case the cosmic ray energy density in
Virgo (and in other clusters of Galaxies) is higher than the local
one. Actually in model E2, it is assumed that the density varies
inversely with distance from the cluster centre.

The results of the calculations are summarised in table I. It is

TABLE 1 Summary of the γ-ray flux predictions for extra-galactic models of cosmic rays origin.

Model	gas (γ-ray isotropic intensity, 10^{-6}cm^{-2}s^{-1}sr^{-1})	Model E1 γ-ray flux from Virgo (10^{-6}cm^{-2}s^{-1}) / γ-ray isotropic intensity, 10^{-6}cm^{-2}s^{-1}sr^{-1}	R	Model E2 γ-ray flux from Virgo (10^{-6}cm^{-2}s^{-1}) / γ-ray isotropic intensity 10^{-6}cm^{-2}s^{-1}sr^{-1}	R
Gas					
1.5 x 10^{13}M$_\odot$ out to 3° radius		0.10	>0.16	5.4	>8.6
2.6 x 10^{13}M$_\odot$ out to 6° radius		0.17	>0.13	6.3	>4.9
Clusters – no halos (5 x 10^{14}M$_\odot$ in average rich cluster)					
Rich clusters only:	0.004	4.5	1	53	12
All clusters:	0.017	19	4	220	51
Clusters with halos (7 x 10^{15}M$_\odot$ in average rich clusters)					
Rich clusters only:	0.056	63	14	140	31
All clusters:	0.069	80	17	310	67
Schramm and Steigman Estimate	0.02 / 0.06	23 / 70	5– / 15		
Intergalactic Gas (for consistency with X-ray background)					
Uniform gas distribution	0.64	700	170		
Gas in rich cluster halos	0.05–0.14	60–160	13–34		

seen that in practically all cases the expected intensities exceed
the observed ones. Taking the case of "all clusters" (which can be
considered as the most likely) we have a ratio of 4. Allowing for
the other processes leading to production of γ-rays we can conclude
that no more than 10% of the local cosmic ray flux could be of
extragalactic origin.

The situation is different for the high energy end of the spectrum.
As was mentioned earlier - the anisotropies of those cosmic rays
indicate their extragalactic origin. A simple model of extragalactic
origin faces serious difficulties caused by the lack of the
expected cut off due to the interaction of the cosmic rays with
the relict radiation (2.7K background). The way out of that
difficulty has been suggested by Giler et al. (1980). The essence
of their model lies in the assumption that the high energy cosmic
rays are produced in the Virgo cluster and they then propagate by
diffusion with an energy dependent diffusion coefficient. With this
model both the energy spectrum and anisotropy can be well re-
constructed for a wide range of parameters. The energy spectrum is
shown in figure 12; the values of the parameters are the same as
those characterising the anisotropy in figure 10.

At the present moment the situation looks to be that the low energy
cosmic rays are of Galactic origin whereas those at the highest
energies are of extragalactic origin. There immediately arises the
problem of the joining point of the Galactic and extragalactic
particles; this question must be regarded as still open. In the
paper by Giler et al. (1980) it was assumed, for simplicity, that
the Galactic cosmic rays disappear around 10^{18}eV, but equally well
they may disappear at a significantly lower energy, perhaps 10^{16}eV
or even 10^{14}eV. Some support for a joining point at 10^{16}eV comes

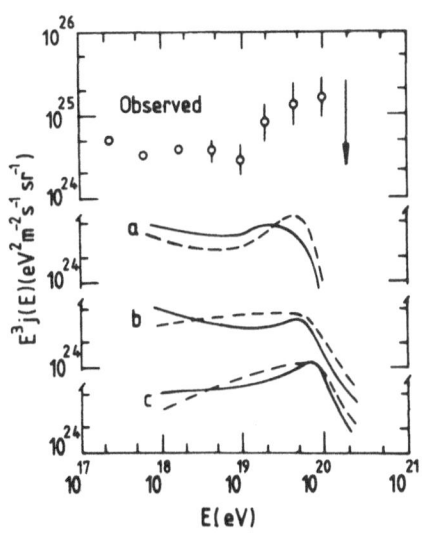

Figure 12

The spectrum of the highest
energy cosmic rays. The energy
spectrum is compared with
predictions of the Virgo origin
model for various assumptions.

from the suggestion of Linsley and Watson (1981) that the average
mass of the primaries decreases when the energy exceeds 10^{16}eV. A
joining at 10^{14}eV would offer some sort of explanation for the
irregular behaviour of the spectrum around 10^{15}eV. Certainly, more
detailed investigations of the EAS region are needed in order to
solve this problem.

4. POSSIBLE COSMIC RAY SOURCES IN THE GALAXY

In general several different types of objects can be considered
as potential cosmic ray sources. Among these the most important
are: supernova remnants, pulsars and shock acceleration in the
interstellar medium. The theories for these sources and for many
others can not avoid significant difficulties. In some models it
is claimed that the difficulties are avoided but the feeling arises
that this is due rather to inappreciation in the knowledge of the
details of the acceleration process. The main contender, the
supernova remnant, gives good energetics but there are serious
difficulties with adiabatic losses; the losses are minimised if it is
assumed that the cosmic rays are produced in later stages of the
expansion of the supernova spell but this leads to restriction of
the possibility of acceleration of high energy particles. The
general difficulty is that the mechanisms are designed to explain
the energy spectra at relatively low energies and they can not
account for the high energy particles.

In the author's opinion in the problem of the cosmic ray origin
investigations of the high energy particles ($\sim 10^{14} - 10^{18}$eV) are
critical. In most theories these particles are neglected simply
because information about them is rather scarce, but a theory which
explains the low energy particles in general would be insufficient
as it is rather unlikely that the particles in this intermediate
region are of an entirely different origin.

Here I would like to describe briefly a phenomenon which may bear
some relation to the problem. Recently it has been shown by
Dzikowski et al.(1980) that there is an excess of cosmic ray
showers from the general direction of the Crab nebula. The most
recent experimental data are summarised in table II. The excess
deduced from the table is statistically very significant as it
amounts to 40.6 ± 10.2. The showers were first interpreted as due
to photons but at the present moment their nuclear origin seems
to be somewhat more likely both by muon content and by indications
that there is no clear point source at the Crab. The nuclear
origin of the showers could be assured by the so called 'neutron
hypothesis'. The life time of neutrons becomes comparable with the
distance of the Crab at energies of the order of a few times 10^{16}eV.
It should be pointed out that if we detect particles from the Crab
the streaming should produce a strong anisotropy as the time from
the date of the explosion is somewhat less than 20% higher than the
transit time for light from the pulsar.

TABLE II Summary of data on arrival direction of cosmic rays with sizes above 10^{16}eV detected in Lodz. The Crab is situated in the middle of the $4^{15} - 6^{45}$ interval.

Sideral time interval (h)	18.12.75 - 22.04.77	23.04.77 -28.11.79	29.11.79 -23.03.81	18.12.75 -23.03.81
$18^{15}-20^{45}$	18	12	29	59
$20^{45}-23^{15}$	15	21	15	51
$23^{15}-1^{45}$	27	22	28	77
$1^{45}-4^{15}$	22	20	16	58
$4^{15}-6^{45}$ CRAB	33	34	37	104
$6^{45}-9^{15}$	27	20	22	69
$9^{15}-11^{45}$	24	16	30	70
$11^{45}-14^{15}$	25	22	23	70
$14^{15}-16^{45}$	21	15	19	55
$16^{45}-19^{15}$	23	14	24	61

5. COSMIC RAYS AND COSMOLOGY

The analysis presented above shows that in the cases where we have some information we find that the presently observed cosmic rays are produced relatively recently, no more than 10^9 years ago. This is the case for the low energy cosmic rays as they are indicated to be of Galactic origin. The average path length of the cosmic rays taken together with data on density of matter in the Galaxy do not allow their life time to be longer than a few times 10^8 years.

The life of the highest energy cosmic rays on the other hand cannot be longer than 10^9 years due to their interaction with the black body radiation. The cosmic rays below 10^{19}eV, in principle, could be much older, as the Universe is transparent for them. If they were much older then those above a few times 10^{19}eV, however, the shape of their spectrum would have to be very different from that observed. In that case the famous black body cut-off would have to exist.

The relation of cosmic rays to cosmology could therefore be only secondary although this statement should not be taken as definite as we still have not proved that any of the known objects could be taken as a definite source of the majority of cosmic rays.

If the sources are unknown we cannot exclude the possibility that cosmic rays are produced by some exotic objects. One of the contenders could be cosmological black holes but the bulk of the low energy cosmic rays cannot be produced by the decay of the Hawking's

type black holes. In such a case the composition of cosmic rays
would have to be dramatically different; the fluxes for antiprotons
and electrons would have to be higher by several orders of magnitude.
Detailed calculations presented earlier by Kiraly (these Proceedings)
confirm that black holes can explain only no more than 10^{-4} of the
local cosmic ray flux. It should be pointed out however, that
energetically black holes could explain the very high energy cosmic
rays. This depends on the mass of black holes at the moment when
they decompose into the massive bosons of the grand unification
theory; the mass would have to be rather high; of the order of
10^{10}g. The attractive feature is that the black holes would
produce a very flat cosmic ray spectrum in agreement with the observation of the reduced rate of fall at the highest energies.

A different approach to the problem has been presented recently by
Dedenko et al. (1981). These workers assume the existence of stable
black holes with the Planck mass, called maximons, and their bound
states (maximonium, maxitronium etc.). Transformation of the
particles may lead to the production of cosmic rays with energies
as high as 10^{28}eV. The spectrum of the particles would be very
flat.

The method for distinguishing between the simple hypothesis, as for
example origin in the Virgo cluster, and the mini (and macro) black
holes would be to investigate the cosmic ray spectrum at energies
above 10^{20}eV. The detection of particles with energies exceeding
10^{21}eV would support the black hole hypothesis as distinct from
Virgo origin, the reason being that the mean free path of the
protons in the black body radiation is shorter than the distance
to the   go cluster. Another point is that in the case of the
black hole hypothesis we can expect at the highest energies the
existence of primary photons as the Universe is more transparent
to them than to protons. If the black holes were concentrated in
the cores of clusters of galaxies - a possible situation - photon
initiated showers could well appear in the energy range 10^{21}-10^{22}eV.

It should be pointed out that the detection of 10^{21} - 10^{22}eV
particles is very difficult because of the very low fluxes involved.
More modern methods based on the detection of radio, scintillation
and Cerenkov light are preferable here.

References

Andrashitov, S.F., et al., 1981, 17th ICRC, Paris, V6, pp 156-159.
Bower, A.J., et al., 1981, 17th ICRC, Paris, V2, pp 113.
Cunningham, G., et al., 1980, Astrophys. J. (Letters), 236,
    pp L71-L75.
Dedenko, L.G., et al., 1981, 17th ICRC, Paris, MN 6-17, late paper.
Dzikowski, T., et al., 1980, IAU Symposium No. 94, Bologna, Italy,
    June 11-14, 1980, pp 327-328.
Dzikowski, T., et al., 1979, 16th ICRC, Kyoto, V8, pp 276-281.

Efimov, N.N., 1981, private communication.
Giler, M., et al., 1980a, J. Phys. G. Nucl. Phys. 6, pp 1561-73.
Giler, M., et al., 1980b, Astron. Astrophys. 84, pp 44-49.
Gregory, J.C., et al., 1981, 17th ICRC, Paris, OG2.1-4, late paper.
Grigorov, N.L., et al., 1971, 12th ICRC, Hobart Vol. 5, pp 1746-51.
Hara, T., et al., 1981, 17th ICRC, Paris, EA 1-29, late paper.
Kakimoto, F., et al., 1981, 17th ICRC, Paris, EA 1-22, late paper.
Kirov, I.N., et al., 1980, Int. Sem. on Cosmic Ray Cascades, Sofia,
   13-19 October, pp 61-74, edited by T.K. Gaisser and T. Stanev.
Linsley, J., and Watson, A.A., 1981, 17th ICRC, Paris, V2, pp 137-40.
Ryan, M.J., et al., 1972, Phys. Rev. Lett. 28, pp 985-88.
Said, S.S., et al., 1981, 17th ICRC, Paris, V1, pp 251-54.
Simon, M., et al., 1979, 16th ICRC, Kyoto Vol. 1, pp 352-57.
Wolfendale, A.W., 1980, IAU Symposium No. 94, Bologna, Italy,
   June 11-14, 1980, pp 309-319.

COSMIC FAR-ULTRAVIOLET BACKGROUND RADIATION

Richard C. Henry

Physics Department, Johns Hopkins University
Baltimore, Maryland 21218   USA

ABSTRACT

It is demonstrated that interstellar dust grains forward-scatter far-ultraviolet radiation extremely strongly: the value of the Henyey-Greenstein scattering parameter g at 1425 Å is shown to be at least 0.75; the actual value is very likely greater than 0.9. Also, observations of the Virgo cluster of galaxies sets a limit $\tau > 2 \times 10^{25}$ sec on the lifetime of 17-20 eV/c^2 heavy neutrinos, if such neutrinos are responsible for the gravitational binding of the cluster.

I.   INTRODUCTION

The simplest and most esthetically satisfying cosmological model is the null model. However, the mere existence of the question "What is the true model of the Universe?" by itself rules out, observationally, the null model. Next most satisfying is the universe as a zero-energy fluctuation, and, happily, the observations of the geometry of the universe (Kristian et al. 1978) suggest that we are at least close to that situation.

It is well known that there is insufficient luminous matter in the universe to provide the gravitational potential energy needed to balance the kinetic energy of expansion of the universe, but invisible matter may be present to do the job: Faber and Gallagher (1979) conclude that "the case for invisible mass in the universe is very strong . . . we think it likely that the discovery of invisible matter will endure as one of the major conclusions of modern astronomy."

One form that invisible matter sufficient to close the universe perhaps could take is extremely hot gas (Field and Henry 1964; Marshall et al. 1980). This requires the re-heating of the universe, a process which would generate unique signatures in the cosmic ultraviolet background (Weymann 1967); signatures that we have searched for (Henry et al. 1978a; Anderson et al. 1979). Closing the universe with baryons is

A. W. Wolfendale (ed.), Progress in Cosmology, 177–188.
Copyright © 1982 by D. Reidel Publishing Company.

difficult, however, because of helium and deuterium abundance problems
(Schramm and Wagoner 1977).

More attractive is closing the universe with massive neutrinos
(Cowsik and McClelland 1972; Schramm and Steigman 1981). Such neutrinos
would decay, producing, again, a characteristic signature in the cosmic
ultraviolet background (Cowsik 1977); a signature that we have also
searched for (Henry 1981a).

Cosmic ultraviolet background radiation has been reviewed recently
by Paresce and Jakobsen (1980), and by Henry (1981a, b). Instead of
giving a general review, therefore, in the present paper I shall present
two new results.

II. SCATTERING PROPERTIES OF INTERSTELLAR GRAINS

The first new result concerns the far-ultraviolet scattering properties
of interstellar dust.

There appears to be general agreement that in the far ultraviolet,
the albedo of the interstellar grains is high; say 0.5 at 1400 Å (see,
for example, Carruthers and Opal 1977). But there is great controversy
concerning the value of the Henyey-Greenstein (1941) scattering parameter
g, with values ranging from 0.95 (Henry et al. 1978b) to 0.25 (Witt
1977-78) being reported. A value of 0.95 corresponds to extremely
strong forward scattering, while 0.25 represents more nearly isotropic
scattering. Knowledge of the correct value is very important, because
if the grains scatter isotropically, a local general glow will be seen
at high galactic latitudes, being the back-scattered light of bright
galactic-plane B stars; this glow would impede the search for ultraviolet
light of cosmological significance, and, incidentally, would limit the
sensitivity of Space Telescope.

I present data which very strongly indicate that $g \geq 0.75$, and that
show in fact that g is probably considerably larger even than this value.

The data are the so-called "Dark South" target observed on Apollo
17 (Henry et al. 1978a). The observed spectrum is shown in Figure 1,
with no correction for stars in the field of view, and no correction for
grating-scattered solar-system Lyman α radiation. The dashed line is at
a level of 1000 photons (cm^2 sec ster Å)$^{-1}$, units which I shall refer to
as "units" from now on. Between 1350 and 1550 Å, these data provide an
absolute upper limit on the cosmic background of about 800 units.

Jura (1979a) has emphasized the model-independence of the value of
g that results from observing dust at high galactic latitudes. The
present target is at $b = -70°$, a latitude at which the dust must be
actually illuminated by the galaxy: we see it; so it sees us, and our
particular region of the galaxy is bright in the ultraviolet (Henry
et al. 1980).

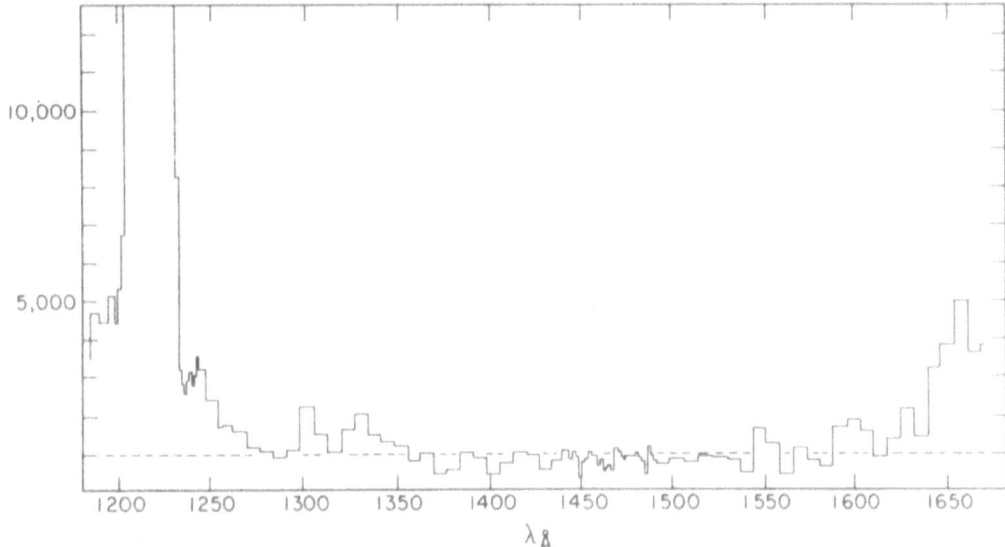

Fig. 1. The spectrum of a region at high galactic
latitudes, as observed on Apollo 17. The intensity
is in photons $(cm^2 \ sec \ ster \ Å)^{-1}$. Lyman α is off-
scale. The data are corrected only for dark current,
which was known precisely. The data provide an
an absolute upper limit, between 1350 and 1550 Å,
of 800 units, on any galactic-plane starlight back-
scattered by dust. Several Rowland ghosts indicate
the presence of grating-scattered Lα radiation, which
shows that the 800 units is actually only an upper
limit on the signal.

How much dust is in the Dark South region of the sky? By good
fortune, the Dark South target is at the *dustiest* high-latitude region
observed by Heiles (1975). Figures 2 and 3 are adapted from Heiles'
paper, and locate the 12° x 12° Dark South target. Figure 17b of Heiles
and Jenkins (1976) shows that most of this hydrogen is in the velocity
range −20 to +20 km sec^{-1}, while Figure 9 of Heiles (1976) shows that
the gas-to-dust ratio is near normal in this region.

There are two additional points that are worth noting regarding
the illumination of this dust. First, the Dark South target is only 70°
away from Orion, one of the most intense sources of ultraviolet illumination in the sky. Second, Sandage (1976) has photographed the optical
light scattered from dust that is located about halfway between Dark
South and Orion.

To deduce limits on the scattering parameter *g*, we use the model
of Jura (1979a, b). This is a conservative choice, in the sense that
the (probably more accurate) numerical integrations of Henry and co-
workers (1978b) predict *more* flux for a given set of parameters (see

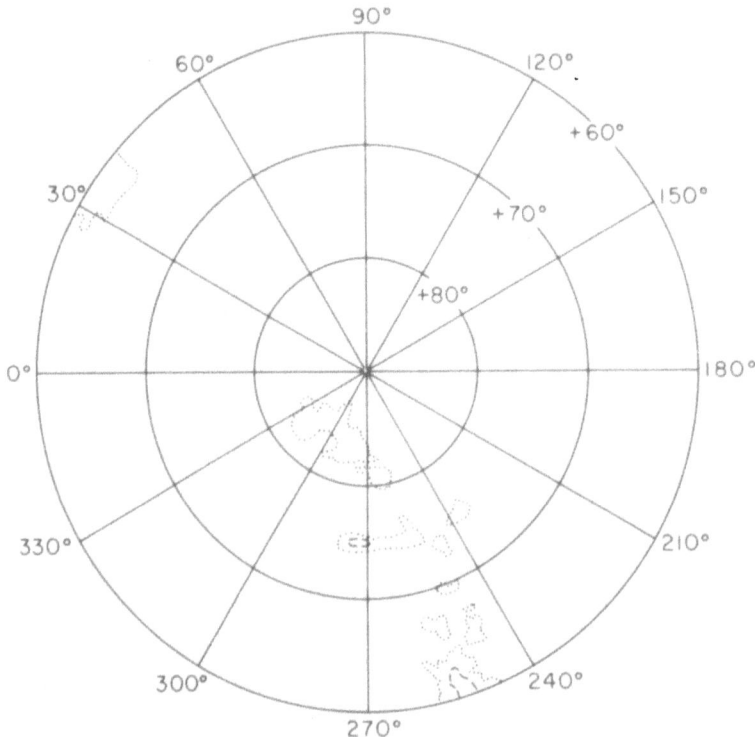

Fig. 2. Regions at high galactic latitudes generally
have only very weak 21 cm emission. This is a 21-cm
map of the north galactic polar cap, adapted from
Heiles (1975). Dotted contours represent 175 Heiles
units, while dashed contours denote 225 Heiles units.
One Heiles unit represents a column density of
2.23×10^{18} hydrogen atoms cm^{-2}.

Figure 4 of Henry 1981c for a comparison of the two models) than does
Jura's model.

In Jura's model, the intensity S of the scattered light is given
by
$$S = \frac{G \tau a}{2} \, H(g, b) \, [1 - E_2 \, (\tau_o)]$$

where
$$H(g, b) = 1 - 1.1 \, g \, (\sin b)^{\frac{1}{2}} = 1 - 1.066 \, g \quad \text{for } b = 70°$$

and
E_2 is the exponential integral (Abramowitz and Stegun 1972)

τ_o is the optical depth in a vertical line through the plane,
= 0.79 at 1400 Å (Jura 1979a);

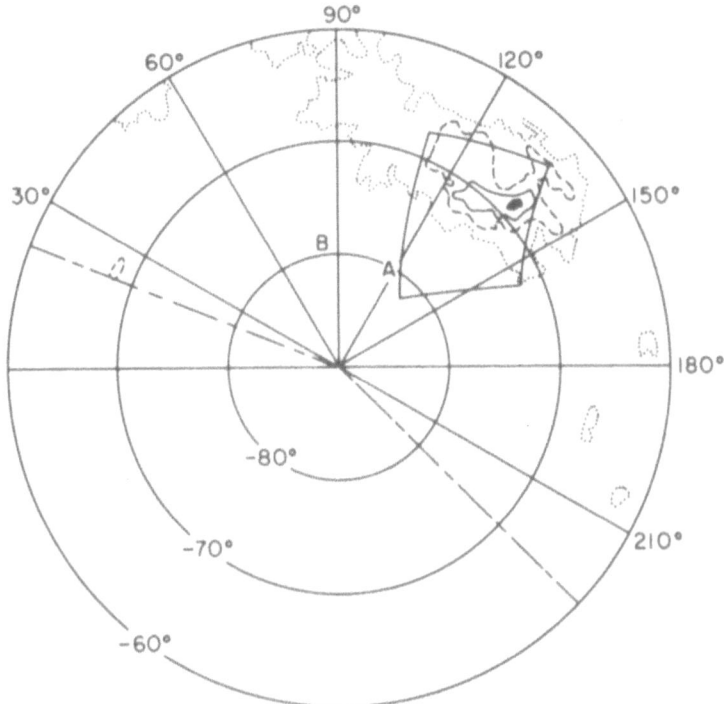

Fig. 3.   The same as figure 2, but for the southern
galactic polar cap.   No 21-cm observations were
made in the lower-left sector bounded by the dashed
lines.   The solid contour represents 275 Heiles units,
while the solid region is > 325 Heiles units.   The
quadrilateral represents the "Dark South" region
observed on Apollo 17 and discussed in the present
paper.   We have observed the dustiest high-galactic-
latitude region observed by Heiles (1975).   Our
upper limit on radiation from this dustiest region
is (figure 1) 800 units.   In contrast, Paresce and
co-workers (1980) report intensities of 1500 units
or greater from positions marked A and B.

G = 3.9 x 10^{-7} ergs cm^{-2} sec^{-1} ster^{-1} Å^{-1} at 1400 Å (Jura 1979a)
  = 27,500 units;

a = grain albedo = 0.5 at 1400 Å;

g = the scattering parameter that we are trying to evaluate;

and τ = the optical depth of the high-latitude cloud.

Jura took τ = 0.31, corresponding to a column density of atomic
hydrogen of 2 x 10^{20} cm^{-2}.   Consulting Figure 3, we see that the average

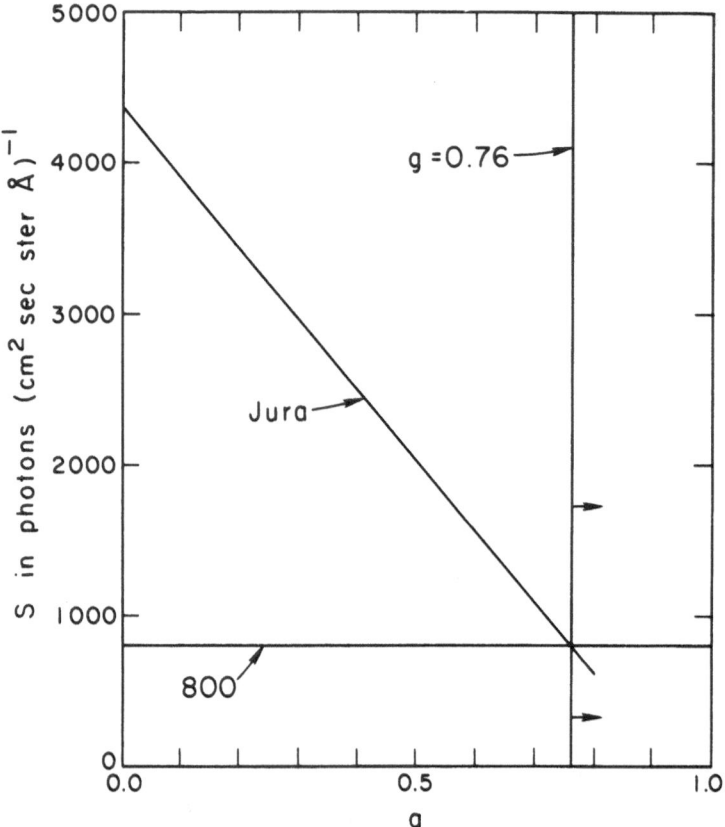

Fig. 4. The sloping line gives the expected high-galactic-latitude intensity, as a function of the scattering parameter g, according to the model of Jura (1979a), using the value of the optical thickness τ appropriate to the cloud we observed. The upper limit on S is, from figure 1, 800 units, so we see that $g \gtrsim 0.75$ (vertical line). This is a lower limit on $\tilde{g}$, for we have not corrected the data for either stars in the field of view, or grating-scattered solar-system Lα radiation.

21-cm column density for the Dark South region is 225 Heiles units, where 1 Heiles unit corresponds to 2.23×10^{18} atoms cm^{-2} (Heiles 1975). So our column density is 5×10^{20} cm^{-2}, and the corresponding value of τ is 0.775. This is the only number we use which is different from that of Jura (1979a).

Thus we obtain S = 4323 (1-1.066 g) units. The computed value for S as a function of g is given in Figure 4, where a horizontal line gives the upper limit of 800 units that was obtained from the data in Figure 1. We conclude that $g > 0.75$. This is the same conclusion as that of Henry

(1981c), but the present result is more secure, because here no corrections have been made for grating-scattered Lα, or for stars in the field
of view.

   This result will resolve some of the "contradictory constraints" on
dust that are noted on pages L181-L182 of Duley and Williams (1980). It
does not by itself, however, guarantee that the flux of 285 ± 32 units
reported at high northern galactic latitudes by Anderson et al. (1979)
is truly extragalactic. Targets 1 and 2 of Anderson et al. are free of
significant stellar flux; grating-scattered Lα is eliminated with a CaF_2
filter, and dark count rate is very low. But Heiles (1975) shows 125
Heiles units of hydrogen at each of these two targets. If there actually
are 800 units of ultraviolet at Dark South (which has 225 Heiles units of
hydrogen), we would expect ∿400 units of ultraviolet at Anderson's
targets, roughly consistent with what he observes.

   But our figure of 800 units for the flux from Dark South was with
no allowance for either stars in the field of view, or grating-scattered
Lα. Those who are of the opinion that a large field of view (the Apollo
17 field of view was 12° x 12°) leads to a large stellar correction,
will obtain a much larger value of g! In fact the stellar correction is
fairly small, particularily at the shortest wavelengths. What is not
small, is the correction for grating-scattered solar system Lα radiation.
Figure 1 shows several features that are certainly Rowland ghosts, while
the laboratory measurements presented in Figure 3 of Fastie and Kerr
(1975) show that grating-scattered Lα is present at all wavelengths. An
estimate of the actual amount of grating-scattered Lα at ∿1425 Å may be
obtained from Figure 13 of Henry (1981a), which shows the amount of such
grating-scattered Lα, as a function of Lα intensity. The Lα intensity
at Dark South was 18.4×10^5 units, which would result in grating-scatter
at 1425 Å of 825 units—which is approximately our total signal.

   If this is so, then g is substantially larger than 0.75; let's say
0.9, and the Anderson et al. data are definitely extragalactic in origin.

   Paresce and co-workers (1980) report a "bright region of intensity
≈1500 units" at position A in Figure 3, and also report an "extended
bright area > 1500 units" at position B. If we are to ascribe the
signal to light scattered from dust, these observations are entirely
incompatible with the present observation of only 800 units or less
from a much dustier region. I have argued elsewhere (Henry 1981a) that
the Paresce and co-workers (1980) results are spurious.

III. DECAY OF HEAVY NEUTRINOS

The second new result that I wish to present concerns the lifetime of
massive neutrinos. Cowsik (1977) set a lower limit of 10^{23} sec on the
lifetime of massive neutrinos, from observations of cosmic background
radiation. New interest in this topic has been stimulated by De Rújula
and Glashow (1980), who set a limit on the muon neutrino decay lifetime

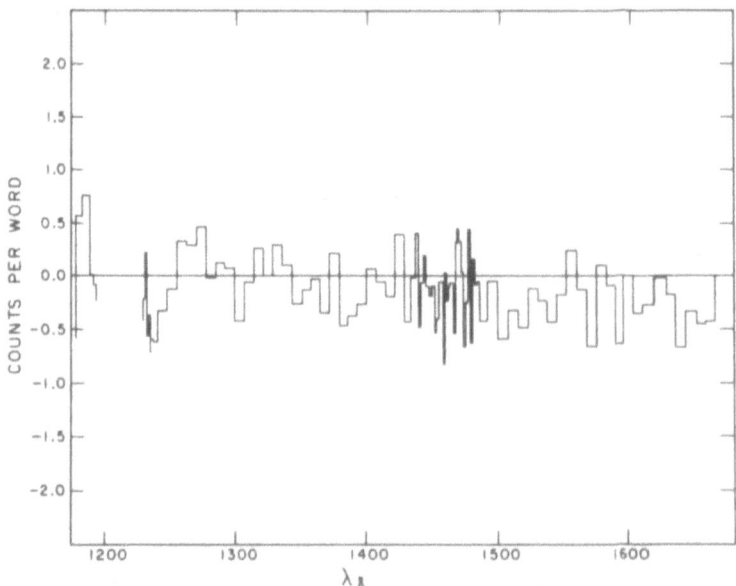

Fig. 5. The far ultraviolet spectrum of the Virgo
cluster of galaxies, observed on the Apollo 17
mission. Wavelengths are in the frame of the spacecraft. A dark region of sky near the earth (the
observations were made about halfway between the
moon and the earth) has been substracted, to remove
dark current and grating-scattered solar system Lα
radiation. One "word" is a 0.1 sec interval. There
is a gap in the data where solar system Lα prevents
meaningful measurement. The average residual
intensity is a shade negative at longer wavelengths,
undoubtedly because of stars in the field of view
in the dark region near the earth. The portion of
sky containing the Virgo cluster is more star-free.
The present observation indicates that the lifetime
of any 16-20 eV/c^2 neutrinos is > 10^{25} sec.

of ∿10^{24} sec, *i.e.*, better than the limit from the ultraviolet background, and who suggest various ultraviolet astronomical observations
as a means of improving the limit. It is possible to use astronomical
observations to improve the limit if, for example, we assume that the
neutrinos are heavily concentrated in rich clusters of galaxies (Shipman
and Cowsik 1981). Henry and Feldman (1981) have presented Apollo 17
data for the Coma cluster of galaxies that provide a limit on the lifetime of any 16-20 eV/c^2 neutrinos of ∿10^{24} sec, *i.e.*, comparable to the
De Rújula and Glashow lifetime, and they state that Apollo 17 data on
the Virgo cluster provide a limit of ∿10^{25} sec.

The Virgo cluster data are displayed in Figure 5. The field of

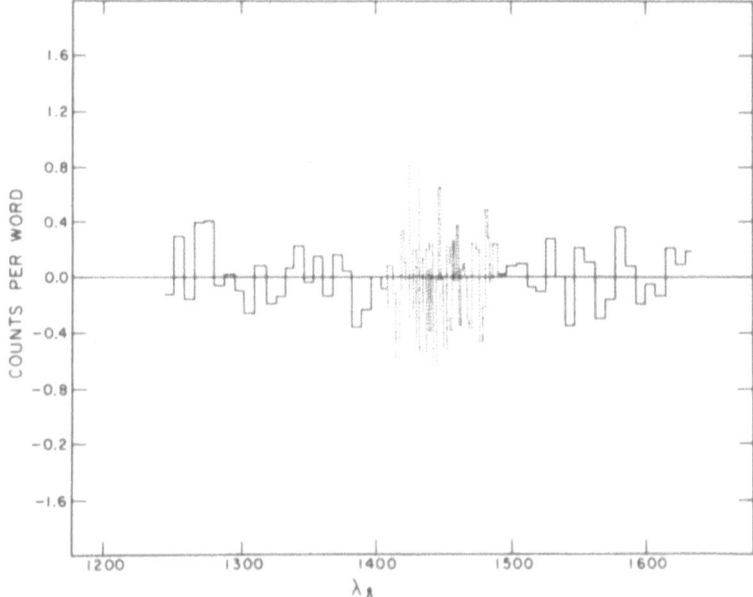

Fig. 6. Combined spectra of the Coma and Virgo
clusters of galaxies, in the laboratory frame.
Spectra in the region 1400-1490 are superposed
rather than averaged. These data set a lower
limit of 2×10^{25} sec on the lifetime of 17-20
eV/c^2 heavy neutrinos, and somewhat lower, lower
limits on 15.5-17 eV/c^2 heavy neutrinos.

view was 12° x 12°, large enough to emcompass even this very large
cluster of galaxies. Spectral resolution was 10 Å, while the velocity
dispersion of the neutrinos in the cluster is only about 5 Å. One "word"
in the figure is a 0.1 sec interval, so the limit on any line feature,
over most of the wavelength range, is about $\frac{1}{4}$ counts per second, corresponding to 180 photons cm^{-2} sec^{-1}, at 1300 Å. The spectrometer
efficiency varies with wavelength. Knowing the mass of the Virgo cluster,
and attributing it predominantly to heavy neutrinos; and knowing the
distance of the Virgo cluster, results, from the observed upper limit on
the ultraviolet flux, in a lifetime > 10^{25} sec, which is better than the
result for muon neutrinos that is deduced from laboratory experiments.
This lower limit on the lifetime, as a function of neutrino mass, is
given in Figure 2 of Henry and Feldman (1981).

A somewhat improved result can be obtained for each cluster, by
combining the result of using different observations of the dark sky as
background corrections. Finally, still further improvement can be
obtained by removing the relative wavelength shift between the average
spectra of the two clusters, and averaging them. The result is shown
in Figure 6. (In the wavelength region near 1470 Å, where the spectrometer cam was programmed to scan more slowly, merging the data proved

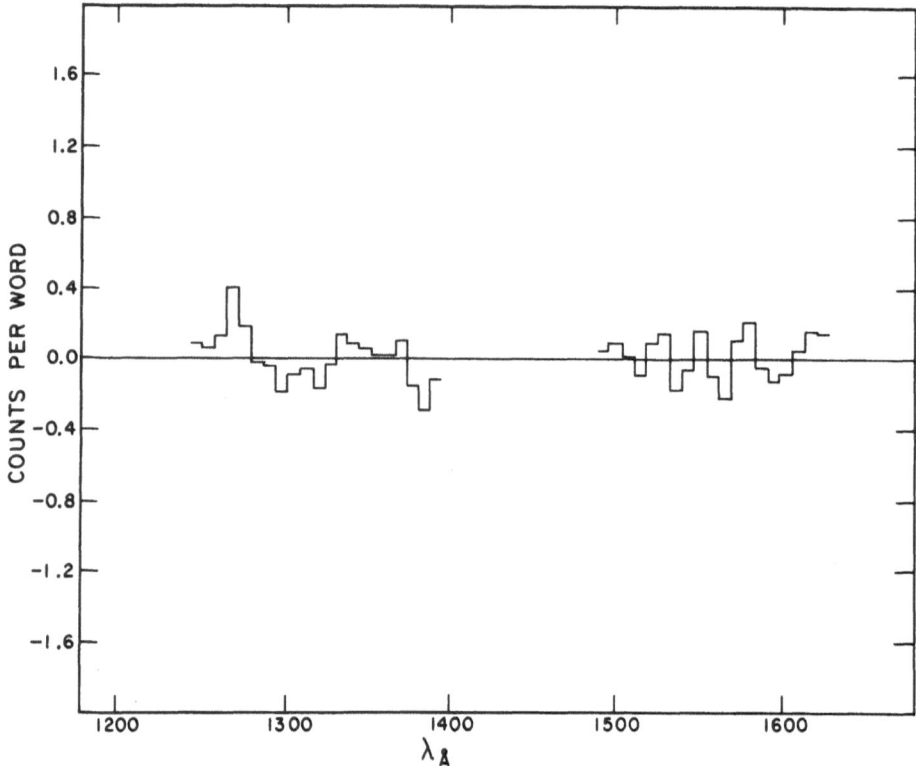

Fig. 7. A running mean of the spectrum that appears
in figure 6.

impractical, and the two spectra are simply superposed.) The result
provides, on average, about a factor two improvement in the limit on the
neutrino lifetime. The probability that the feature at 1275 Å is due to
chance is 20%. This 1275 Å "blip," for what it is worth, shows up more
clearly (Figure 7) when a running mean of two wavelength bins is taken.
Also, it should be noted that the noise is expected to be larger in the
1500-1600 Å region than in the 1250-1400 Å region, because the spectrometer quantum efficiency is about a factor two lower, at the longer
wavelengths.

The present observations are heavily contaminated by grating-
scattered Lα, which is removable by the subtraction process used here,
but which increases the statistical noise. Insertion of a CaF_2 filter
would completely eliminate all Lα, including grating-scattered Lα. Thus,
a very much more sensitive version of the Apollo 17 spectrometer—which
was otherwise ideal for observation of clusters of galaxies—is feasible.
The Space Shuttle itself provides adequate pointing, without any separate
experiment pointing. An early flight of such an experiment would be very
desirable.

IV.   CLOSING THOUGHT

If it is true that the universe is a zero-energy fluctuation, we will never know whether the universe is open or closed. For if the universe is a zero-energy fluctuation, $q_o = 1/2$ exactly, and the universe is "on a knife-edge" between being open or closed. Whether it is the one way or the other presumably would be due to a "fluctuation of a fluctuation," and beyond the scope of any conceivable experiment to decide.

ACKNOWLEDGEMENTS

I thank W. G. Fastie, Principal Investigator on the Apollo 17 ultraviolet spectrometer experiment, and I thank A. Pevsner for originally drawing my attention to the problem of neutrino decay. This work was supported by NASA contract NAS 9-11528, and NASA grant NGR 21-001-001, to the Johns Hopkins University.

REFERENCES

Abramowitz, M. and Stegun, I. A.:   1972, *Handbook of Mathematical Functions*, Washington: U. S. Government Printing Office.
Anderson, R. C., Henry, R. C., Brune, W. H., Feldman, P. D. and Fastie, W. G.:   1979, Astrophys. J. 234, pp. 415-426.
Carruthers, G. R., and Opal, C. B.:   1977, Astrophys. J. (Letters) 212, pp. L27-L31.
Cowsik, R.:   1977, Phys. Rev. Lett. 39, pp. 784-787.
Cowsik, R., and McClelland, J.:   1972, Phys. Rev. Lett. 29, pp. 669-670.
De Rújula, A., and Glashow, S. L.:   1980, Phys. Rev. Lett. 45, pp. 942-944.
Duley, W. W., and Williams, D. A.:   1980, Astrophys. J. (Letters) 242, pp. L179-L182.
Faber, S. M., and Gallagher, J. S.:   1979, Ann. Rev. Astron. Astrophys. 17, pp. 135-187.
Fastie, W. G., and Kerr, D. E.:   1975, Applied Optics 14, pp. 2133-2142.
Field, G. B., and Henry, R. C.:   1964, Astrophys. J. 140, pp. 1002-1012.
Heiles, C.:   1975, Astron. Astrophys. Suppl. 20, pp. 37-55.
Heiles, C.:   1976, Astrophys. J. 204, pp. 379-402.
Heiles, C., and Jenkins, E. B.:   1976, Astron. Astrophys. 46, pp. 333-360.
Henry, R. C.:   1981a, in *Sixteenth Recontre de Moriond*, in press, ed. Tran Thanh Van, J.
Henry, R. C.:   1981b, in *Tenth Texas Symposium on Relativistic Astrophysics,* in press, ed. Ramaty, R., and Jones, F. C.
Henry, R. C.:   1981c, Astrophys. J. (Letters) 244, pp. L69-L72.
Henry, R. C., Anderson, R. C., and Fastie, W. G.:   1980, Astrophys. J. 239, pp. 859-866.
Henry, R. C., Anderson, R., Feldman, P. D., and Fastie, W. G.:   1978b, Astrophys. J. 222, pp. 902-908.
Henry, R. C., and Feldman, P. D.:   1981, Phys. Rev. Lett. 47, pp. 618-619.

Henry, R. C., Feldman, P. D., Fastie, W. G., and Weinstein, A.:   1978a,
    Astrophys. J. 223, pp. 437–446.
Henyey, L. G., and Greenstein, J. L.:   1941, Astrophys. J. 93, pp. 70–83.
Jura, M.:   1979a, Astrophys. J. 231, pp. 732–735.
Jura, M.:   1979b, Astrophys. J. 227, pp. 798–800.
Kristian, J., Sandage, A., and Westphal, J. A.:   1978, Astrophys. J.
    221, pp. 383–394.
Marshall, F. E., Boldt, E. A., Holt, S. S., Miller, R. B., Mushotsky,
    R. F., Rose, L. A., Rothschild, R. E., and Serlemitsos, P. J.:
    1980, Astrophys. J. 235, pp. 4–10.
Paresce, F., and Jakobsen, P.:   1980, Nature 288, pp. 119–126.
Paresce, F., McKee, C. F., and Bowyer, S.:   1980, Astrophys. J. 240,
    pp. 387–400.
Sandage, A.:   1976, Astron. J. 81, pp. 954–957.
Schramm, D. N., and Steigman, G.:   1981, Astrophys. J. 243, pp. 1–7.
Schramm, D. N., and Wagoner, R. V.:   1977, Ann. Rev. Nuclear Sci. 27,
    pp. 37–74.
Shipman, H. L., and Cowsik, R.:   1981, Astrophys. J. (Letters) 247,
    pp. L111–L114.
Weymann, R.:   1967, Astrophys. J. 147, pp. 887–900.
Witt, A. N.:   1977–1978, Publ. Astron. Soc. Pacific 89, pp. 750–757.

PROSPECTS FOR X-RAY OBSERVATIONS OF COSMOLOGICAL SIGNIFICANCE

K. A. Pounds

X-ray Astronomy Group, University of Leicester, U.K.

ABSTRACT

The cosmic X-ray background (at least in the energy band $\sim$2-10 keV) shares with the microwave background the property of originating at a high redshift ($z_x \sim 1-5$; $z_m \sim 10^3$). Thus, studies of the structure, spectrum and origin of the X-ray background are potentially important cosmologically. Existing measurements of the background isotropy and deductions made therefrom are reviewed and seen to provide interesting limits on the matter distribution on scales larger than that of super-clusters. Source counts from the Einstein Observatory and earlier sky survey experiments show a significant (and possibly dominant) component of the X-ray background to arise from a strongly evolving population of high redshift QSO's. However, the present X-ray data do not yield definitive cosmological data, and it is concluded that the realisation of this potential must await the deep all-sky survey of ROSAT (in $\sim$1987) and, more particularly, the 1.2 metre AXAF X-ray telescope ($\sim$1990?) with its capability to study many types of X-ray source to redshifts $z \gtrsim 1$.

1. INTRODUCTION

The large scale isotropy of the cosmic X-ray background (XRB) radiation (at E $\gtrsim$ 3 keV, where the emission of most stellar sources and of the hot component of the interstellar medium becomes negligible) demonstrates its probable extragalactic origin; the small scale isotropy and source counts show the XRB arises mainly at a redshift $z \gtrsim 1$. Thus it is immediately evident that X-ray observations are potentially important for cosmology.

In section 2 the present limits on the isotropy of the ($\sim$2-10 keV) XRB are reviewed and their implications discussed. In particular, it is noted that X-ray observations already yield the most sensitive constraints to density inhomogeneities in the Universe on scale lengths

189

A. W. Wolfendale (ed.), Progress in Cosmology, 189–202.

of 100 - 1000 Mpc.

Section 3 briefly summarises the observations of individual extragalactic X-ray sources, which support the view that the XRB originates
largely at z ⪆ 1 and show that a substantial fraction - perhaps most -
of the background radiation is the integrated emission of an evolving
distribution of QSO's.

Section 4 suggests that marginal further advances may be expected
from the analysis of existing data (from the HEAO-1 and Einstein
Observatory spacecraft) and from EXOSAT, due for launch next year.
The overall conclusion is that definitive cosmological data must await
the next generation of X-ray satellites, viz. ROSAT, a German 0.8 metre
imaging telescope (due for launch in 1987), and the NASA Advanced X-ray
Astronomy Facility (AXAF), hopefully to be launched early in the 1990's.
Finally, in section 5, the potential of several cosmological tests in
the ROSAT/AXAF era is reviewed.

2.   THE ISOTROPY OF THE X-RAY BACKGROUND

Maps of the X-ray sky in the ~ 2 - 10 keV band have been produced
from each of the earlier all sky survey experiments, Uhuru (Schwartz,
1980), Ariel-5 (Warwick et al, 1980) and HEAO-1 A2 (Boldt, 1981). They
are all in agreement in showing a residual galactic plane excess at the
few percent level, with the North Galactic Pole being brighter than
the South Galactic Pole. The origin of this galactic asymmetry is not
clear, but it does conform to the known differences in optical galaxy
counts and in the soft X-ray (E ⪅ 1 keV) background. Several results
of 'cosmological interest' have also been derived from the above X-ray
maps. These are summarised below.

2.1 After removal of the galactic component from the Ariel-5 sky map,
Warwick et al (1980) found a statistically significant, oblate, 12-hour
(quadrupole) variation in the 2 - 18 keV background. Moreover, the
pole position of this effect ($l \sim 245°$, $b \sim 53°$) was coincident, within
rather coarse errors, with the direction of the 24-hour (dipole)
anisotropy in the cosmic microwave background (CMB) discovered by Smoot
et al (1977). Figure 1 illustrates a possible explanation of this
effect proposed by Fabian and Warwick (1979). Their idea is that the
gravitational effect of a large density perturbation, of mass δM (lying
beyond at least a substantial fraction of the XRB), causes a 'tidal
shear' which reduces the apparent X-ray intensity towards and directly
away from the perturbation. The discussion of measured source counts
to high redshift (section 3) suggests the observed 12-hour anisotropy
(if real) requires a mass perturbation at z ⪆ 1. Warwick et al also
point out that a large matter perturbation at z ⪅ 3 should also be
detectable by other means, such as radio source and QSO counts; however,
current limits on large-scale anisotropies in these counts are only at
the ~ 10% level. In fact, the Ariel-5 12-hour variation still requires
experimental confirmation and was not found in an analysis of the Uhuru

XRB map (Protheroe et al, 1980). Systematic errors, particularly the adequate removal of the galactic variation, are admitted by both groups to be significant. Nevertheless, the reported 12-hour effect does provide an interesting preview of the potential of XRB measurements in the future.

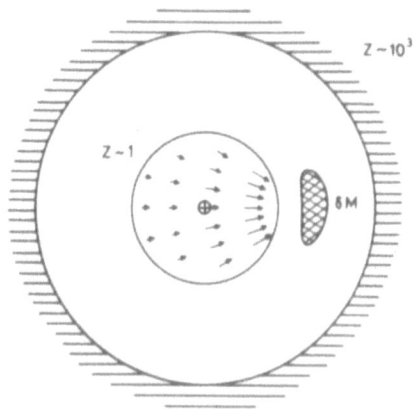

Figure 1.   The gravitational effect of a large density perturbation of mass δM causing a 'tidal shear' and a resulting 12-hr anisotropy in the intensity of the X-ray background radiation (after Fabian and Warwick, 1979)

2.2 The same Ariel-5 analysis produced a global upper limit of ∼ 1% to any 24 hour (dipole) anisotropy. In particular, this upper limit corresponds to a peculiar velocity of the Earth with respect to the microwave background of $v_{pec} \simeq$ 475 km s^{-1} (90% confidence), just consistent with the reported positive effect, on the CMB, of 390 $\pm$ 60 km s^{-1} (Smoot et al, 1977). In a similar analysis of the Uhuru 2 - 7 keV background, Protheroe et al (1980) derive a Compton-Getting velocity in the direction given by Smoot et al of 560 $\pm$ 230 km sec^{-1}. Although no positive 24-hour intensity variation has been detected in the XRB to date, Wolfe (1970) and later authors have pointed out that <u>in principle</u> this waveband is a factor ∼ 3.5 times more sensitive to a given than is the CMB. This is so because the Compton-Getting effect on the background intensity due to the velocity of the Earth relative to the source of background radiation (of spectrum $J_\nu \propto \nu^\alpha$) has amplitude $\Delta I/I = (3-\alpha)v_{pec}/c$ and $\alpha \simeq$ - 0.5 for the 2 - 20 keV XRB, compared with $\alpha \simeq$ 2 for the (Rayleigh-Jeans) microwave spectrum.

2.3 Wolfe (1970) and others have also noted the potential of measurements of the XRB (and CMB) anisotropy in outlining the large scale structure of the Universe at the current epoch. On the (reasonable)

assumption that large scale ($\gtrsim$ 50 Mpc) fluctuations in the matter
distribution will be accompanied by variations in the total (discrete
and diffuse) X-ray emissivity, the nearest matter fluctuations of
amplitude $\delta\rho/\rho$ and co-moving scale λ ($\ll$ present Hubble radius λ_H) will
cause an XRB anisotropy of

$$\Delta I / I \simeq \delta\rho/\rho \times \lambda / \lambda_H$$

A recent assessment of this effect by Rees (1980) confirms the earlier
view that the present $\sim$1% limits of the XRB on intermediate angular
scales (tens of degrees) provides the <u>best</u> current limits of the
homogeneity of the Universe, on scales of $\sim$ 100 Mpc to $\sim$ 1000 Mpc
(Figure 2). On still larger scales the CMB (originating at a much
earlier epoch than the XRB) is likely to remain the best indicator of
Universal structure. Note also that (as Rees, 1980, has pointed out)
a comparison of fluctuations in the XRB and the CMB due to the same
large scale 'clumping' yield, in principle, a value of Ω_0.

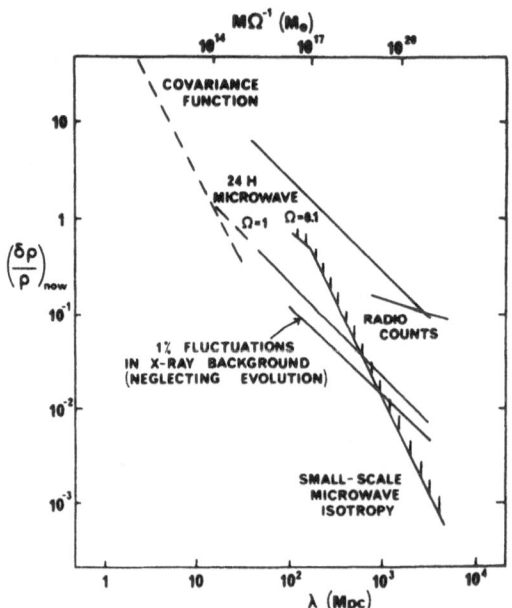

Figure 2.   Constraints on density perturbations ($\delta\rho/\rho$) on various
             length scales λ (after Rees, 1980). In particular, note
             that the 1% limits on the XRB isotropy yield the lowest
             limits on $\delta\rho/\rho$ on scales $\sim$ 100 - 1000 Mpc.

3.   X RAY-SOURCE COUNTS

     The Uhuru and Ariel-5 all-sky surveys showed a majority of the brightest X-ray sources at a galactic latitude $\gtrsim 20^{\circ}$ to be clusters or active galaxies, (23% and 30% respectively, of the sources in the '3A' high latitude catalogue, and 27% and 35% of those optically identified; McHardy et al, 1981). The X-ray luminosity functions of these 'local' objects have been used to predict contributions to the (2 - 10 keV) XRB of $\sim 20\%$ and 5% respectively. The present 1% small-scale ($\sim 5 - 10$ degrees) fluctuations in the XRB limit the total contribution from rich clusters to $< 50\%$ of the background (2 - 10 keV), however they evolve. In fact, Einstein observations of a few high redshift clusters (Perrenod and Henry, 1981; White et al, 1981) suggest weak <u>negative</u> evolution (ie clusters were less luminous and probably cooler in the past) and therefore lead to the conclusion that the cluster contribution to the XRB (at least for E $\lesssim$ 20 keV) is small. On the other hand, the detection of many QSO's in Einstein medium- and deep- exposure fields suggests that they evolve in a strong, positive sense at high z.

Figure 3 summarises the published extragalactic source counts from Uhuru, Ariel-5 and the Einstein Observatory. These range from the Uhuru/Ariel-5 limit of $\sim 2 \times 10^{-11}$ erg cm^{-2} sec^{-1} (2 - 10 keV) down to $\sim 2 \times 10^{-14}$ erg cm^{-2} sec^{-1} (1 - 3 keV), the approximate limit of the Einstein deep survey. It is seen that the Einstein medium-survey point falls below the Euclidean extrapolation from Uhuru/Ariel-5, and has been interpreted as the effect of the local cosmological expansion of a non-evolving source sample (Maccacaro et al, 1981). Between the Einstein medium- and deep-survey source counts, however, significant positive evolution is indicated. Relating this to the individual source identifications in the Einstein surveys suggests this evolution is dominated by the QSO's (Zamorani et al, 1981). As Dr Zamorani points out in an accompanying paper, the observed QSO's (to $m_{\theta} \sim 20$) account for $\sim 30\%$ of the (1 - 3 keV) XRB, and this could rise to $\sim 80\%$ with the inclusion of optically fainter QSO's. It is also evident from Figure 3 that the QSO (or any other extragalactic) source counts are formally constrained by the observed intensity of the XRB. Thus, the source counts integral must flatten shortly below the Einstein deep survey limit.

In summary, despite the proven ability of the Einstein Observatory to detect the most luminous QSO's to z $\sim$ 3, the lack of complete samples (and of X-ray derived luminosity functions) severely limits attempts to obtain precise estimates of the contributions of discrete sources (and hence of any diffuse component) to the XRB. Further, the considerable uncertainty in the density or luminosity evolution of any class of extragalactic source currently prevents the derivation of useful cosmological results. The prospects for the future (over 10 - 15 years) look more promising.

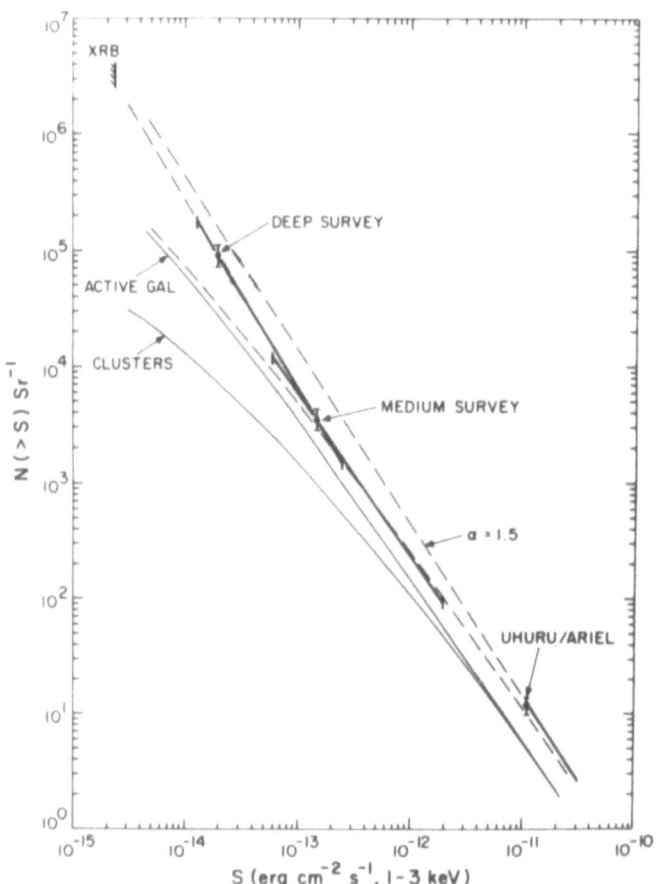

Figure 3.   Log N v. Log S for extragalactic X-ray sources. The Uhuru/
              Ariel-5 sky survey result is shown as well as the deep and
              medium survey results from Einstein. Also shown are the
              expected source counts for clusters and active galaxies
              (assuming no evolution and q_0 = 0) and their sum. The
              X-ray background intensity would be reached by extrapolation
              of the source counts to 'XRB'.

4.   PROSPECTS FOR RELEVANT X-RAY OBSERVATIONS OVER THE NEXT DECADE

     In the near future, a modest extension in the statistics of high
redshift sources will come from the analysis of the remaining Einstein
data (factor ∿2) and from the somewhat smaller imaging telescopes on
EXOSAT, due for launch in late 1982 (Pounds, 1981). Neither mission is
likely to alter the conclusion towards the end of section 3. The study
of the isotropy of the XRB, however, may be advanced more significantly

when the analysis of data from the HEAO-1 A2 experiment (originally designed for the study of the XRB, of course) is completed in the near future. The A2 experiment has good photon statistics on the XRB and an order of magnitude increase in the number of independent sky cells compared with Ariel-5 or Uhuru.

Beyond EXOSAT, the next relevant X-ray astronomy mission now being planned is the German spacecraft, ROSAT, carrying a fourfold-nested, 0.8 metre diameter X-ray telescope ($\sim$3 times the Einstein sensitivity) with $\sim$5 arc sec resolution (Trümper, 1981). During the first 6-9 months in orbit, following the presently scheduled launch in 1987, ROSAT is planned to carry out an all-sky survey to a sensitivity limit of $\sim 10^{-13}$ erg cm^{-2}s^{-1}(Figure 4). This is comparable to the medium sensitivity surveys of Einstein and should yield $\sim 10^{5}$ sources, including greatly increased numbers of extragalactic X-ray source, e.g. clusters, Seyferts, BL Lacs, QSO's, etc. In addition, the chances must be good of discovering new and relatively rare types of X-ray source.

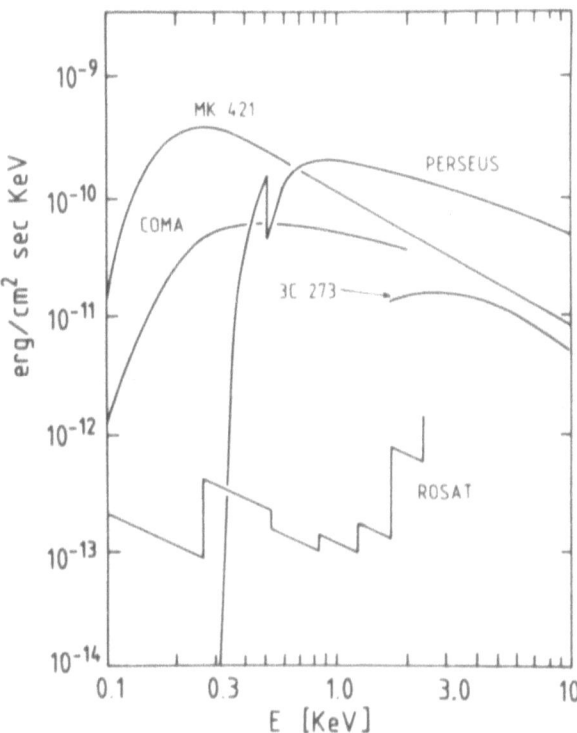

Figure 4.   Energy spectra of selected extragalactic X-ray sources compared with the sensitivity limit reached in a uniform all-sky survey of 6 months with the ROSAT 0.8 metre X-ray telescope.

Such complete (X-ray derived) samples of the different classes of
extragalactic source should provide accurate luminosity functions of
each class and hence yield more directly their individual contributions
to the XRB (at least at the local epoch). From the cosmological viewpoint, however, it seems unlikely that the ROSAT survey will have
sufficient sensitivity to detect enough sources at high redshift (most
luminous QSO's apart) to determine their evolution and hence yield
usefully precise values of q_o or H_o.

The successor to the Einstein Observatory in the NASA programme,
the Advanced X-ray Astronomy Facility (AXAF) appears, in contrast, to
be the first X-ray mission with a clear 'cosmological potential'.
Defined over the period 1978-80, AXAF (NASA Tech. Memo. TM-78285)
incorporates 6 nested Wolter-I X-ray mirrors with an outer element of
diameter 1.2 metre and a photon collection area ~ 4 times Einstein
(Figure 5). Its major advance, however, will be in angular resolution
(Figure 6), with a specified on-axis resolution of ≲ 0.5 arc sec, and a
consequential further substantial gain in point-source sensitivity
compared with Einstein; overall, the point-source sensitivity of AXAF
is expected to be ~ 50 times that of Einstein. A second crucial advantage

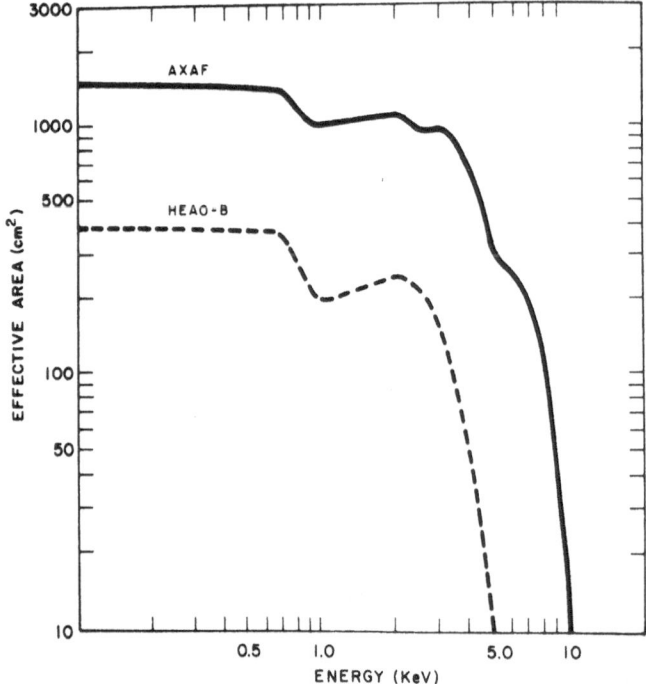

Figure 5.   AXAF photon collection area compared with that of the
              Einstein Observatory.

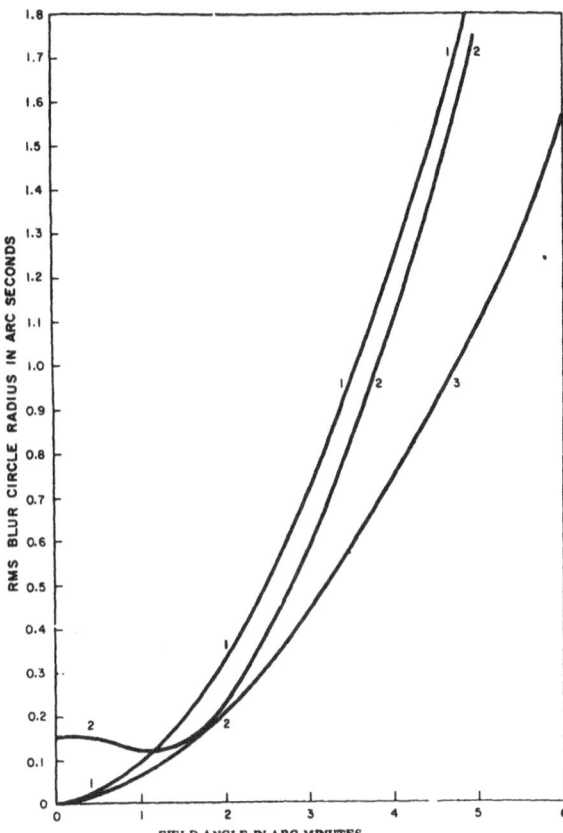

Figure 6.   Off-axis resolution of the AXAF 1.2 metre X-ray telescope,
              assuming a perfect mirror system. Within 3 arc min of
              the optical axis fabrication tolerances will degrade the
              image quality to 0.5 arc sec. Curve (1) represents the
              Gaussian focal plane, (2) the focal plane displaced 0.15mm
              toward the mirrors, and (3) the optimum focal surface.

offered by AXAF is the extension of a useful response to $\sim$ 7 keV,
allowing good sensitivity for detection of the characteristic iron lines
(and thereby making possible direct, X-ray determined redshifts in many
cases).

      Figure 7 summarises in graphical form the sensitivity limits of
three envisaged focal plane detectors on AXAF, referred to point-like
and extended sources at high redshifts. The potential is obvious, with
the ability to detect sources over a wide range of X-ray luminosity to
$z \gtrsim 1$ and a source of the luminosity of 3C 273 ($\sim 10^{46}$ erg sec^{-1}) to $z \gtrsim 10$,
if they exist at such an early epoch. It is apparent from present

Einstein results (e.g. see Figure 3) that strong evolutionary effects
must occur in this redshift interval and - quite likely - that the
formation of active galaxies may be directly observed. Very little is
known about clusters at z ≥1 and their direct observation by X-ray
mapping promises to be a very powerful tool for investigating their
early evolution. Since the most likely source of the hot gas producing
the diffuse cluster X-ray emission is the galaxies themselves, stripped
by mutual collision and by the ram pressure on the intracluster gas,
the development of the X-ray emission (intensity and spatial distribution)
should be very instructive in comparison with the galaxy clustering
itself.

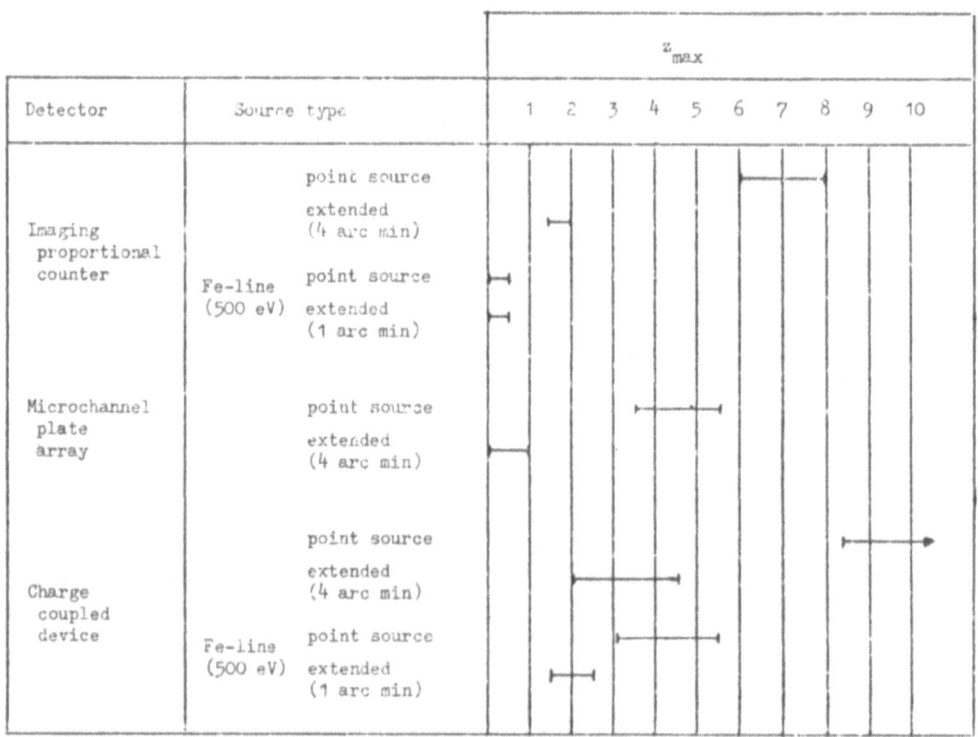

Figure 7.   The maximum observable redshift for three AXAF focal
plane detectors. Source luminosity $L_x = 10^{45}$ erg s^{-1} and
observing time of 10^5 sec are assumed. The range of
z_{max} covers a range of density parameters $\Omega_o = 0.1 - 1.0$

5.   COSMOLOGICAL TESTS USING X-RAY DATA

To first order the principal cosmological aim of observation is the determination of the expansion and deceleration parameters, whose present values are the Hubble constant H_o and the deceleration constant q_o. As noted earlier, no useful limits of H_o and q_o have been yielded by X-ray observation to date, although measurements of the isotropy of the X-ray background have yielded some interesting limits on the large scale structure of the Universe. In this final section the prospect for a more positive outcome from X-ray observations with the planned ROSAT and AXAF missions is briefly considered.

(a)   Apparent magnitude v. redshift.

This classical cosmological test depends on the variation of apparent magnitude m with z for sources of a given luminosity. The rate of variation uniquely determines the deceleration parameter q_o, but application of this test has to date been bedevilled mainly by the uncertainty in determining M v. z, i.e. the lack of a 'standard candle'. The observed cut-off in the luminosity function of X-ray binaries at several times 10^{38} erg s^{-1} seems likely to be due to Eddington-limited accretion. If so, the observation of X-ray binaries out to the Virgo cluster with AXAF may yield a direct measurement of the luminosity distance and, in comparison with the observed redshifts, an improved value of H_o. It _may_ be that the ROSAT sky survey, combined with AXAF observations to high z, will sufficiently well define the evolution of some class of luminous extragalactic X-ray emitter (perhaps involving Eddington-limited accretion onto a 'standard' massive black hole) to allow the m - z test to yield a useful value of q_o but this is at present unclear.

(b)   Limiting X-ray variability.

Variability is found to be common in the X-ray emission of active galactic nuclei (Marshall et al, 1981) and occurs on timescales so short as to offer the potential of a measurement of luminosity distance on a much larger scale than from the study of the X-ray binaries. This possibility arises from the fact that the minimum timescale for variability (relativistic beaming apart) in a source of luminosity L_x occurs for an electron scattering optical depth ~ 1. Given a maximum efficiency for mass-to- energy conversion of ~ 0.1, variability of time-scale Δt will be related to the amplitude ΔL_x by the relation (Cavallo and Rees, 1978),

$$\Delta L_x \lesssim 2 \times 10^{41} \, \Delta t \ \text{erg} \ s^{-1}$$

This method has already been applied to the $\Delta t \sim 100$ s variability detected in the low luminosity Seyfert NGC 6814. The calculated luminosity gives, in comparison with the measured recession velocity of 1400 km s^{-1}, a minimum H_o of 50 km s^{-1} Mpc^{-1}. The sensitivity of ROSAT and repeated observations over several hours allow many high luminosity active galaxies to be examined (to z $\gtrsim 1$) on the critical timescales for the luminosity distance to be determined. Provided sufficient high redshift objects are seen to vary rapidly, this method could provide the statistical basis of a precise determination of H_o and q_o.

(c)   Angular diameter (Θ) v. redshift.

This second classical test uses the non-linearity in $(\Delta\Theta)^{-1}$ - z
to derive q_o and is again severely limited by uncertainties in the
evolution of the objects used (e.g. optical galaxies, and clusters). If
future X-ray observations can determine the evolution of the X-ray
clusters (i.e. core radius, temperature) to z $\gtrsim$ 1, this test should be
feasible. Again, however, the precision of q_o obtainable will be strongly
dependent on the uncertainties in the evolutionary pattern so determined.
One reason for optimism is that the X-ray surface brightness of a rich
cluster (at least at the present epoch) is substantially greater than
the general XRB, making distant clusters easier to detect than in the
optical band. Against this, the (negative) evolution of X-ray emission
may be stronger than that of the optical galaxy clustering itself.
(d)   Differential source counts v. redshift.

Schwartz (1976) has pointed out that evolutionary effects are less
serious and line-of-sight inhomogeneities unimportant on scales $<$ tens
of Mpc if the differential number of objects as a function of z is
measured, rather than the redshift relations for apparent magnitude or
size. He further suggested that X-ray clusters would provide a sensitive
means for applying this test and estimated, for example, that at z $\sim$ 1
the ratio of differential source counts for q_o = 0 and q_o = 1 is $\sim$ 3.8
(the comparable ratio for tests (a) and (c) above is $\sim$ 2.2 and $\sim$ 1.5
respectively). An assessment of this test, post-Einstein, confirms its
promise, since rich clusters do stand out as extended sources with a
surface brightness substantially greater than the XRB. Furthermore,
the evolution determined near z $\sim$ 0.5 is only weakly negative and there
are reasons (e.g. cluster lifetimes) for supposing that this will remain
so to z $\sim$ 1. Nevertheless, very few X-ray clusters have been detected
by Einstein and it remains questionable whether the ROSAT sky survey
will satisfy the requirement of the Schwartz test, viz. to detect all
the clusters in a well-defined (luminosity) class, and yield a usefully
precise value of q_o. Again, however, AXAF should certainly do so over
specified areas of the sky.
(e)   A promising (new) cosmological test, again using the X-ray emission
from rich clusters, has been discussed by several authors (e.g. Silk
and White, 1978; Boynton, 1981). Briefly, this involves tha mapping of
the X-ray cluster, to determine the emission measure and core radius (r_c)
and to compare this with the microwave shadowing effect, caused by
scattering of the CMB photons by the hot electrons (T_e) in the cluster.
Thus,

$$L_x \sim n_e^2 \, T_e^{-0.5} \, r_c^3 \qquad\qquad \text{(XRB)}$$

and $$\Delta T/T \sim \Delta I/I \sim n_e \, T_e \, r_c \qquad\qquad \text{(CMB)}$$

which combine to give

$$L_x \sim (\Delta I/I)^2 \, T_e^{-2.5} \, r_c \sim (\Delta I/I)^2 \, T_e^{-2.5} \, D_A \, \theta_c$$

where θ_c is the angular size of the cluster and D_A its angular size
distance. Relating the angular size and luminosity distance (D_L) in

the usual way, the luminosity distance is then given by five observable quantities, the X-ray flux (F_x), the microwave decrement, the electron temperature and the angular size and redshift of the cluster.

$$D_L \simeq C_L \, F_x \, (\Delta I / I)^2 \, T_e^{-2.5} \, \theta_c \, / \, (1 + z)^2$$

where C_L is a coefficient depending weakly on the element abundance of the cluster gas.

Since also, $D_L = c / H_0 \, (z + (1 - q_0) z^2 / 2 + \quad \ldots \text{higher terms in z})$

for Lemaitre-Friedmann models, q_0 can be derived from the observation of a single cluster (i.e. no knowledge of cluster evolution is required). In principle, from the study of two clusters, covering a range of z, both H_0 and q_0 may be obtained. Provided luminous X-ray clusters do occur at z ~1, AXAF should certainly provide the necessary X-ray data for this test. The limiting accuracy on q_0 and H_0 is in fact likely to rest with the difficulty of mapping the CMB shadowing, since $\Delta T / T$ for nearby rich clusters is quite small (~10^{-4})and will be still smaller for distant clusters if they are indeed cooler than those at the local epoch.

6. REFERENCES

Boldt, E., 1981, Washington Acad. Sci., 71, pp. 24-44

Boynton, P.E., 1981, in X-ray Astronomy with the Einstein Satellite. Ed. Giacconi, pp. 297-310. D. Reidel Pub. Co.

Cavallo G. and Rees, M.J., 1978, Mon. Not. R. astr. Soc. 183, pp. 359-366

Fabian, A.C. and Warwick, R.S., 1979, Nature, 280, pp. 39-40

Maccacaro, T., Feigelson, E.D., Fener, M., Giacconi, R., Gioia, I.M., Griffiths, R.E., Murray, S.S. and Zamorani, G., 1981, Astrophys. J. in press.

Marshall, N., Warwick, R.S. and Pounds, K.A., 1981, Mon. Not. R. astr. Soc. 194, pp. 987-1002

McHardy, I.M., Lawrence, A., Pye, J.P. and Pounds, K.A., 1981, Mon. Not. R. astr. Soc.

Perrenod, S.C. and Henry, J.P., 1981, Astrophys. J., 247, pp. L1-L4

Pounds, K.A., 1981, Washington Acad. Sci., 71, pp. 104-114

Protheroe, R.J., Wolfendale, A.W. and Wdowczyk, J., 1980, Mon. Not. R. astr. Soc., 192, pp. 445-454

Rees, M.J., 1980, in X-ray Astronomy, Ed. Giacconi and Setti, pp. 377-384. D. Reidel Pub. Co.

Schwartz, D.A., 1976, Astrophys. J., 206, pp. L95-L98

Schwartz, D.A., 1980, Physica Scripta, 21, pp. 644-649

Silk, J. and White, S., 1978, Astrophys. J., 226, pp. L103-L106

Smoot, G.F., Gorenstein, M.V. and Muller, R.A., 1977, Phys. Rev. Lett. 39, pp. 898-901

Trümper, J., 1981, Washington Acad. Sci. 71, pp. 114-124

Warwick, R.S., Pye, J.P. and Fabian, A.C., 1980, Mon. Not. R. astr. Soc. 190, pp. 243-260

White, R.A., Sarazin, C.L., Quintana, H. and Jaffe, W.J., 1981,
    Astrophys. J., 245, pp. L1-L4
Wolfe, A.M., 1970, Astrophys. J., 159, pp.L61-L67
Zamorani, G., Henry, J.P., Maccacaro, T., Tananbaum, H., Soltan, A.,
    Avni, Y., Liebert, J., Stocke, J., Strittmatter, P.A., Weymann, R.J.,
    Smith, M.G. and Condon, J.J., 1981, Astrophys. J., 245, pp. 357-374

THE ORIGIN OF THE X-RAY BACKGROUND

G. Zamorani
Istituto di Radioastronomia CNR, Bologna, Italy

Despite the hundredfold increase in sensitivity achieved with the launch
of the soft X-ray Einstein Observatory, the problem of the origin of the
X-ray background (XRB) is not completely resolved yet, so that the old
controversy between supporters of different theories (XRB as due to discrete sources or to truly diffuse processes) is likely to continue.
On the other hand, the results obtained with the Einstein Observatory
have the capability of constraining quite strongly the allowed range of
possibilities. Two quite different and independent approaches have
been followed in order to estimate the contribution of discrete sources
to the XRB. The first (Einstein Deep and Medium Surveys) has extended
the observed log N - log S down to fluxes of $\sim 3 \times 10^{-14}$erg cm^{-2}s^{-1}
providing a firm lower limit of $(26 \pm 11)\%$ for the contribution of discrete sources to the XRB. The second (observations of several hundreds
of previously known extragalactic objects) has clearly shown that quasars,
as a class, are the most significant contributors to the XRB. An estimate of the allowed range of their contribution is here presented,
taking into account both the existence of selection effects in the observed sample of quasars and the large uncertainty about the exact shape
of the optical log N-m at faint magnitudes.

1.   INTRODUCTION

1.1. The first measurements

     The existence of a diffuse X-ray background was discovered almost
twenty years ago (Giacconi et al. 1962). Since then, a large number of
experiments on board of rockets, balloons and satellites have been carried out with the aim of measuring the intensity, spectral shape and
isotropy of this radiation. It was very soon established that, away
from the galactic plane, this radiation is isotropic to a level of a
few percent at least. This isotropy was the first proof in favour of an
extragalactic origin of the X-ray background (XRB) above ~ 2 keV.

     As for the intensity and spectral shape, the information available

A. W. Wolfendale (ed.), Progress in Cosmology, 203–214.
Copyright © 1982 by D. Reidel Publishing Company.

at the beginning of the seventies was summarized in two review papers by
Silk (1973), and Schwartz and Gursky (1974). As stressed by the authors,
the experimental errors and the scatter of the data obtained at comparable energies were such that the shape of the XRB in the range 1-100 eV
was equally well fitted by either a thermal spectrum $I(E) \propto \exp(-E/kT)$
with $kT \sim 35$ keV or two power law spectra

$$I(E) \propto \begin{cases} E^{-0.4} & \text{for } 1 \leq E \leq 20 \text{ keV} \\ E^{-1.4} & \text{for } 20 \leq E \leq 100 \text{ keV} \end{cases}$$

1.2 Proposed theories for the origin of the XRB

Quite a number of theories were proposed to explain the origin of
the XRB. They can be divided into two main groups: those which invoke
truly diffuse mechanisms and those which explain the XRB as the summed
contribution of discrete sources.

The favoured mechanisms for diffuse emission were both thermal
(X-ray emission from hot intergalactic gas: Field, 1972; Field and
Perrenod, 1977) and non-thermal (Inverse Compton of relativistic electrons
on the photons of the cosmic microwave background: Brecher and Morrison,
1969).

As for the contribution of discrete sources to the XRB, it was soon
established that known classes of objects, without evolution, can account
for $\sim 20\%$ of the observed background (Setti and Woltjer, 1973; Rowan
Robinson and Fabian, 1975). Moreover, on the basis of a one-object
(3C 273) statistics, it was suggested (Setti and Woltjer, 1973) that
"QSOs may well suffice to account for most of the observed background".
Almost exactly the same conclusion can be reached today using the wealth
of data on X-ray properties of QSOs obtained with the Einstein Observatory (see section 3.5).

2. THE HEAO 1 RESULTS

2.1 The spectrum of XRB from 3 to 50 keV

The HEAO 1 A2 experiment, designed to provide detailed spectral
information on the XRB up to ~ 50 keV, represents a real quality jump with
respect to all previous results. In fact, for the first time a single
experiment offers data with small errors over a large energy range.
Figure 1b (adapted from Marshall et al., 1980) shows that a thermal
bremsstrahlung model with a temperature of 40 keV gives an excellent
description of the data over the entire range of energy; the incident
spectrum corresponding to the 40 keV model is shown in Figure 1a. While
no single power law spectrum fits the observations, the combined emission
from power law sources can provide a very good fit of the same HEAO 1-A2
data (De Zotti et al., 1981). However, the required properties (mean
spectral index and energy cutoff) appear to be different from those of

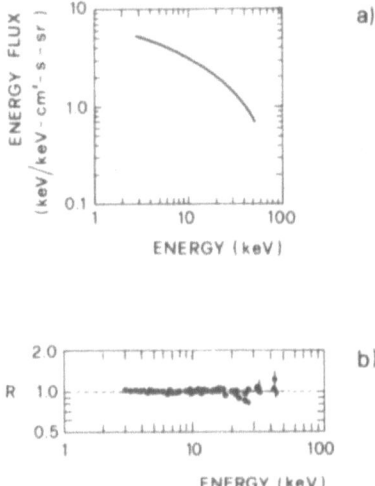

Fig. 1. (a) The incident thermal bremsstrahlung spectrum for kT = 40 keV. (b) The ratio of the observed counts for the XRB to those predicted by the model. (Adapted from Marshall et al. 1980).

the few active galactic nuclei (Seyfert galaxies; Mushotzky et al., 1980) for which spectral information is available. Even more restricted (limited to a couple of objects) is our knowledge of the X-ray spectra of quasars. In the near future EXOSAT will offer the opportunity to obtain such information on some of the brighter X-ray quasars.

2.2 The HEAO 1 – A2 all sky survey

In the scanning mode the HEAO 1-A2 experiment has performed an all-sky survey down to a limiting flux of ∿3.1 x 10^{-11} erg $cm^{-2}s^{-1}$ (2-10 keV). In the 8.2 steradians with galactic latitude b^{II} > 20°, 86 sources have been detected; 61 have been identified with extragalactic objects, 17 are galactic and 6 are still unidentified. These data have been presented and discussed by Piccinotti et al. (1981). The very high percentage of identifications has allowed them to compute X-ray luminosity functions for different classes of objects, i.e. clusters (29 objects; slope of the luminosity function = 2.05 ± 0.15) and active galactic nuclei (AGN, 23 objects; slope = 2.75 ± 0.15). The integration of the luminosity functions gives an estimate for the contribution to the 2-10 keV XRB of ∿4% for clusters and ∿18% for AGNs. These estimates are obtained assuming no evolution both for clusters and AGNs.

3.   THE EINSTEIN RESULTS

3.1 The EINSTEIN Surveys

With the launch of the Einstein Observatory a completely new way to investigate the problem of the contribution of discrete sources to the soft XRB became possible. The imaging capabilities and the hundred-fold increase in sensitivity of the Einstein Observatory made it possible to obtain deep X-ray exposures in selected fields counting discrete sources down to $\sim3 \times 10^{-14}$ erg cm^{-2}s^{-1} (1-3 keV). To this limiting flux, Giacconi et al. (1979) found a surface density of 19 $\pm$ 8 sources per square degree. Assuming that the X-ray log N-log S relation can be fitted by a power law with slope 1.5, the total contribution of discrete sources brighter than 2.6×10^{-14} erg cm^{-2}s^{-1} has been estimated to be $\sim(26 \pm 11)\%$ of the diffuse background. This direct measurement gives a firm lower limit for the contribution of discrete sources to the XRB. Unfortunately, very few of the X-ray sources found in the deep surveys have been identified with optical counterparts so far. However, a guess about the nature of the sources which are dominant at these faint flux levels can be obtained using the information from the medium sensitivity survey (Maccacaro et al. 1981). This survey which covers ~50 square degrees at high galactic latitude with sensitivities in the range $7 \times 10^{-14} - 5 \times 10^{-12}$ erg cm^{-2}s^{-1} (0.3-3.5 keV), resulted in the detection of 48 extragalactic sources. From these sources, the number-flux relation for the extragalactic population has been derived, yielding a best fit power law slope of 1.53 $\pm$ 0.16 (Figure 2). The big advantage of this sample is that the identification of these sources is almost complete. Table 1 shows the identification content of the Einstein Medium Survey compared with that expected on the basis of the HEAO 1$^-$A2 survey.

TABLE 1

Class of Objects	Einstein Medium Survey	Expected Composition from HEAO 1 - A2 data
AGNs	37	19
Clusters	9	24
Others	2	5

It is clear that the two major classes of extragalactic sources (i.e. clusters of galaxies and AGNs) are represented in a different proportion in the two samples (the significance of the difference is greater than 3σ): AGNs, which include both Seyfert galaxies and QSOs, become dominant at lower X-ray fluxes. Extrapolating this trend to even lower fluxes, we can reasonably conclude that most of the sources (>77%) in the Deep Survey are AGNs and derive a lower limit of $\gtrsim 20\%$ for the percentage contribution of this class of objects to the XRB.

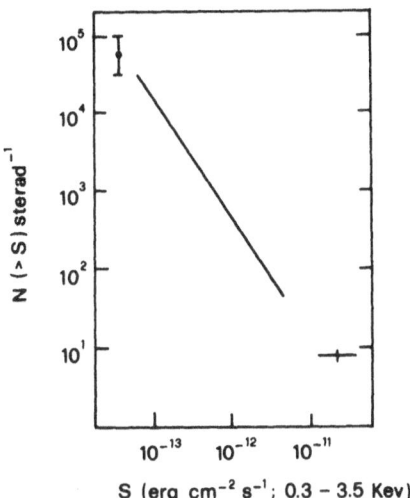

Fig. 2. The integral log N - log S relation for the extragalactic X-ray
sources (solid line). The data point at high flux level is from the
HEAO 1 -A2 study of Piccinotti et al. (1981). The data point at low
flux level is from the Einstein deep survey of Giacconi et al. (1979).
(Adapted from Maccacaro et al. 1981).

3.2 The EINSTEIN Observations of QSOs

     The second, more traditional way to estimate the contribution of
discrete sources to the XRB is through the use of the X-ray observations
of previously known extragalactic objects. As soon as the first Einstein
observations became available, it was immediately clear that quasars, as
a class, are significant contributors to the XRB (Tananbaum et al. 1979).
Today, at the end of the Einstein mission, a few hundred previously
known quasars have been observed and data for more than two hundred are
already available in the literature (Tananbaum et al. 1979; Ku, Helfand
and Lucy, 1980; Zamorani et al. 1981). Despite this wealth of data the
direct estimate of the contribution to the soft XRB due to QSOs is not
straightforward.

     First, no X-ray observations of "complete" samples are available
yet. As a consequence, no X-ray luminosity function can be derived and
the best way to estimate the contribution of quasars to the XRB is still
to use the optical counts plus the observed values of the ratio between
X-ray and optical luminosities. In this way, the uncertainty associated
with the optical counts, which is quite substantial at magnitudes
fainter than ∿20, affects also the estimate of the contribution of QSOs
to the background.

     Second, a number of correlations among X-ray, optical and radio luminosities have been found which must be taken into account properly.

The importance of the associated uncertainties can be realized by considering that the existing estimates range from ∿30% to ∿100% (Bonoli et al. 1980; Cavaliere et al. 1981; Cheney and Rowan-Robinson 1981; Kembhavi and Fabian 1981; Zamorani 1981). The X-ray data used by the various authors are essentially the same; the different conclusions are due either to approximations introduced in the use of X-ray data or to different assumptions on the optical counts at faint magnitudes. In the following we derive a new estimate of the contribution of QSOs to the XRB, making full use of the available X-ray data and applying an evolutionary model which gives a very good fit to the best data on the optical counts.

3.3 Correlations among X-ray, optical, and radio luminosities in QSOs

At least three correlations have to be taken into account if one wants to have an unbiased estimate of the average ratio between X-ray and optical luminosities ($<l_x/l_o>$).
i) The average value of l_x/l_o is a function of optical luminosity (l_x/l_o decreasing with increasing l_o). This effect is shown in Figures 3 and 4 where the nominal spectral index between optical and X-ray frequencies (α_{ox}) is plotted versus the monochromatic luminosity at 2500 Å for radioloud quasars (Figure 3), and for radioquiet quasars and Seyfert 1 galaxies (Figure 4). The straight lines in both figures represent the best fit obtained considering both the detections and the X-ray upper limits following the method of Avni (1981). In deriving the two fits we have assumed that α_{ox} has a gaussian distribution around the best straight line relation. The data used here are taken mostly from Zamorani et al. (1981), complemented with a few unpublished measurements obtained at the Center for Astrophysics. The best fit line for the radioloud QSOs (75 objects; 72 X-ray detections) is:

$$\alpha_{ox} = (0.110 \pm 0.025) \log l_o - 2.064, \text{ i.e. } l_x \propto l_o^{0.71 \pm 0.07}$$

for the radioquiet QSOs (45 objects; 20 detections) is:

$$\alpha_{ox} = (0.123 \pm 0.029) \log l_o - 2.224, \text{ i.e. } l_x \propto l_o^{0.68 \pm 0.08}$$

In both cases (radioloud and radioquiet QSOs) the slope of the best fit line is different from a slope zero at a significance level >4σ. This result shows that, on the average, the X-ray luminosity increases less rapidly than the optical luminosity. As a consequence, one cannot use a unique value for $<l_x/l_o>$, when estimating the contribution of QSOs to the XRB.

Note that in Figure 4 the best fit line has been derived using only the QSOs and not the Seyfert 1 galaxies. Yet, the data for Seyferts 1 (taken from Kriss, Canizares and Ricker, 1980) are perfectly consistent, at a lower absolute luminosity, with the correlation found for QSOs, providing a new continuity argument between the two classes of objects.

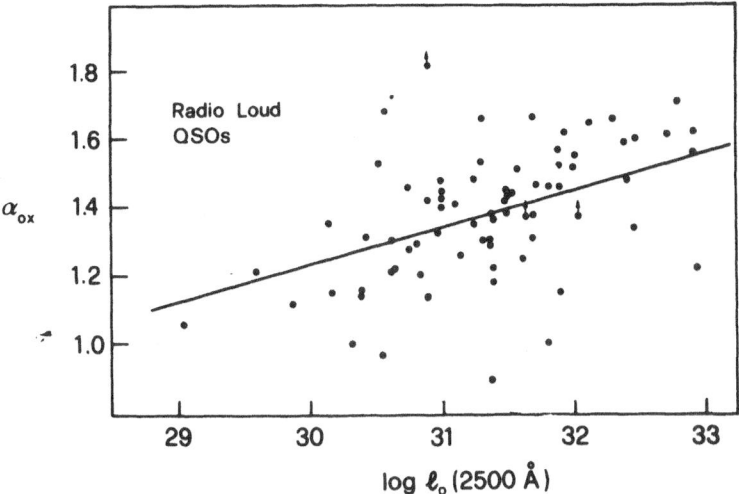

Fig. 3. The nominal spectral index between optical and X-ray frequencies (α_{ox}) versus the monochromatic luminosity at 2500 Å for radioloud quasars. The solid line is the best fit line.

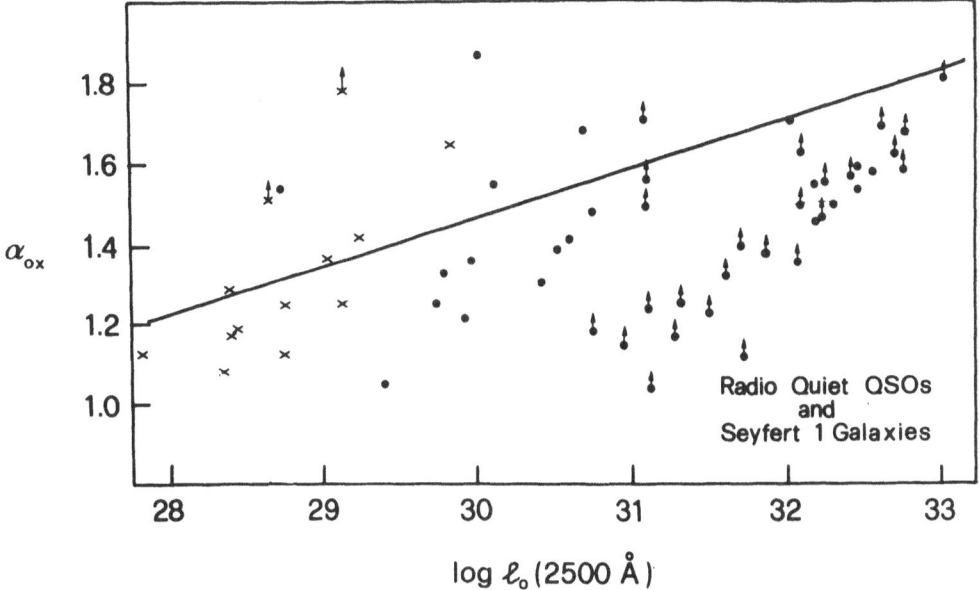

Fig. 4. The same as Fig. 3 for radio quiet quasars and Seyfert 1 galaxies. The solid line is the best fit line derived using only the radioquiet quasars.

ii) The radioloud QSOs are, on the average, stronger X-ray emitters than the radioquiet QSOs with the same optical luminosity. This effect,

already found by Ku, Helfand and Lucy (1980) and Zamorani et al. (1981),
is clearly seen by looking at Figures 3 and 4: the displacement between
the two, almost parallel, best fit lines corresponds to a difference of
a factor ~ 3 in l_x/l_o. This implies that the two classes of QSOs (radioloud and radioquiet) must be considered separately in estimating their
contribution to the background.

iii) The ratio of l_x to l_o in radioloud QSOs is a function of the ratio
of radio to optical luminosities $R = l_R/l_o$ (see Figure 4 in Zamorani et
al., 1981). But the R distribution of the QSOs observed in X-ray is <u>not</u>
representative of the R distribution of QSOs in the sky ($\Psi(R)$). This
fact implies that the $<l_x/l_o>$ for radioloud QSOs has to be weighted with
the $\Psi(R)$ distribution obtained from radio observations of optically
selected QSOs (see, e.g., Sramek and Weedman 1980).

3.4 The evolutionary model

    In Figure 5 the best existing data points for the integral counts
of QSOs are shown as a function of the apparent magnitude. (For a
critical, completely updated discussion on the log N-m of QSOs, see
Setti and Woltjer 1981).

    Because of the existence of the correlations discussed in Section
3.3, we are not allowed to assume an average value of l_x/l_o and to in-
tegrate along the observed log N-m using this value. In fact, at different magnitudes we expect different distributions of intrinsic optical
luminosities, which, in turn, implies different distributions of l_x/l_o.
Hence, we have to use a model for the evolution of QSOs.

    As shown by Braccesi et al. (1980) a pure density evolution law is
not capable of reproducing both the steep slope of the optical counts up
to $m_B \simeq 20$ and the flattening which occurs at $m_B \simeq 21$. For this reason
we have chosen to adopt a pure luminosity evolution model $L(z) = L(0)e^{c\tau}$,
where τ is the look-back time ($\tau = z/(1+z)$ for a $q_o = 0$ Universe).

    In order to find the model parameters (local luminosity function
and amount of evolution), we have used Table 8 of Braccesi et al. (1980),
which gives the space density of quasars as a function of redshift **and** magnitude, derived from complete optical samples. The best representation
of these data is obtained with $c = 8$, the curve in Figure 5 shows the
log N-m computed by integrating up to $z_{Max} = 3.5$ the luminosity function
corresponding to $c = 8$. The agreement with the experimental data is
excellent.

    Figure 6 gives a schematic description of the model in the plane
redshift-optical luminosity. The solid lines represent the lower and
upper limits of integration of the luminosity function; the dashed lines
represent the approximate upper bounds corresponding to different limiting magnitudes. From this figure, one can see clearly the effect of
taking into account the anticorrelation between l_x/l_o and l_o: if the
redshift cutoff at $z \simeq 3.5$ is real, at magnitudes fainter than ~ 18-19,
one is essentially picking up objects from the lower end of the optical

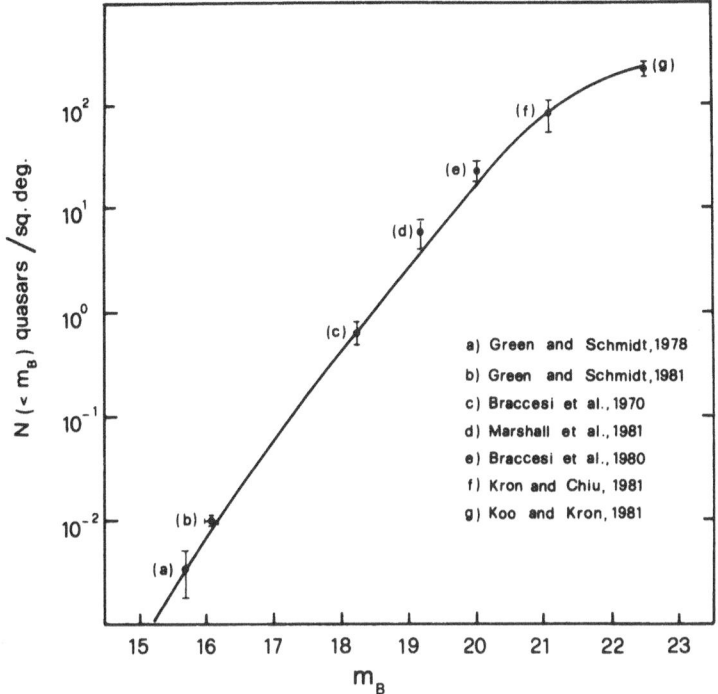

Fig. 5. Optical source counts of quasars. The number of quasars per
square degree brighter than a given magnitude is plotted versus blue
magnitude. The solid line represents the expected counts on the basis
of a pure luminosity evolution model.

luminosity function, which implies higher average value of l_x/l_o.

It is important to point out that the lower end of the local (z=o)
luminosity function which has been derived and used in this model over-
laps quite substantially with the range of luminosity for which the
luminosity function of Seyfert 1 galaxies has been determined (Veron,
1979). In this range of luminosities the two luminosity functions are
in very good agreement with each other both in slope and normalization.
This fact and the continuity in the X-ray properties of Seyferts and
QSOs allows us to give an estimate of their contribution to the XRB which
includes both classes of objects at the same time.

3.5 The contribution to the XRB of Seyferts and QSOs (AGNs)

From Sections 3.3 and 3.4 we have all the information which is
needed to estimate the contribution of AGNs to the soft XRB, namely:
a) an optical luminosity function and an evolution law which reproduce
the optical counts;
b) a relation between X-ray and optical luminosity for radioquiet and

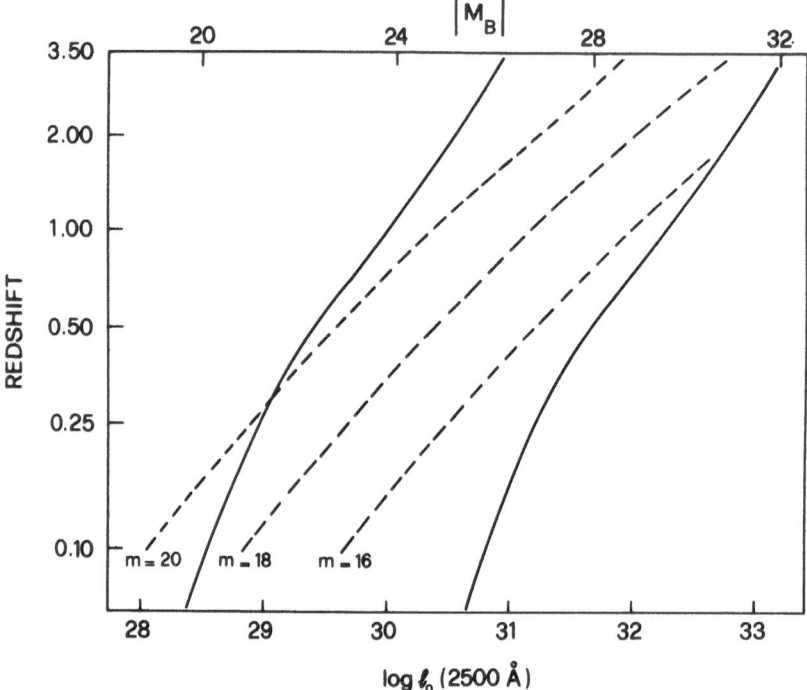

Fig. 6. Schematic description, in the plane redshift-optical luminosity, of the luminosity evolution model described in the text. The solid lines represent the lower and upper limits of the luminosity function. The dashed lines represent the approximate upper bounds corresponding to different magnitudes.

radioloud QSOs.

From a) and b) we can derive, separately, the X-ray luminosity function for radioquiet and radioloud QSOs and estimate their contribution to the XRB integrating the two luminosity functions up to $z_{Max} = 3.5$. From the collection of radio data on optically selected quasars assembled by Sramek and Weedman (1980), we estimate that radioloud quasars are ∿8.6% of the total number of QSOs. Assuming that this percentage does not change with redshift we estimate that radioquiet QSOs contribute ∿52% of the XRB and radioloud QSOs contribute ∿9%.

4. CONCLUSIONS

We have shown that a model which gives a very good representation of the optical counts of QSOs and makes full use of the available information on the X-ray properties of Seyferts and QSOs gives the best estimate of ∿60% for their contribution to the soft X-ray background.

Unfortunately, the uncertainty on this number is still large. Part of it is due to the relatively small number of radioquiet QSOs which have been detected with the Einstein Observatory (see Figure 4). This situation cannot improve until 1986, when new observations will be possible with the German X-ray satellite ROSAT. On the other hand, there is the realistic possibility of reducing at least part of the uncertainty by means of optical works. Two observational programs presently in progress will lead to important results with respect to the problem of the XRB. The first one, identification of a large number of faint candidate quasars, will allow us to obtain a considerably better determination of the log N-m relation at magnitudes fainter than $m \simeq 20$, where most of the contribution to the XRB is coming from. The results of the second project, determination of the redshift distribution of a complete sample of quasars at $m \simeq 20$, will provide strong constraints on different evolutionary models.

Acknowledgments: It is a pleasure to thank Y. Avni, G. Setti and H. Tananbaum for many helpful discussions and comments.

REFERENCES

Avni, Y.: 1981, in preparation.
Bonoli, F., Braccesi, A., Marano, B., Merighi, R., and Zitelli, V.: 1980, Astron.Astrophys.(Letters) 90, L10.
Braccesi, A., Formiggini, L., and Gandolfi, E.: 1970, Astron.Astrophys. 5, p. 264.
Braccesi, A., Zitelli, V., Bonoli, F., and Formiggini, L.: 1980, Astron. Astrophys. 85, p. 80.
Brecher, K., and Morrison, P.: 1969, Phys.Rev. Letters 23, p. 802.
Cavaliere, A., Danese, L., De Zotti, G., and Franceschini, A.: 1981, Astron.Astrophys. 97, p. 269.
Cheney, J.E., and Rowan-Robinson, M.: 1981, Monthly Notices Roy. Astron. Soc., in press.
De Zotti, G., et al.: 1981, Astrophys.J., in press.
Field, G.B.: 1972, Ann.Rev.Astron.Astrophys. 10, p. 227.
Field, G.B., and Perrenod, S.C.: 1977, Astrophys.J. 215, p. 717.
Green, R.F., and Schmidt, M.: 1978, Astrophys.J.(Letters) 220, L1.
Green, R.F., and Schmidt, M.: 1981, private communication.
Giacconi, R., Gursky, H., Paolini, F., and Rossi, B.: 1962, Phys. Rev. Lett. 9, p. 439.
Giacconi, R., et al.: 1979, Astrophys.J.(Letters) 234, L1.
Kembhavi, A.K. and Fabian, A.C.: 1981, Monthly Notices Roy. Astron. Soc., in press.
Koo, D.C. and Kron, R.G.: 1981, preprint.
Kriss, G.A., Canizares, C.R., and Ricker, G.R.: 1980, Astrophys.J. 242, p. 492.
Kron, R.G., and Chiu, L.T.G.: 1981, Pub. A.S.P. 93, p. 397.
Ku, W.H., Helfand, D.J., and Lucy, L.B.: 1980, Nature 288, p. 323.
Maccacaro, T., et al.: 1981, Astrophys.J., in press.

Marshall,, F.E., et al.: 1980, Astrophys.J. 235, p. 4.

Marshall, H.L., et al.: 1981, in preparation.

Mushotzky, R.F., Marshall, F.E., Boldt, E.A., Holt, S.S., and
  Serlemitsos, P.J.: 1980, Astrophys.J. 235, p. 377.

Piccinotti, G., Mushotzky, R.F., Boldt, E.A., Holt, S.S., Marshall, F.E.,
  and Serlemitsos, P.J.: 1981, Astrophys.J., in press.

Rowan-Robinson, M., and Fabian, A.C.: 1975, Monthly Notices Roy. Astron.
  Soc. 170, p. 199.

Schwartz, D., and Gursky, H.: 1974, in "X-Ray Astronomy", pp. 359-388,
  ed. Giacconi, R. and Gursky, H.

Setti, G., and Woltjer, L.: 1973, in "X- and Gamma-Ray Astronomy", pp.
  208-211, ed. Giacconi, R. and Bradt, H.V.

Setti, G., and Woltjer, L.: 1981, Proceedings of the Vatican Study Week
  on "Cosmology and Fundamental Physics".

Silk, J.: 1973, Ann. Rev. Astron. Astrophys. 11, p. 269.

Sramek, R.A., and Weedman, D.W.: 1980, Astrophys.J. 238, p. 435.

Tananbaum, H., et al.: 1979, Astrophys.J.(Letters) 234, L9.

Veron, P.: 1979, Astron.Astrophys. 78, p. 46.

Zamorani, G.: 1981, in the Proceedings of the "X-Ray Symposium", pp. 61 -
  71, ed. Davis Philip, A.G.

Zamorani, G., et al.: 1981, Astrophys.J. 245, p. 357.

RELATIVISTIC ELECTRONS IN GALAXIES AND THEIR NONTHERMAL RADIATION FROM
RADIO TO GAMMA RADIATION

R. Schlickeiser[1,2]
1) Max-Planck-Institut für Radioastronomie, Auf dem Hügel 69,
   D-5300 Bonn 1, West Germany (permanent address)
2) Institute of Astronomy, University of Cambridge, Madingley
   Road, Cambridge CB3 0HA, UK.

Abstract. Relativistic electrons are responsible for most of the non-
thermal radiation in galaxies. Their dynamics is determined by interactions with cosmic magnetic, radiation and matter fields. After
summarizing the main radiation processes we discuss recent radio and
gamma ray observations of our Galaxy and derive observational constraints
on the propagation parameters for electrons. In particular, we concentrate on the nature of relativistic electron sources and the nature
of gamma ray point sources. A discussion of nonthermal radiation from
selected extragalactic objects which includes its contribution to the
extragalactic diffuse gamma ray background completes the review.

I. INTRODUCTION

        Although relativistic electrons (RES) - positrons e^+ and
negatrons e^- - have no direct impact on cosmology, they are of vital
importance since being responsible for most of the nonthermal radiation
in galaxies which very often dominates the thermal component. Cosmologically one may distinguish between two classes of REs: (i) those
being produced before galaxy formation e.g. by $p - \bar{p}$ collisions, (ii)
those being produced in galaxies by an acceleration mechanism linked
more or less to stellar objects (pulsar, supernova explosion, shock
wave acceleration, secondary $e^{\pm}$ - decay in interstellar matter). Here
I want to concentrate on the latter component although a study of the
existence of pregalactic REs may be very interesting.
        Most of our knowledge of the dynamics of REs comes from
galactic observations. REs have been detected in the primary cosmic
ray flux in 1961 by Meyer and Vogt (1961) and Earl (1961). At high
energies (E > 10 GeV) the ratio of interstellar REs to cosmic ray protons
is about 1 percent. Studies of their nonthermal synchrotron radiation
in cosmic magnetic fields, their inverse Compton and nonthermal
bremsstrahlung radiation allow us to draw conclusions on their energy
spectrum $N(E,\underline{r})$ in different regions of the Galaxy and on their spatial
distribution $\bar{n}_e(\underline{r}) = \int dE N(E,\underline{r})$.
        In section II we will shortly review their main radiation

A. W. Wolfendale (ed.), Progress in Cosmology, 215–232.

processes, whereas section III is devoted to recent observations of REs and their nonthermal radiation, and the current theoretical interpretation is discussed. Section IV indicates the importance of X- and gamma-ray emission on the total emission from extragalactic objects and discusses the nature of the extragalactic diffuse gamma-ray background radiation.

II. MAIN RADIATION PROCESSES OF REs

REs produce non-thermal continuum radiation by
(a) Synchrotron radiation in cosmic magnetic fields.
(b) Inverse Compton radiation of ambient photon gases.
(c) Nonthermal electron bremsstrahlung in the Coulomb fields
    of atoms and molecules of the interstellar gas.

Table 1 summarizes the main features and predictions of these radiation processes under the assumption of optically thin emission, if the spatially isotropic intensity of REs is given by

$$I(E,\underline{r}) \equiv \frac{c}{4\pi} \; N(E,\underline{r}) = \times(\underline{r}) \; E^{-p} \quad cm^{-2} \; s^{-1} \; ster^{-1} \; eV^{-1} \tag{1}$$

where $1 < p = $ const.

TABLE 1: Continuum radiation processes of relativistic electrons
          (optically thin case)

Process	Energy Flux $\propto$	Remarks
Synchrotron radiation	$\nu^{(1-p)/2} \int\limits_0^\infty d\ell \times(\underline{r}) \; H_\perp^{(p+1)/2}(\underline{r})$	polarization $<\nu>(MHz) \cong 16 H_\perp (\mu G) E^2 (GeV)$
Inverse Compton scattering	$\nu^{(1-p)/2} \int\limits_0^\infty d\ell \times(\underline{r}) \; n_{ph}(\underline{r})$	for $\varepsilon E << (mc^2)^2$ "Thomson Limit" $<E_\gamma> \cong \varepsilon E^2/(mc^2)^2$
	$\nu^{-p} \int\limits_0^\infty d\ell \times(\underline{r}) \; n_{ph}(\underline{r})$	for $\varepsilon E >> (mc^2)^2$ "Klein-Nishina" cut-off $<E_\gamma> \cong E$
Nonthermal bremsstrahlung	$\nu^{1-p} \int\limits_0^\infty d\ell \, n_{gas}(\underline{r}) \times(\underline{r})$	for $E > 10$ MeV, $<E_\gamma> \cong 0.5 \; E$

$n_{ph} (\varepsilon,\underline{r})$: target photon distribution; $H_\perp (\underline{r})$: magnetic field strength; isotropic relativistic electron distribution assumed:
$$I(E,\underline{r}) = \times(\underline{r}) \; E^{-p}$$

The interstellar magnetic field strength is of the order $H_\perp \cong 3\mu G$ so that the synchrotron radiation of REs in the Galaxy produces diffuse continuum radio emission in the MHz frequency range. In objects with larger magnetic fields REs synchrotron radiation may be responsible for continuum infrared, optical, ultraviolet and even X-ray emission. By measuring the degree of polarization this has indeed be confirmed for a number of sources.

The mean energy of a bremsstrahlung photon from REs is $\langle E_\gamma \rangle \cong 0.5\ E$, so that this radiation may be detected in the gamma ray range (see section III.2). In inverse Compton interactions of REs we have to consider two cases: (i) the Thomson limit $E \ll (mc^2)^2/\varepsilon$ where ε is the photon energy <u>before</u> collision, (ii) the extreme Klein-Nishina range, $E \gg (mc^2)^2/\varepsilon$. In the latter case the typical energy of the photon <u>after</u> collision $\langle E_\gamma \rangle \cong E$, so this radiation again is detectable only in the gamma ray regime. However, in the Thomson limit all photons (depending on E and ε) from radio to gamma ray photons can be produced in this interaction. Physically, in the Klein-Nishina range recoil effects in the interaction become important, lowering the cross section value which results in the so-called "Klein-Nishina" cutoff in the inverse Compton radiation spectrum. The detection of such a continuous steepening of the gamma ray energy spectrum with increasing energy would reveal the inverse Compton origin of this radiation (see also section III.4).

As a result of these radiation processes and the energy loss of REs by ionizing and exciting atoms and molecules of the interstellar gas, the "source spectrum" of REs $q(e,\underline{r},t)$ is changed into the "equilibrium spectrum" of REs $N(E,\underline{r},t)$ given in the simplest approximation by the solution of a continuity equation in energy space (see e.g. Blumenthal and Gould 1970).

$$\frac{\partial N(E,\underline{r},t)}{\partial t} + \frac{\partial}{\partial E}\left[N(E,\underline{r},t) \cdot \frac{dE}{dt}(E,\underline{r})\right] + \frac{N(E,\underline{r},t)}{T} = q(E,\underline{r},t) \qquad (2)$$

where the energy loss of a single RE is

$$-\frac{dE}{dt}(E,\underline{r}) \cong [3 \cdot 10^{-7} + 8 \cdot 10^{-16}E]\left(\frac{n_{gas}(\underline{r})}{cm^{-3}}\right) + 10^{-25}E^2\left\{\frac{W_{ph}(\underline{r})}{0.7 eVcm^{-3}} + \frac{3}{16\pi}\left(\frac{H_\perp(\underline{r})}{3\mu G}\right)^2\right\} eVs^{-1} \qquad (3)$$

The term $\frac{N}{T}$ represents the zeroth approximation of leakage losses out of the confinement region considered. Equation (2) is a simple version of the continuity equation in phase space neglecting completely spatial gradients. Solutions of equation (2) for different circumstances have been given by Kardashev (1962).

In the steady-state case, $\frac{\partial N}{\partial t} \equiv 0$, the solution of equation (2) with the boundary condition $N(E = \infty,\underline{r}) = 0$ is given by

$$N(E,\underline{r}) = \left[\frac{dE}{dt}(E,\underline{r})\right]^{-1} \int_E^\infty dx\, q(x,\underline{r}) \exp\left\{\frac{1}{T}\int_E^x \frac{dy}{\frac{dy}{dt}(y,\underline{r})}\right\} \qquad (4)$$

For example: if $q(E,\underline{r}) = q_o(\underline{r}) E^{-p}$ and $\frac{dE}{dt} = a(\underline{r}) E^2$, one finds

$$N(E,\underline{r}) \cong \begin{cases} q_o(\underline{r})T\ E^{-p} & \text{for } E << \frac{1}{aT} \\ \dfrac{q_o(\underline{r})\ E^{-(p+1)}}{a(\underline{r}).(p-1)} & \text{for } E >> \frac{1}{aT} \end{cases} \qquad (5)$$

which enables an easy physical interpretation. For energies smaller than $E = (aT)^{-1}$ the leakage time is smaller than the radiation loss time, so that particles leak out the confinement region before loosing their energy in radiation processes. In the opposite case $E >> (aT)^{-1}$, the radiation time scale $\tau_r = (aE)^{-1} << T$, so that the spectrum steepens.

Finally, we mention that $e^+- e^-$ pair annihilation produces the 511 keV line radiation. The detection of this line from the galactic center with strong time variability has been recently reported by Riegler et al. (1981). We refer the reader to this work for further references and interpretation of this result.

III. OBSERVATIONS OF REs AND THEIR INTERPRETATION

1. Direct Measurements Near the Solar System

The energy spectrum between 30 GeV and 10^3 GeV has been measured to follow a power law in energy (Nishimura et al. 1980 and references therein):

$$I(E) \propto E^{-p} , \quad p = 3.2 \pm 0.2 \qquad (6)$$

In this energy range, the ratio of RE intensity to that of cosmic ray protons amounts about 1 percent.

Below 10 GeV the outstreaming solar wind strongly modifies the measured intensities. After "demodulation" one finds again a power law intensity distribution (6) with p = 2.2 ± 0.4 for 0.3 GeV<E<2 GeV (Cummings et al. 1973a, Daugherty et al. 1975).

2. Measurements of the Frequency Spectrum of Nonthermal Radiation from
   REs

As can be seen from Table 1, the synchrotron energy flux from a given direction (ℓ^{II}, b^{II}) is proportional to $\nu^{(1-p)/2}$, if the form of the RE energy spectrum does not change along the line of sight. By measuring the frequency spectrum of the diffuse background synchrotron radiation from the anticenter direction, the value of p has been determined by various authors (e.g. Goldstein et al. 1970, Cummings et al. 1973b, Webber et al. 1980). For energies larger than 0.3 GeV this method confirms the results of section III.1. However, at lower electron energies, E<0.3 GeV, corresponding to synchrotron frequencies ν < 10 MHz (see Table 1), the absorption of the nonthermal radiation by free-free transitions and by the Razin-effect has to be taken into account. The poor knowledge of the spatial variation of the magnetic field strength $H_\perp(r)$ and the ionized gas density $w_{IG}(r)$ leads to large uncertainties in the absolute values of the electron flux. Also, a major part of these low-energy (E<0.1 GeV) REs may be of Jovian origin (McDonald and Trainor 1976, Eraker and Simpson 1981).

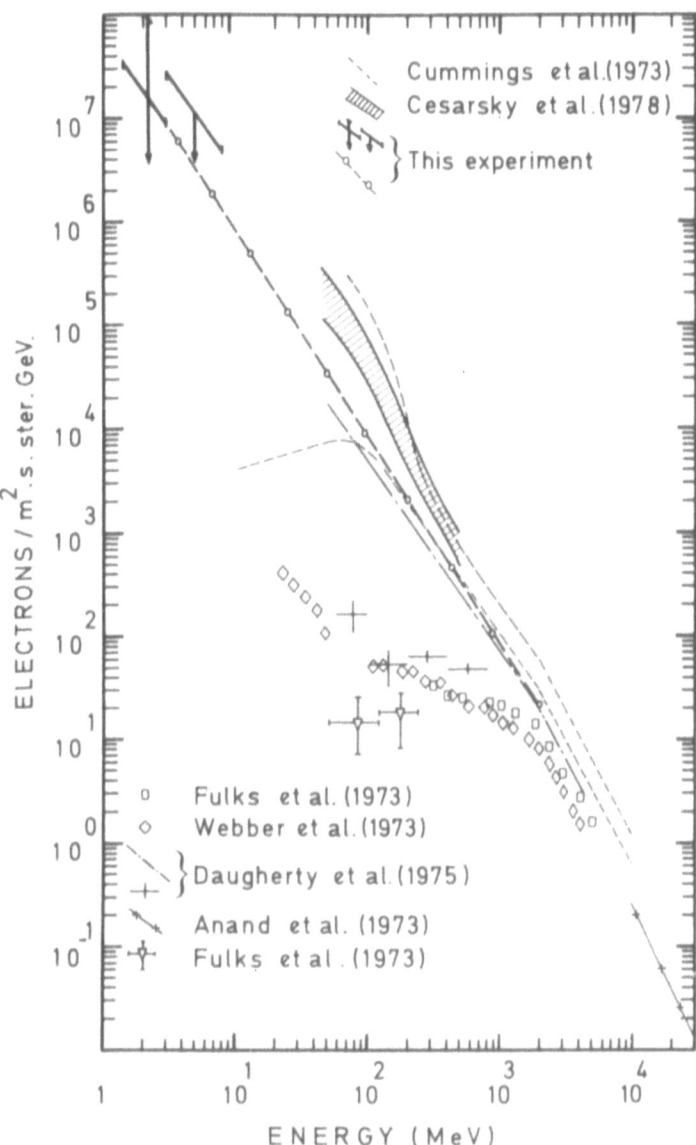

Figure 1. Calculated and measured cosmic ray electron spectrum. Reproduced from Mandrou et al. (1980) by courtesy of the Astrophysical Journal.

        Fortunately, we may estimate the interstellar RE energy
spectrum below 0.3 GeV from observations of the galactic gamma ray
energy spectrum below E_γ = 100 MeV. These gamma rays are mainly due
to nonthermal RE bremsstrahlung (Fichtel et al. 1976, Schlickeiser and
Thielheim 1978, Cesarsky et al. 1978 - see also section III 3c). Figure
1 shows the results. Cesarsky et al. (1978) have used COS-B measurements
of gamma rays with energies between 70 MeV and 5000 MeV (Paul et al.
1978) to derive the interstellar electron spectrum at energies E > 70
MeV. In the electron energy range below 10 MeV only upper limits can
be derived from observations of the galactic disk $|\ell^{II}|$ < 45° in the
gamma ray energy range 80 keV - 8 MeV by Mandrou et al. (1980). These
authors note that a RE spectrum $I(E) \propto E^{-2}$ in the range from 10^{-3} to
2 GeV is consistent with the COS-B observations above 0.07 GeV and their
upper limits at lower energies.
        COS-B has measured gamma ray energy spectra from four different
regions of the Galaxy (Paul et al. 1978). Since their shape does not
vary with position, this indicates that the shape of the RE energy
spectrum is independent of position, i.e. p = const. in the galactic
disk.
        Following the argument of Weinberg and Silk (1973) recently
Protheroe and Wolfendale (1980) estimated the diffuse galactic X-ray
and gamma ray background produced by interactions of REs. According to
their results the source spectrum of REs is a single power law in energy
$q(E) \propto E^{-2.14}$.
        Summarizing this section and section III.1 we note that the
RE intensity distribution in the Galaxy follows a power law $I(E,\underline{r})$ =
$x(\underline{r})\ E^{-P}$ where
        (i)   p = 3.2 ± 0.2   for E > 30 GeV
        (ii)   p = 2.2 ± 0.4   for 0.1 < E < 1 GeV
        (iii)   only upper limits exist on the flux of REs below 0.1
                GeV. A value of p ≅ 2.0, however, is consistent with
                these limits.
The "break" in the electron energy spectrum around 10 GeV is interpreted
with simple lifetime arguments (5) - see e.g. Hartmann et al. (1977),
Prince (1979).

3. Spatial Distribution of REs

        There exist complete sky surveys of the nonthermal synchrotron
radiation at 150 MHz (Landecker and Wielebinski 1970) and 408 MHz (Haslam et al. 1981). They display the Milky Way as strong emitter, reveal
local features (North galactic spur, loops), point sources and other
galaxies (e.g. Cen A, LMC, Andromeda nebula). Hence these surveys
indicate directly that REs originate in galaxies. An indirect proof
based on the overall presence of the 3K microwave background radiation
and the non-detected X-ray background due to inverse Compton scattering
by REs has been formulated earlier by Fazio et al. (1966).
        Assuming cylindrical geometry ($\underline{r}$ = (R,φ,z) one may derive
from these surveys the radial (R) and latitudinal (z) distribution of
the synchrotron emissivity.

(a) <u>Radial distribution</u>. Taking the longitudinal profile at $b^{II}=0°$ of the 408 MHz - survey and assuming some symmetry, Philipps et al. (1981) unfolded the radio profile to derive the radial distribution of the averaged synchrotron emissivity

$$< x(R)H_{\perp}^{\frac{p+1}{2}}(R) > \quad .$$

However, interpretations in terms of $<x(R)>$ and/or $<H_{\perp}(R>$ is difficult due to the effect of turbulence enhancement (Cowsik and Mitteldorf 1974).

(b) <u>Distribution perpendicular to the galactic plane</u>.

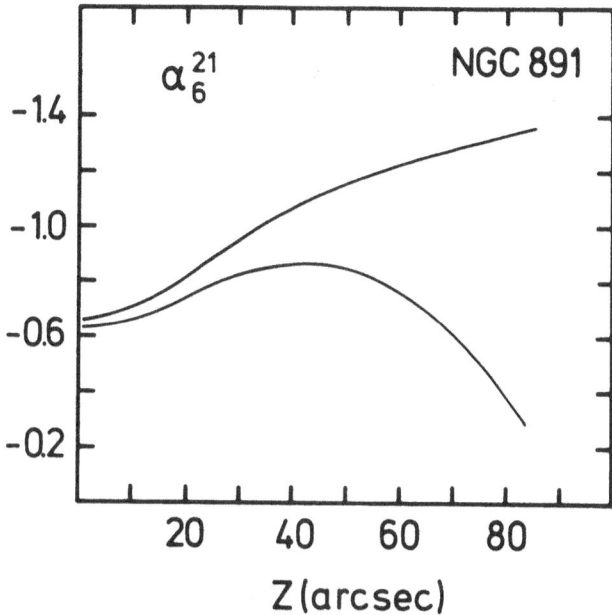

Figure 2. Measured radio spectral index between $\lambda=21$ cm and $\lambda=6$ cm ($S(\nu) \sim \nu^{\alpha}$) as a function of distance from the galactic plane for NGC 891 (Strong and Allen 1981).

Due to the location of the solar system inside the galactic disk the latitude radio profile of synchrotron emission of our own Galaxy is difficult to analyse with respect to the z-distribution of the synchrotron emission. Therefore one uses observations of external galaxies seen edge-on where the variation perpendicular to the galactic plane directly corresponds to the z-variation of the synchrotron emissivity. Measurements of NGC 4631 (Ekers and Sancisi 1977) and NGC 891 (Allen et al. 1978, Beck et al. 1979, Strong and Allen 1981) at different frequencies indicate that the radio distribution perpendicular to the plane is significantly broader than the optical image. The extent of this radio halo is frequency dependent; observations so far indicate that the extent decreases with increasing frequency. As a consequence, the radio spectral index $\alpha = \frac{1-p}{2}$ increases with distance from the ga- lactic plane as is shown for the case of NGC 891 in Figure 2. For a

number of arguments summarized by Webster (1978) our Galaxy shows a
similar behaviour. Lerche and Schlickeiser (1980; 1981a, b, c) have
studied this phenomenon in detail and explained the spectral gradient
with an energy dependent transport mechanism of REs. Diffusion in
partially irregular magnetic fields, convection and adiabatic deceleration in galactic winds and energy loss processes compete in determining
the dynamics of relativistic electrons.

(c) <u>Gamma ray sky map (70 MeV - 5 GeV)</u>. The spatial distribution of
REs may be also inferred from the COS-B gamma ray sky survey, which now
is available for latitudes $|b^{II}| < 25°$ (Mayer-Hasselwander et al. 1981).
Besides the concentration of gamma ray emission towards the galactic
plane, the survey indicates the occurrence of several point sources
which will be discussed in detail in section 3d (see Figure 4).

High-energy (>10 MeV) gamma rays result from interactions of
cosmic ray nucleons of energies 1-30 GeV/nucleon ($\pi°$-decay, $\pi^{\pm}$ decay
into $e^{\pm}$ and subsequent electron bremsstrahlung) and REs (nonthermal
bremsstrahlung, inverse Compton scattering of ambient photon gases).
A review of gamma ray production processes can be found in Schlickeiser
(1981a). We want to emphasize here two points:

(i) Cosmic ray nucleons produce high-energy gamma rays in
inelastic collisions with the interstellar gas atoms and molecules
giving rise to neutral <u>and</u> charged pions. The $\pi°$-meson immediately
decays into two gamma rays whereas the $\pi^{\pm}$-mesons decay into $e^{\pm}$. If the
confinement time of these secondary electrons in the Galaxy is longer
than their bremsstrahlung loss time, these secondary electrons give
rise to nonthermal bremsstrahlung. Marscher and Brown (1978) and
Schlickeiser (1981b) have shown that the contribution from secondary
electrons is significant, especially at gamma ray energies smaller than
100 MeV as can be seen from Figure 3. The combined production rate
gives a good fit to the COS-B spectral measurements and explains why
pure $\pi°$-decay gamma ray spectra have not been observed. In this picture high-energy gamma rays (E_{γ} > 70 MeV) are mainly due to cosmic ray
nucleons.

(ii) If one uses the <u>observed</u> electron spectrum to calculate
the gamma ray production rates one finds that the $\pi°$-decay and electron
bremsstrahlung dominate the galactic production of gamma rays. The
basic question, whether high energy gamma rays entirely originate from
cosmic ray nucleons, reduces to the question: are REs in our Galaxy
mainly secondaries (from cosmic ray nucleons) or dominated by primary
REs sources?

The integral gamma ray intensity (E_{γ} > 100 MeV) is given by

$$I_{\gamma}(>100\text{MeV}; \ell^{II}, b^{II}) = \frac{(1.9\pm0.2)10^{-25}}{4\pi} \int_{0}^{\infty} d\ell \ x_{p}(\underline{r})n_{g}(\underline{r})\text{cm}^{-2}\text{s}^{-1}\text{ster}^{-1} \ , \quad (7)$$

enabling to find the spatial distribution of cosmic ray nucleons ($x_{p}(\underline{r})$)
if the distribution of interstellar matter as inferred from 21-cm and
CO-line observations is used (e.g. Stecker 1975).

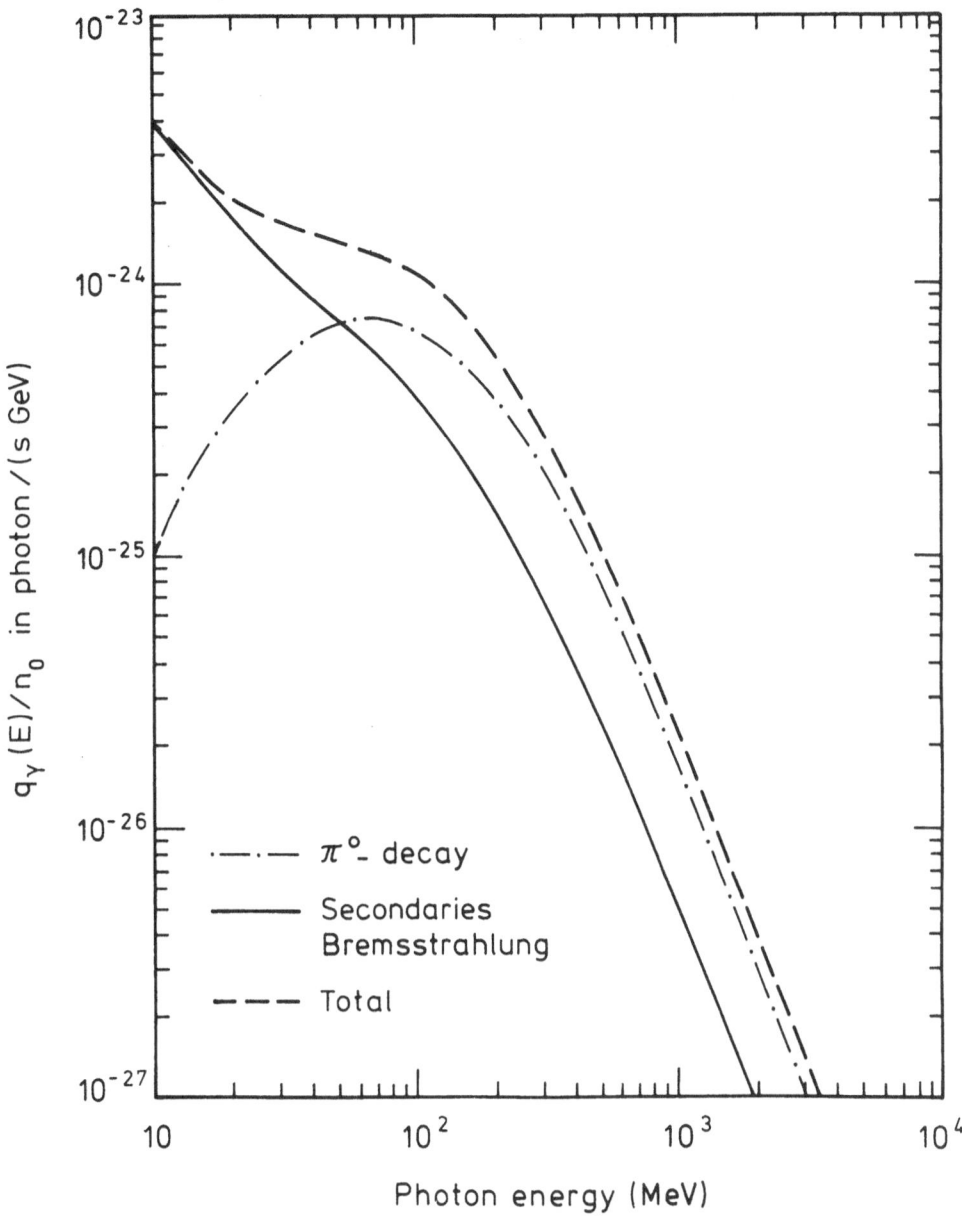

Figure 3. Differential production rate of gamma rays per
interstellar gas atom from π°-decay and electron bremsstrahlung from cosmic ray nucleon secondaries (from Schlickeiser
1981b).

4. Gamma Ray Point Sources - REs Sources?

COS-B has revealed the existence of 25 high-energy gamma ray point sources: two pulsars (NP0532, PSR0833-45), one quasar (3C273) and two interstellar gas clouds (ρ Oph, Orion) (Swanenburg et al. (1981)). The remaining 20 sources are still unidentified with known astrophysical objects. Due to the poor angular resolution of the COS-B detector (∼2°), some sources may be extended objects with angular size up to 2° (Hermsen 1980, Li and Wolfendale 1981). Figure 4 shows the location of the discovered point sources in the sky. The strong concentration of these sources to the galactic plane is apparent. This concentration excludes that the sources are either closeby or very far away favouring typical distances 0.5 kpc $\lesssim$ D < 4 kpc. The typical integral (>100 MeV) flux is of the order 2.10^{-6} photons cm^{-2} s^{-1} resulting in a luminosity estimate in the range of $(0.1 - 6).10^{35}$ erg s^{-1}, assuming their emission is isotropic.

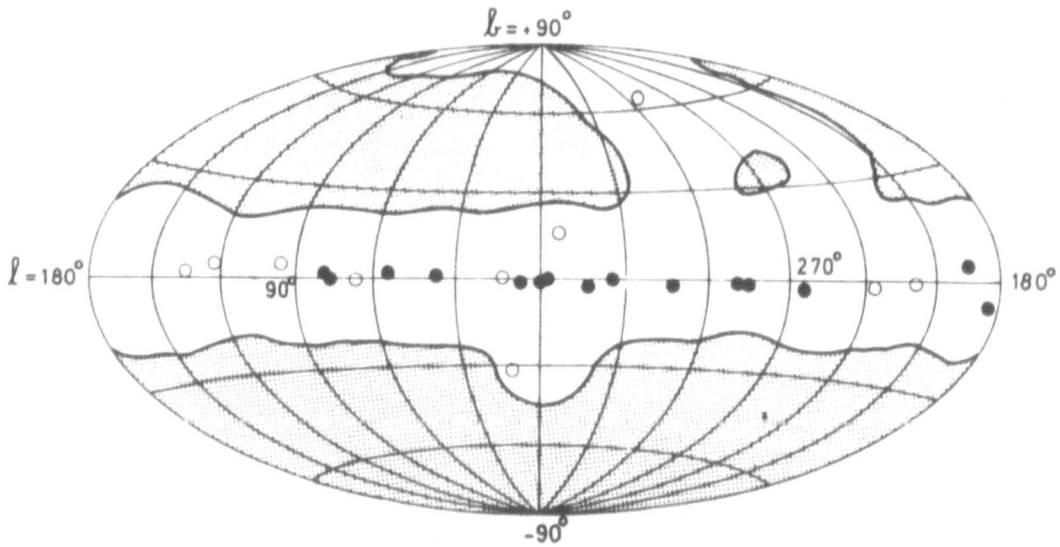

Figure 4. Gamma ray sources contained in the 2CG catalogue. Filled circle: sources with flux > $1.3x10^{-6}$ photons cm^{-2} s^{-1}. Open circles: weaker sources. Unshaded area: region of the sky searched. Reproduced from Swanenburg et al. (1981) by courtesy of The Astrophysical Journal.

The nature of the unidentified sources is still unknown. In the following we shortly review some proposals. Of course, there may be different classes of sources as has been already verified for the identified ones (pulsars, quasars, gas clouds).

(a) Pulsars. Since already two gamma ray sources are pulsars, it is obvious to propose that some of the unidentified sources are also

pulsars. Assuming that the energy reservoir of pulsars is rotation of
a neutron star of constant moment of inertia $I \cong 10^{45}$ g cm², the rate
of rotational energy loss of known radio pulsars is calculated from the
period P and its derivative $\dot{P}$:

$$\dot{E}_{rot} = - 4\pi \ I \ \dot{P}/P^3 \tag{8}$$

Values of $\dot{E}_{rot}$ range from $4.7.10^{38}$ erg s^{-1} for the Crab pulsar and
7.10^{36} erg s^{-1} for the Vela pulsar to a minimum around 10^{30} erg s^{-1};
for most of the known pulsars $\dot{E}_{rot}$ is in the range $10^{32} < E_{rot} < 10^{34}$
erg s^{-1} (Buccheri et al. 1977). Even if their efficiency of converting
rotational energy into high-energy (>100 MeV) gamma rays is much higher
than for the Crab ($\eta \cong 3.10^{-4}$) and Vela pulsar ($\eta \cong 6.10^{-3}$), known radio
pulsars seem not to be luminous enough to account for the observed gamma
ray sources.

However, since the gamma ray luminosity of pulsars seems to
decrease with age, Buccheri (1980, 1981) suggested that young pulsars,
so far undetected in radio astronomy, have to be associated with the
detected gamma ray sources; but, the number of pulsars younger than
Vela pulsar is highly uncertain (see e.g. Helfand 1981).

(b) Supernova Remnants, OB-stars, dense interstellar clouds.
Pinkau (1979) has pointed out that the narrow z-distribution of discrete
gamma ray sources resembles that of O-stars, supernova remnants, H_2-
clouds and HII-regions indicating a link between gamma ray sources and
young galactic objects. This is supported by observational evidence
that six gamma ray sources are associated with optically visible supernova remnants (van den Bergh 1979).

In line with these findings, Montmerle (1979) has proposed
the SNOB-model of gamma ray sources: OB-stars are cosmic ray sources,
injecting suprathermal particles into the surrounding medium by flares
which then are boosted to relativistic energies by shock waves of
neighbouring supernova remnants and produce gamma rays mainly by non-
thermal RE bremsstrahlung although contributions from π°-decay cannot
be excluded (Montmerle and Cesarsky 1979). Although energetically
possible, the details of shock wave acceleration are very uncertain, in
particular its efficiency in converting mechanical energy of the shock
into accelerating REs and cosmic ray nucleons.

Following the suggestion of Black and Fazio (1973) that individual massive clouds irradiated by cosmic ray particles are gamma ray
point sources, recently Issa and Wolfendale (1981) have argued that 60
percent of the COS-B sources can be explained in this way. The integral
(>100 MeV) gamma ray luminosity of a cloud of mass M irradiated by an
intensity I of cosmic ray nucleons is derived from the integral production rate due to π°-decay and secondaries bremsstrahlung as

$$L(>100 \text{ MeV}) \cong 4.10^{28} \ f_{c-r} (M/M_\odot) \text{ erg s}^{-1} \tag{9}$$

where $f_{c-r} = I/I_0$ is the ratio of the cosmic ray nucleon flux in the
cloud to that measured near the solar system. Support for this model
is the fact that it agrees with the measured gamma ray luminosities of
the Orion (Caraveo et al. 1980) and ρ-Oph (Issa et al. 1981) cloud

complexes with $f_{c-r} \cong 1$ if the radioastronomical mass estimates are used. There is agreement that this model explains nearby (distance smaller than 1 kpc gamma point sources. However, the luminosities for "passive" clouds ($f_{c-r} \cong 1$) are too small to make these objects visible at large distances with the COS-B sensitivity. Giant molecular clouds have typical masses of $(1-5)\ 10^5\ M_\odot$ (Blitz 1980, Solomon and Saunders 1980), so that for $f_{c-r} \cong 1$ according to (9) the gamma ray luminosity is smaller than $L(>100\ \text{MeV}) < 2.10^{34}\ \text{erg s}^{-1}$; one order of magnitude too small to account for bright gamma ray point sources. It has been speculated that there are clouds with $f_{c-r} \gg 1$ ("active" clouds) where the enhancement of cosmic rays may be due to e.g. acceleration of particles by stellar winds from active stars in the complex (Casse and Paul 1980). However, the efficiency of such acceleration models is very uncertain.

Another proposed model for gamma ray sources is associated with extended low-density HII-regions heated by young hot OB-stars. Mezger et al. (1981) have recently shown that 80 percent of the far-infrared/submm emission of our Galaxy comes from dust embedded in extended low-density (ELD) HII-regions. The energy density of infrared and ultraviolet photons in these objects is very high suggesting to consider the inverse Compton interactions of these photons with ambient interstellar REs. Here we follow a simple approach and consider only the interactions of the infrared photons. More detailed calculations are given in Biermann et al. (1981). According to Mezger et al. the infrared luminosity of the Galaxy results from the superposition of ELD HII-regions, so that

$$0.8\ L_{IR}^{Gal} = 5.10^9\ L_\odot \cong \sum_{i=1}^{N} 4\pi\ c w_i r_i^2 = 3c \sum_{i=1}^{N} V_i W_i r_i^{-1} \approx 3c\ f\ V_{Gal} W\ R^{-1} \tag{10}$$

where f is the filling factor of ELD HII-regions ($f \cong 10^{-3}$), W their mean infrared energy density, R their mean (assumed spherical) radius and $V_{Gal} \cong 10^{66}\ \text{cm}^3$ the galactic volume. Hence we find $W \cong 3.10^{-13}\ R(\text{pc})$ erg cm^{-3}. The inverse Compton gamma ray luminosity is (Schlickeiser 1981c)

$$L(>E_\gamma) \cong \sigma_T\ c\ V\ W\ \frac{E_\gamma}{\varepsilon_i}\ N\ (E> mc^2\ (E_\gamma/\varepsilon)^{\frac{1}{2}})\ \text{erg s}^{-1} \tag{11}$$

where σ_T is the Thomson cross section, $V = (4\pi/3)R^3$ the volume of an individual ELD HII-region and $\varepsilon_i \cong 10^{-2}$ eV the mean energy of the infrared photons. Expressing the integral number density of REs in the HII-region as $N(>50\ \text{GeV}) = g_e\ N_\odot(>50\ \text{GeV})$, where $N_\odot(>50\ \text{GeV}) \cong 10^{-13}\ \text{cm}^{-3}$ is measured near the solar system, we find from (11)

$$L(>100\ \text{MeV}) \cong 6.10^{26}\ (\frac{R}{pc})^4 g_e\ \frac{\text{erg}}{\text{sec}} < 10^{32} g_e\ \frac{\text{erg}}{\text{sec}} \tag{12}$$

if we use 20 pc as an upper limit to the radius of an ELD HII-region. For a flux of high-energy (>50 GeV) REs as measured near the solar system, $g_e = 1$, this is at least two orders of magnitude smaller than the required gamma ray luminosity. Enhancements of the order $g_e > 100$ may be achieved either by a nearby supernova explosion which leads essentially to the SNOB-model discussed already, or by a closeby REs source e.g. a pulsar forming a binary system with the OB-star (Jackson 1972, Schlickeiser 1981d).

The last hypothesis is supported by the analysis of the energy spectrum of the unidentified gamma ray source CG 195+4. It has been found that an inverse Compton origin of this radiation fits the observations best (Schlickeiser 1981d); the apparent steepening of the gamma ray spectrum is attributed to the Klein-Nishina cut-off (see Table 1) and allows us to estimate the energy of the target photons in CG195+4: $2eV < \varepsilon < 20$ keV. This suggests that CG195+4 also emits radiation in the XUV-frequency range. And indeed: observations with the Einstein X-ray telescope have revealed the existence of an X-ray source inside the $0.4° \times 0.4°$ error box which also has been optically identified (Bignami 1981). So CG195+4 might be an example of a gamma ray source consisting of a RE source (pulsar?) and a source of XUV-photons (OB-star?).

Summarizing our discussion on unidentified galactic gamma ray sources, we note that these objects probably are related to a young galactic population. Estimates on the basis of conservative assumptions on the flux of relativistic particles yield luminosities which are one or two orders of magnitude smaller than observed. Future gamma ray experiments with tenfold improved sensitivity should therefore reveal a dramatic increase in the number of gamma ray sources. It is possible that the COS-B gamma ray sources are the upper tail of a luminosity distribution function centered at $\sim 10^{33}$ erg s^{-1}, so that the high luminosity of these sources can be explained by adequate fluctuations in the physical parameter values. The discussion has shown that several source mechanisms are possible.

IV. EXTRAGALACTIC OBJECTS AND THE DIFFUSE EXTRAGALACTIC GAMMA RAY BACKGROUND

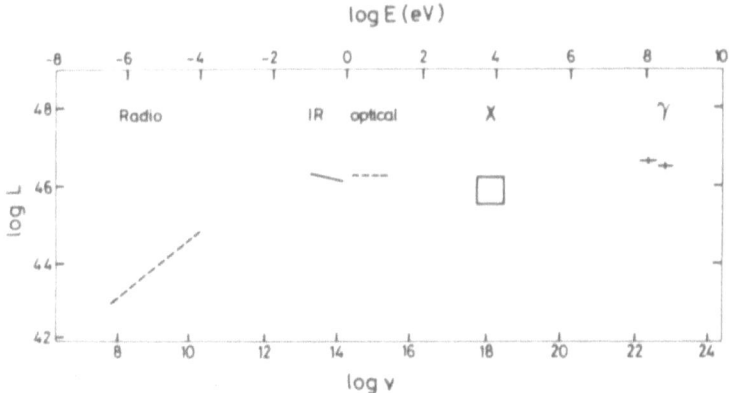

Figure 5. The energy distribution of 3C273 per decade of frequency. Reproduced from Ulrich (1981) by courtesy of Space Science Review.

The ratio of the gamma ray to the total luminosity of our
Galaxy, $L(>100 \text{ MeV})/L_{tot} \cong 10^{-6}$ indicates that in normal galaxies gamma
rays are of minor importance for the energy budget of these objects.
This seems to be remarkably different for active galaxies. The quasar
3C273 has been detected as point source of high-energy gamma rays in
the COS-B experiment (Swanenburg et al. 1978, Bignami et al. 1981). As
can be seen from Figure 5, its gamma ray luminosity accounts for at
least 10 percent of the total luminosity. For the Seyfert galaxies
NGC4151 (Perotti et al. 1979) and MCG8-11-11 (Perotti et al. 1981) luminosity ratios $L(>100 \text{ keV})/L_{tot} \approx 10^{-2}$ have been measured. In active
galaxies nonthermal radiation accounts for a significant fraction of
the total radiation output.

By comparing upper limits on the gamma ray flux (>35 MeV) from
active galaxies measured in the SAS-2 experiment with their X-ray
spectra, Bignami et al. (1979) noted that in these objects a significant
steepening of the energy spectrum from X-rays to gamma rays occurs.
The exact location of this break is not known, however there are indications from the measurements of NGC4151 and MCG8-11-11 that the break
occurs around ∿1 MeV. This immediately prompted several authors (e.g.
Grindlay 1978, Schönfelder 1978, Bignami et al. 1978) to explain the
bump in the extragalactic isotropic gamma ray background between 1 and
10 MeV (see Figure 6) by a superposition of unresolved sources like
NGC4151, MCG8-11-11 and 3C273.

Observations of gamma rays from high galactic latitudes have
revealed the presence of a diffuse, at energies smaller than 40 MeV
highly isotropic component, the energy spectrum of which is shown in
Figure 6. It has been pointed out (Schlickeiser and Thielheim 1976)
that, because the solar system lies inside the galactic disk, at high
energies (>100 MeV) most of the gamma ray flux arriving from large
latitudes is of local galactic origin giving rise to a steeper spectrum
for the true extragalactic gamma rays. This prediction has been confirmed by a careful analysis of the SAS-2 gamma ray arrival directions
(Özel 1980). Due to instrumental background problems, the COS-B
experiment is unable to measure the extragalactic diffuse gamma ray
flux. Hence, models for the diffuse extragalactic gamma ray background
have to explain both, the bump between 1 and 10 MeV and the steep
spectrum above 100 MeV. Stecker (1978) has noted that most models proposed to explain the bump had difficulties with reproducing the steep
spectrum. We are left with two alternatives for the origin of the
diffuse extragalactic gamma ray background:
(i) matter-antimatter annihilation in the baryon-symmetric big bang
    (Stecker 1978, 1981) where the peak from the π°-decay at 70 MeV is
    redshifted to lower energies to explain the bump.
(ii) superposition of many unresolved external sources. Schönfelder
    (1978) has emphasized that such an interpretation is possible if the
    spectrum of NGC4151 is typical for all external objects.

However, we want to point out that the gamma ray measurements
of external galaxies between 0.5 - 40 MeV are very poor mainly due to
the difficult experimental technique in this energy range. Published
energy spectra are very controversial: e.g. for the case of NGC4151
several experiments (Meegan and Haymes 1979, White et al. 1980) have

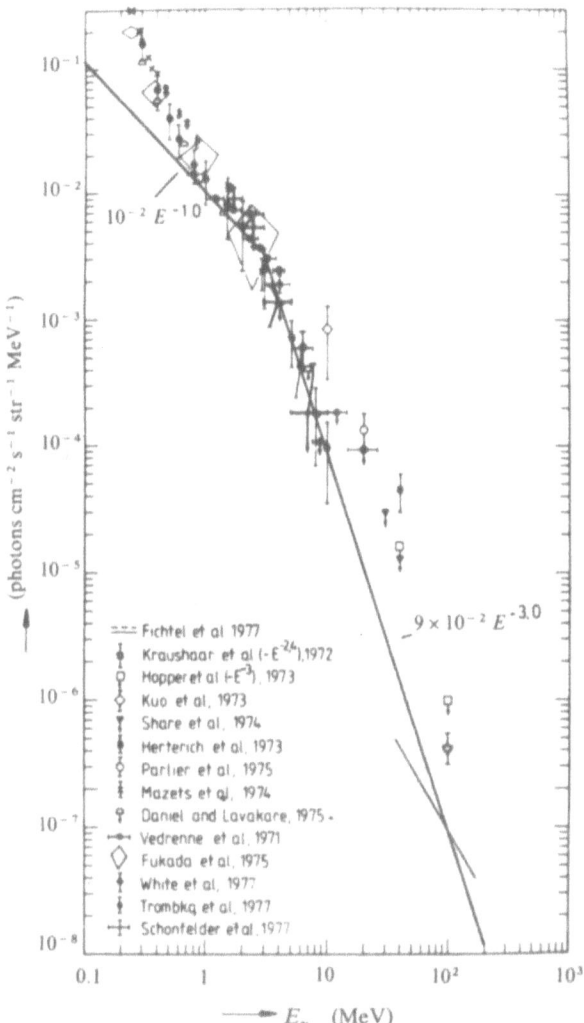

Figure 6.   Energy spectrum of the diffuse gamma ray background
radiation from high latitudes. Reproduced from Schönfelder
(1978) by courtesy of Nature.

found no evidence for a gamma ray flux around 1 MeV from this object,
whereas Perotti et al. (1979) have confirmed Schönfelder's (1978)
detection who in the meantime has withdrawn his results. It is questionable whether this controversy is due to time variability of the source
NGC4151. We have to wait for future more sensitive measurements.

The energy range around 1 MeV is astrophysically very important; not only to answer the cosmologically interesting question of the
origin of the diffuse extragalactic gamma ray background. The threshold

energy for photon-photon annihilation is 1.02 MeV (Herterich 1974),
low-energy cut-offs in the energy spectrum of REs (Schlickeiser 1980)
are visible too as are deviations from the interaction cross sections
derived under the assumption of relativistic RE energies (E>>0.511 MeV).
Also, gamma rays from Penrose powered black holes (Kafatos 1980), the
$e^+ - e^-$ annihilation line and nuclear gamma ray lines (Ozernoy and
Aharonian 1979) can be observed. As so often in physics, the merits
are high where the experimental technique is difficult.

V. ACKNOWLEDGEMENTS

        I thank Drs. P. Biermann, R. Güsten, I. Lerche, P. Mezger,
M. Rees and A. Wolfendale for helpful discussions. My stay in Cambridge
is supported by the Deutsche Forschungsgemeinschaft (Schl 201/2-1).
        I acknowledge the help of Mrs. A. Julier and Mr. R. Sword in
preparing this manuscript.

References

Allen, R.J., Baldwin, J.E., Sancisi, R., 1978, Astr. Ap. 62, 397.
Beck, R., Biermann, P., Emerson, D.T., Wielebinski, R., 1979, Astr. Ap.
        77, 25.
Biermann, P., Güsten, R., Mezger, P.G., Schlickeiser, R., 1981, in pre-
        paration.
Bignami, G.F., 1981, private communication.
Bignami, G.F., Lichti, G.G., Paul, J.A., 1978, Astr. Ap. 68, L15.
Bignami, G.F., Fichtel, C.E., Hartman, R.C., Thompson, P.J., 1979,
        Ap. J. 232, 649.
Bignami, G.F., Bennett, K., Buccheri, R., Caraveo, P.A., Hermsen, W.,
        Kanbach, G., Lichti, G.G., Masnou, J.L., Mayer-Hasselwander,
        H.A., Paul, J.A., Sacco, B., Scarsi, L., Swanenburg, B.N.,
        Wills, R.D., 1981, Astr. Ap. 93, 71.
Black, J.H., Fazio, G.G., 1973, Ap. J. 185, L7.
Blitz, L., 1980, in: Giant Molecular Clouds in the Galaxy, eds. P.M.
        Solomon & M.G. Edmunds, Pergamon Press: Oxford, p.1.
Blumenthal, G.R., Gould, R.J., 1970, Rev. Modern Phys. 42, 237.
Buccheri, R., 1980, in: Non-Solar Gamma Rays, eds. R. Cowsik & R.D.
        Wills, Pergamon Press: Oxford, p.17.
Buccheri, R., 1981, in: Pulsars, IAU Symp. 95, eds. W. Sieber & R.
        Wielebinski, Reidel: Dordrecht, p.241.
Buccheri, R., D'Amico, N., Massaro, E., Scarsi, L., 1977, Proc. Int.
        School of Gen. Rel. Effects, MPI-PAE/Astro. 138, 376.
Caraveo, P.A., Bennett, K., Bignami, G.F., Hermsen, W., Kanbach, G.,
        Lebrun, F., Masnou, J.L., Mayer-Hasselwander, H.A., Paul, J.
        A., Sacco, B., Scarsi, L., Strong, A.W., Swanenburg, B.N.,
        Wills, R.D., 1980, Astr. Ap. 91, L3.
Casse, M., Paul, J.A., 1980, Ap. J. 237, 236.
Cesarsky, C.J., Paul, J.A., Shukla, P.G., 1978, Ap. Space Sci. 59, 73.
Cowsik, R., Mitteldorf, J., 1974, Ap. J. 189, 51.

Cummings, A.C., Stone, E.C., Vogt, R.E., 1973a, Proc. 13th Int. Cosmic
        Ray Conf. (Denver), Vol. 1, 335.
Cummings, A.C., Stone, E.C., Vogt, R.E., 1973b, Proc. 13th Int. Cosmic
        Ray Conf. (Denver), Vol. 1, 340.
Daugherty, J.K., Hartman, R.C., Schmidt, P.J., 1975, Ap. J. 198, 493.
Earl, J.A., 1961, Phys. Rev. Letters 6, 125.
Ekers, R.D., Sancisi, R., 1977, Astr. Ap. 54, 953.
Eraker, J.H., Simpson, J.A., 1981, Proc. 17th Int. Cosmic Ray Conf.
        (Paris), Vol. 3, 279.
Fazio, G.G., Stecker, F.W., Wright, J.P., 1966, Ap. J. 144, 611.
Fichtel, C.E., Kniffen, D.A., Thompson, D.J., Bignami, G.F., Cheung,
        C.K., 1976, Ap. J. 208, 211.
Goldstein, M.L., Ramaty, R., Fisk, L.A., 1970, Phys. Rev. Lett. 24, 1193.
Grindlay, J.E., 1978, Nature 273, 211.
Hartmann, G., Müller, D., Prince, T., 1977, Phys. Rev. Lett. 38, 1368.
Haslam, C.G.T., Salter, C.J., Stoffel, H., Wilson, W.E., 1981, Astr.
        Ap. Suppl., in press.
Helfand, D.J., 1981, in: Pulsars, IAU Symp. 95, eds. W. Sieber & R.
        Wielebinski, Reidel: Dordrecht, p.249.
Hermsen, W., 1980, PhD thesis, Univ. Leiden.
Herterich, K., 1974, Nature 250, 311.
Issa, M.R., Strong, A.W., Wolfendale, A.W., 1981, J. Phys. G7, 565.
Issa, M.R., Wolfendale, A.W., 1981, Nature 292, 430.
Jackson, J.C., 1972, Nature Phys. Sci. 236, 39.
Kafatos, M., 1980, Ap. J. 236, 99.
Kardashev, N.S., 1962, Sov. Astron.- AJ 6, 317.
Landecker, T.L., Wielebinski, R., 1970, Aust. J. Phys. astrophys.
        Suppl. 16. 1.
Li, T.P., Wolfendale, A.W., 1981, Astr. Ap. 100, L26.
Lerche, I., Schlickeiser, R., 1980, Ap. J., 239, 1089.
Lerche, I., Schlickeiser, R., 1981a, Ap. J. Suppl. 47, in press.
Lerche, I., Schlickeiser, R., 1981b, Ap. Letters, 22, 161.
Lerche, I., Schlickeiser, R., 1981c, Astr. Ap., in press.
Mandrou, P., Bui-Van, A., Vedrenne, G., Niel, M., 1980, Ap.J. 237, 424.
Marscher, A.P., Brown, R.L., 1978, Ap. J. 221, 588.
Mayer-Hasselwander, H.A., Bennett, K., Bignami, G.F., Buccheri, R.,
        Caraveo, P.A., Hermsen, W., Kanbach, G., Lebrun, F., Lichti,
        G.G., Masnou, J.L., Pauli, J.A., Pinkau, K., Sacco, B.,
        Scarsi, L., Swanenburg, B.W., Wills, R.D., 1981, Astr. Ap.,
        in press.
McDonald, F.B., Trainor, J.H., 1976, in: Jupiter, ed. T. Gehrels, Un.
        of Arizona Press: Tucson, p.961.
Meegan, C.A., Haymes, R.C., 1979, Ap. J. 233, 510.
Meyer, P., Vogt, R.E., 1961, Phys. Rev. Letters, 6, 193.
Mezger, P.G., Mathis, J., Panagia, N., 1981, Astr. Ap., submitted.
Montmerle, T., 1979, Ap. J. 231, 95.
Montmerle, T., Cesarsky, C., 1979, Proc. 16th Int. Cosmic Ray Conf.
        (Kyoto), Vol. 1, p.191.
Nishimura, J., Fujii, M., Taira, T., Aizu, E., Hiraiwa, H., Kobayashi,
        K., Niu, K., Ohta, I., Golden, R.L., Koss, T.A., Lord., J.J.,
        Wilkes, R.J., 1980, Ap. J. 238, 394.

Özel, M.E., 1980, Astr. Ap. 81, 33.
Ozernoy, L.M., Aharonian, F.A., 1979, Astrophys. Space Sci. 66, 497.
Paul, J.A., Bennett, K., Bignami, G.F., Buccheri, R., Caraveo, P.,
        Hermsen, W., Kanbach, G., Mayer-Hasselwander, H.A., Scarsi,
        L., Swanenburg, B.N., Wills, R.D., 1978, Astr. Ap. 63, L31.
Perotti, F., Della Ventura, A., Sechi, G., Villa, G., DiCocco, G.,
        Baker, R.E., Butler, R.C., Dean, A.J., Martin, S.J., Ramsden,
        D., 1979, Nature 282, 484.
Perotti, F., Della Ventura, A., Villa, G., DiCocco, G., Butler, R.C.,
        Carter, J.N., Dean, A.J., 1981, Nature 292, 133.
Phillips, S., Kearsey, S., Osborne, J.L., Haslam, C.G.T., Stoffel, H.,
        1981, Astr. Ap. 98, 286.
Pinkau, K., 1979, Nature 277, 17.
Prince, T., 1979, Ap. J. 227, 676.
Protheroe, R.J., Wolfendale, A.W., 1980, Astr. Ap. 92, 175.
Riegler, G.R., Ling, J.C., Mahoney, W.A., Wheaton, W.A., Willetti, J.B.,
        Jacobsen, A.S., Prince, T.A., 1981, Ap. J. 248, L13.
Schlickeiser, R., 1980, Ap. J. 240, 636.
Schlickeiser, R., 1981a, Fortschritte der Physik 29, 95.
Schlickeiser, R., 1981b, Astr. Ap., submitted.
Schlickeiser, R., 1981c, Astr. Ap. 94, 229.
Schlickeiser, R., 1981d, Astr. Ap. 94, 57.
Schlickeiser, R., Thielheim, K.O., 1976, Nature 261, 478.
Schlickeiser, R., Thielheim, K.O., 1978, Mon. Not. Roy. astr. Soc.,
        182, 103.
Schönfelder, V., 1978, Nature 274, 344.
Solomon, P.M., Saunders, D.B., 1980, in: Giant Molecular Clouds in the
        Galaxy, eds. P.M. Solomon & M.G. Edmunds, Pergamon Press:
        Oxford, p.41.
Stecker, F.W., 1975, Phys. Rev. Letters 35, 188.
Stecker, F.W., 1978, Nature 273, 493.
Stecker, F.W., 1981, this symposium.
Strong, A.W., Allen, R.J., 1981, Proc. 17th Int. Cosmic Ray Conf.
        (Paris), Vol. 2, 248.
Swanenburg, B.N., Bennett, K., Bignami, G.F., Caraveo, P., Hermsen, W.,
        Kanbach, G., Masnou, J.L., Mayer-Hasselwander, H.A., Paul,
        J.A., Sacco, B., Scarsi, L., Wills, R.D., 1978, Nature 275,
        298.
Swanenburg, B.N., Bennett, K., Bignami, G.F., Buccheri, R., Caraveo,
        P., Hermsen, W., Kanbach, G., Lichti, G.G., Masnou, J.L.,
        Mayer-Hasselwander, H.A., Paul, J.A., Sacco, B., Scarsi, L.,
        Wills, R.D., 1981, Ap. J. 243, L69.
Ulrich, M.H., 1981, Space Sci. Rev. 28, 89.
Van den Bergh, S., 1979, Astr. J. 84, 71.
Webber, W.R., Simpson, G.A., Cane, H.V., 1980, Ap. J. 236, 448.
Webster, A.S., 1978, Mon. Not. Roy. astr. Soc. 185, 507.
Weinberg, S.L., Silk, J., 1973, Ap. J. 183, 49.
White, R.S., Dayton, B., Gibbons, R., Long., J.L., Zanrosso, E.M., Zych,
        A.D., 1980, Nature 284, 608.

COSMIC GAMMA RAYS

A.W. Wolfendale,
Physics Department,
University of Durham,
Durham DH1 3LE

1. Introduction

   The view of the Universe through the gamma ray window is without
doubt still a hazy one but nevertheless there is interesting
information to be gained already and as time goes on and improvements
in technique are made very important advances can be confidently
predicted.

   The present paper which complements that by Dr. Schlickeiser,
concerns just two topics in the field which are of contemporary
interest : the nature of the γ-ray 'sources' and the information
contained in the apparently extragalactic γ-ray flux.

2. Gamma ray sources

   Both the SAS II experiment (Fichtel et al., 1975) and the later
COS B experiment (Swanenburg et al., 1978) have provided data from
their satellite-borne spark chambers which point to the existence of
γ-ray 'sources' as well as a rather smooth continuum. Although
some of these 'sources' can be identified with known objects :
specifically the CRAB and VELA pulsars and, probably, 3C273 and the
molecular cloud associated with ρ Oph, most can not and there is
thus a problem. This problem has relevance for Cosmology for a
number of reasons, principally as follows.

   (i) If the 'source' contribution is small then the bulk of the
        Galactic γ-radiation can be attributed to cosmic ray inter-
        actions in the ISM and, since the distribution of target
        nuclei is known, to moderate precision, the spatial distri-
        bution of cosmic rays can be examined. Such a distribution
        is clearly related the important cosmological problem : where
        do cosmic rays come from?

A. W. Wolfendale (ed.), Progress in Cosmology, 233–237.
Copyright © 1982 by D. Reidel Publishing Company.

(ii)   There may be amongst the unidentified sources objects which
       are important cosmologically, both on Galactic and extra-
       galactic scales.

       The approach taken·by the author's group has been the conservative
one, conditioned by previous experience in cosmic ray research : to
see to what extent the sources can be explained in terms of the
interaction of cosmic rays with the clumpy ISM.   Inevitably there
has been conflict with those starting from other standpoints who
are tempted to claim rather more exotic origins for the intentisy
peaks.

       It is tempting to draw analogies with other parts of the electro-
magnetic spectrum.   For example, in the γ-ray region the fraction of
the Galactic flux not due to discrete sources is very small indeed,
in fact it is not clear whether a diffuse X-ray flux has been
detected or not.   A similar situation exists for the extragalactic
X-ray flux.   One might be tempted to think that the true situation
in the X-ray regime is the same and that the only reason why so few
sources has been seen is the very poor angular resolution of the
detectors ($\sigma_\theta \sim 2$-3°).   However, I would counter this by drawing
attention to the situation with the synchrotron radiation in the
region of 100's of MHz.   Here, it is generally agreed that the bulk
of the radiation is not from discrete sources but is due to cosmic
ray electrons interacting with the magnetic field in the ISM.   Those
same electrons (together with protons) interacting with the nuclei
of the ISM can generate the bulk of the detected γ-rays.   In fact,
Haslam et al. (1981) have drawn attention to the very similar profile
of γ-ray and radio intensities with longitude along b = 0° thus
adding weight (but not proving) the contention.

       It must be admitted that the arguments favouring a sizeable
fraction of the γ-ray sources (specifically, those listed in the 2CG
catalogue of Swanenburg et al.) being irradiated molecular clouds
is circumstantial.   It is as follows:

(a)   Li and Wolfendale (1981a) argue that the angular sizes of the
claimed sources could well be quite large.   Specifically they
claim that sources of diameter 4° could not be distinguished
from 'point' sources and indeed that there are some sources which
have actual sizes above this limit.   Many distant molecular clouds
would subtend angles of this order.

(b)   Statistical studies (Li and Wolfendale, 1981b) using
published data on the distribution of gas in the Galaxy, i.e.
hydrogen in both atomic and molecular forms (plus heavier atoms),
and the size distribution of molecular clouds, indicate that about
60% of the sources with $E_\gamma > 100$ MeV can be explained by the cosmic
ray interaction mechanism.   This figure still seems reasonable at
the present time.

(c) Detailed examination of local (within ∿ 2 kpc) molecular
clouds has led Issa and Wolfendale (1981) to conclude that these
clouds at least do provide γ-rays by virtue of cosmic ray
irradiation at about the expected level. For one cloud at least,
that associated with Orion, there seems to be general agreement
on this point (Caraveo et al., 1980; Wolfendale, 1980). It would
be surprising if cosmic ray irradiation were not a common
situation.

Turning to those unidentified sources which are unlikely to be
explained in the above fashion there are presumably a variety of
likely explanations. Undetected pulsars must rank high on the list,
unusual stars and supernova remnants associated with molecular clouds,
where the shock is accelerating cosmic rays in the cloud (Montmerle
1979) are also possibilities. To these should be added high mass-loss
stars within molecular clouds again accelerating protons and
electrons (Montmerle, 1981) and the possibility of some clouds
accelerating pre-existing cosmic rays as they collapse. Improved
resolution will surely allow an eventual distinction between these
mechanisms.

At present the author is of the view that the contribution to
the Galactic γ-ray intensity from unresolved sources which do not
involve ambient cosmic rays interacting with the ISM is perhaps only
about 20%. Proceeding then to analyse the remainder there seems no
objection to the conclusion reached by Dodds et al. (1975) and
spelled out in more detail by Issa et al. (1981) that there is
evidence for a radial gradient of cosmic ray intensity in the Galaxy
(for both protons and electrons) of such a form as might be expected
for a Galactic origin of the particles. Although the energies of
the particles are low ($\gtrsim$ 10 GeV for protons and $\gtrsim$ 1 GeV for electrons),
whereas the proton component of cosmic rays extends right up to
10^{20}eV and perhaps beyond, it is an important energy range because
this is where the bulk of the energy of the cosmic radiation lies.
Thus, if the arguments about neglecting sources are correct, γ-ray
Astronomy has already brought forward an important answer, viz that
most cosmic ray protons (below 10 GeV) are not of extragalactic origin,
as was previously a possibility, but are generated in the Galaxy.

Turning now to the question of unidentified sources which might
have individual significance it is important not to persist too
strongly with the conservative view. The main reason for this remark
is that the second brightest source, 2CG195 + 04, with a flux above
100 MeV of 4.8×10^{-6}cm^{-2}s^{-1}, has not yet been identified. The object
is, at a latitude of + 4.5°, sufficiently far out of the Galactic
plane to be in a region of good visibility; further, there are no
identified molecular clouds in the vicinity. Davies et al. (1978) have
suggested identification with the (postulated) neutron star companion
of a nearby star : γ Geminorum, but this is of course, not proven.

There are clearly surprises in store in this area.

3. Extragalactic gamma rays

The evidence favouring a finite flux of diffuse γ-rays of extragalactic origin comes from the SAS II experiment (Fichtel et al., 1977) although these authors would probably admit that the evidence favouring strict isotropy is not strong. In fact, it is just conceivable that the bulk of the flux comes from the Galactic halo and this is another reason why the present author does not propose to examine the cosmological models which have been put forward to explain the 'isotropic γ-ray background'.

Notwithstanding the above remarks the 'extragalactic' γ-ray flux is of great value in allowing a limit to be set on the contribution of extragalactic cosmic rays to the cosmic ray flux, viz the problem considered in some detail in §2. There need be no excuse for another, independent, examination of this important problem.

The advance that has made progress possible is the determination of the mass of gas in galaxy clusters, a development which has followed from the measured X-ray fluxes, the X-rays originating in hot inter-cluster gas. Said et al. (1982) have recently investigated the problem in some depth and they conclude that for all reasonable models of the distribution of cosmic rays in the Universe in which the observed cosmic rays are entirely extragalactic the predicted γ-ray flux (E_γ > 100 MeV) is higher than the measured upper limit. The only model which would even remotely survive is one in which the cosmic ray intensity is the same everywhere rather than, as would have been expected, peak somewhat in galaxy clusters. An additional requirement is that the gas in clusters is confined to rich clusters only.

Said et al. conclude, and this conclusion seems to be still valid, that a limit of about 10% can be set to the fraction of protons in the (1-10)GeV energy band which have come from extragalactic sources. Thus, although there are undoubtedly significant fluxes of cosmic rays in extragalactic space it is unlikely that their energy density exceeds about 0.1eV cm^{-3} (the Galactic energy density is $\sim$ 1eV cm^{-3}). Hence cosmic rays are probably not a dominant force in cosmological processes except perhaps in restricted regions.

References

Davies, R.E., Fabian, A.C., and Pringle, J.E., 1978, Nature 271, 634.

Caraveo, P.A., Bennett, K., Bignami, G.F., Hermsen, W., Kanbach, G., Lebrun, F., Masnou, J.L., Mayer-Hasselwander, H.A., Paul, S.A., Sacco, B., Scarsi, L., Strong, A.W., Swanenburg, B.N., and Wills, R.D., 1980, Astron. Astrophys. 91, L3.

Dodds, D., Strong, A.W., and Wolfendale, A.W., 1975, Mon. Not. R. astron. Soc. 171, 569.

Fichtel, C.E., Kniffen, D.A., and Thompson, D.J., 1975, Astrophys. J., 198, 163.

Fichtel, C.E., Hartman, R.C., Kniffen, D.A., Thompson, D.J., Ogelman, H.B., Ozel, M.E., and Tumer, T., 1977, Proc. 12th ESLAB Symp. Frascati, 191.

Haslam, C.G.T., Kearsey, S., Osborne, J.L., Phillipps, S., and Stoffel, H., 1981, Phil. Trans. R. Soc. Lond. A301, 573.

Issa, M.R., and Wolfendale, A.W., 1981, Nature, 292, 430.

Issa, M.R., Riley, P.A., Strong, A.W., and Wolfendale, A.W., 1981, J. Phys. G, 7, 973.

Li, T.P., and Wolfendale, A.W., 1981a, Astron. Astrophys., 100, L26; 1981b, Astron. Astrophys., 103, 19.

Montmerle, T., 1979, Astrophys. J., 231, 95; 1981, Phil. Trans. R. Soc. Lond. A301, 505.

Said, S.S., Wolfendale, A.W., Giler, M., and Wdowczyk, J., 1982, J. Phys. G. (in the press).

Swanenburg, B.N., Bennett, K., Bignami, G.F., Caraveo, P., Hermsen, W., Kanbach, G., Masnou, J.L., Mayer-Hasselwander, H.A., Paul, J.A., Sacco, B., Scarsi, L., and Wills, R.D., 1978, Nature, Lond. 275, 298.

Wolfendale, A.W., 1980, IUPAP/IAU Symp. 94, Bologna, Origin of Cosmic Rays (Reidel, Dordrecht, Holland), 309.

ORIGIN OF STARS AND GALAXIES

W.H. McCrea.
Astronomy Centre, University of Sussex, Brighton BN1 9QH
England.

With due tentativeness, arguments are presented in support of the
following proposition:
A galaxy is produced by a collision between two portions of raw material
moving under their mutual gravitation; shocked material produces primary
condensations in which short-lived exploding massive stars produce the
first heavy elements; a condensation then becomes a globular cluster of
the first normal stars, the entire galaxy being thus initially a 'cluster'
of such clusters; tidal dissipation - and possibly other effects to be
mentioned - converts most of these into a halo stellar population; subsequent evolution depends upon whether residual diffuse material falls
mainly into a disk or into a nucleus (or perhaps expands into a larger
halo).

1. INTRODUCTION

The problem to be considered here is, how were galaxies/stars first
formed? This will be treated from the standpoint that, if we are able
to infer what processes took place in the formation of these bodies, we
should know something about the state of the universe - or some part of
it - immediately before their formation. We should then know at least
what we ought to ask about how that state was produced.

In broadest possible terms the exercise is to form a gravitationally
bound system from material not initially so bound. Therefore there has
to be a process of *accumulation* of the required material, and a process
of *energy-loss* by this material. Obvious as this may be, the guiding
principle in all such work must be to seek for such processes . And
obvious as the general principle may be, the actually operative processes
have not been firmly identified in the case of any astronomical body -
astronomers are not agreed as to how any gravitationally bound astronomical body has been formed.

Although it may still not be known how, in particular, an individual star has been formed, there is a good deal of agreement about the

239

A. W. Wolfendale (ed.), Progress in Cosmology, 239–257.

essential features surrounding the formation of stars in the galactic
disk (Population I), which is taking place all the time in the Galaxy
as we know it. We can assert:- Stars in the disk are formed from
material of interstellar *clouds* that is involved in a *shock* produced
by a *collision* between clouds; an 'open' *cluster* of typically a few
hundred stars is formed in what may be viewed as any one operation; a
supernova outburst may be a significant element in the operation;
clusters so formed are gradually dispersed by *tidal effects* to produce
the disk population as a whole; there is nothing to contradict a claim
that essentially the whole of this population has arisen in this way.

Clearly the conditions under which the first stars were formed must
have been different. But stars are stars and it is natural to expect
the essential physical mechanism of formation to be always fundamentally
the same.

2.   GLOBULAR CLUSTERS

2.1 Properties

In the light of what was said above about one type of stellar
cluster, it is inevitable that we should contemplate the only other
recognized species. The following are some properties of *globular
clusters*:-

1.   Globular clusters exist as parts of our Galaxy; in the first
      instance we confine attention to these.

2.   Globular clusters follow a generally standard pattern, having
      masses and diameters that range over not much more than an order
      of magnitude.

3.   At the same time globular clusters are found to differ from one
      another considerably in the relative abundances of chemical
      elements heavier than hydrogen and helium.

4.   The stars in globular clusters are among the oldest known.

5.   Globular clusters must be gradually dispersed by tidal action
      of the rest of the Galaxy; it is generally inferred that they
      were much more numerous in the past.

6.   Nothing contradicts the hypothesis that all the oldest stars now
      in the Galaxy (Population II or halo stars) came from dispersed
      globular clusters.

7.   A globular cluster - in at anyrate some cases - contains near
      its centre (but how near is not yet known) an X-ray source,
      indicating a site of past violent activity of some sort, such
      as a supernova outburst.

8.   All properties imply that a globular cluster was formed all at
      once as a whole- in the course of its existence it may lose stars
      or other material, but there is no indication that in the course
      of its existence one has acquired fresh matter in any form.

9. Amongst globular clusters belonging to the Galaxy there is no indication of a significant spread of ages.

10. Other 'nearby' galaxies are observed to possess globular clusters having generally similar, though not identical, properties.

Probably the most contentious of these statements is number 6. In support, one may quote the important finding of Kinman (1959) that the mean angular momentum per unit mass of globular clusters is the same as for the rest of the Galaxy. Frenk and White (1980) have comprehensively and critically reviewed existing knowledge of the kinematics and dynamics of the galactic globular cluster system, but their discussion seems not to conflict with this feature of Kinman's work. Again Woolley (1978) has obtained important properties of the kinematics of RR Lyrae stars in the Galaxy in the neighbourhood of the Sun. If such work could be extended to larger regions of the Galaxy, and perhaps to other sorts of stars, it should ultimately be of much value in the present context. But it is difficult to see that the existing results can be so utilized. The work by Fall and Rees (1977) on the survival and disruption of galactic substructure certainly shows that there can be alternative explanations of some of the properties considered here, but one cannot see that in the present state of knowledge it rules out the line of thought here pursued.

2.2   Inferences

     Accepting for the moment the validity of the foregoing properties we may fairly securely draw the following inferences, not now with any special reference to our own Galaxy:-

1.   A globular cluster originated from what we shall call a *primary condensation* of the raw material of galaxy formation.

2.   A galaxy consisted initially of a set of primary condensations forming a gravitationally bound system, presumably along with some uncondensed raw material.

3.   A primary condensation consisted initially of raw material containing no heavy elements; later it consisted of stars containing such elements. So we infer that some of the raw material must have formed massive stars which in exploding produced heavy elements and dispersed them into the remaining raw material of the condensation.

4.   The explosions could have been part of the mechanism for forming normal stars - in the way that some astronomers suggest that supernova outbursts play a part in forming stars from interstellar material in the disk of the present Galaxy. Also they could have left behind collapsed objects possibly to take part in X-ray production.

5.   There could be chemical differences between globular clusters,

and between stars of the same age in a galactic halo, traceable to simple differences between primary condensations in mass or density, which could cause differences in the explosions and their consequences.

6.   It appears much more natural for the first heavy elements to have been made *in situ* in primary condensations, and the first normal stars to have been formed there forthwith, these stars being subsequently gradually dispersed through a halo, rather than for the heavy elements to have been made elsewhere in the galaxy concerned, and then dispersed over great distances before any normal stars could be made any-where - the latter being apparently the widely accepted opinion at the present time.

The conjectured massive stars formed from the raw material seem to be in some respects similar to the Population III stars that have been discussed by recent writers, particularly M. Rowan-Robinson in this volume. But here they are going to arise as one step in what one hopes is a reasonably coherent wider scheme of star-formation.

Without any particular theory, the existence and known properties of globular clusters thus lead persuasively to the far-reaching inferences here sketched out. We shall see now that a simple theoretical model leads to the same inferences and renders them quantitative. The inferences of the present work need not therefore depend upon the foregoing presen-tation. It has been given because the following theory may appear unduly simplified, but it may prove more acceptable when it is seen to agree apparently so well with what may be called the 'detective story inferen-ces' from the evidence in hand. That is to say, the two approaches will serve to support each other.

3.   THEORY: PRELIMINARIES

3.1 Raw Material

We accept the general features of simple big-bang cosmology, accor-ding to which the formation of galaxies and stars takes place not before the epoch of decoupling of matter and radiation. Thus the raw material available for the first formation of such bodies, at cosmic time t_0, say, is required to satisfy the initial conditions:- It consists effectively of only hydrogen and helium in the proportion yielding mean molecular weight

$$\mu \approx 1.2 \qquad\qquad (3.1)$$

At t_0 the density ρ_0 is uniform, and we write

$$\rho_0 = n_0 H, \qquad H = \text{mass of hydrogen atom}, \quad (3.2)$$

so that n_0 is the density expressed as the equivalent number-density of hydrogen atoms. We expect ρ_0 to be less than the mean density within the main part of a typical galaxy. This is because a galaxy is a condensation in the material so that certainly at t_0 a galaxy must have mean density greater than ρ_0, and it is usually assumed that the mean density of a galaxy does not change much once it is formed. For various reasons this may be an oversimplification, but at anyrate it gives a rough indication that

$$n_0 \lesssim 1 \text{ cm}^{-3} \tag{3.3}$$

At t_0 the temperature T_0 is uniform, and since t_0 is later than the epoch of decoupling

$$T_0 < T_{dec} \approx 3000 \text{ K,} \tag{3.4}$$

this last being the currently accepted estimate of the decoupling temperature.

If anything happens to the material to heat some of it locally above T_{dec} then, losing energy by radiation, the temperature of that material must fall back rapidly to a value close to T_{dec}; it will take much longer to cool below T_{dec}. This is in fact almost the defining property of T_{dec}. It implies that, whenever the temperature T of the material is relevant in the following discussion, for the most part we may take it to be near to T_{dec}. It occurs always in the combination T/μ, so in applications we shall use for simplicity

$$T/\mu = 3000 \text{ K.} \tag{3.5}$$

3.2 Gravitational contraction and collapse

We assume that, if gaseous material of density ρ, temperature T, be brought to a configuration in which it can readily fragment into portions that can collapse gravitationally, then it will actually so fragment. For the material as stated the 'Jeans' length and mass are

$$\lambda_J = (\pi RT/\mu G\rho)^{\frac{1}{2}}, \qquad m_J = \rho\lambda_J^3 \tag{3.6}$$

where R = gas constant, G = gravitation constant. We suppose a body of the material of roughly spherical shape and mass m to be such a fragment provided

$$m \gtrsim m_J \tag{3.7}$$

This is the property of gravitational instability as commonly assumed in such work. As it is usually employed, all that can be inferred is that the stated fragment will start to contract. In the present case, however, we have the additional property that by radiation loss the temperature T must remain at the value given by (3.5) so that, if contraction starts, it will pass over into actual gravitational collapse.

This seems to be a fair inference from the work of Ebert, 1955 (see also McCrea, 1957).

3.3   Activity

We are studying what is necessarily a transitory phase of the raw material described in 3.1. According to the simple big-bang picture, at any cosmic epoch before t_0 , the material fills the universe uniformly. By definition of t_0 , at any later epoch much of this material is condensed into stars and galaxies. We are attempting to discover what happens in between. Now perfectly uniform material cannot turn itself into non-uniform material. Simple big-bang cosmology thus lacks some essential feature that can cause the transition. Without some modification of its postulates, that model will never produce stars and galaxies. Here we are not proposing to discuss those postulates as such. It is, however, necessary to introduce some feature that *causes action* to occur in the material. The simplest possible is to suppose that at epoch t_0 the material is as we have described it but that it happens to be *broken into pieces of all sizes.*

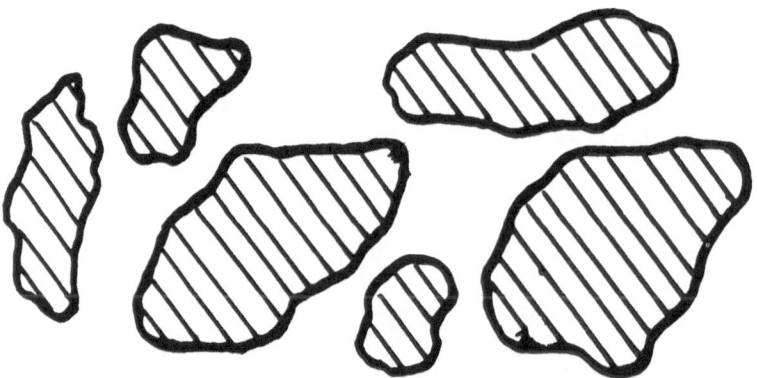

Figure 1.   Raw material at epoch t_0 broken into clouds.

We shall call the pieces of material 'clouds'. Later on we shall discuss the significance of our supposition somewhat farther. Meanwhile, apart from it, we keep as closely as possible to big-bang cosmology. Thus the overall tendency of the clouds must be to move apart in accordance with the Hubble expansion. But some neighbouring clouds will tend to fall together under their mutual gravitation. Also small clouds will tend to dissipate, while large clouds will tend to contract under self-gravitation. However, there need be no essential difference between an irregular cloud falling in upon itself and two neighbouring clouds falling upon each other.

4.   CLOUD COLLISIONS

4.1   Descriptive account

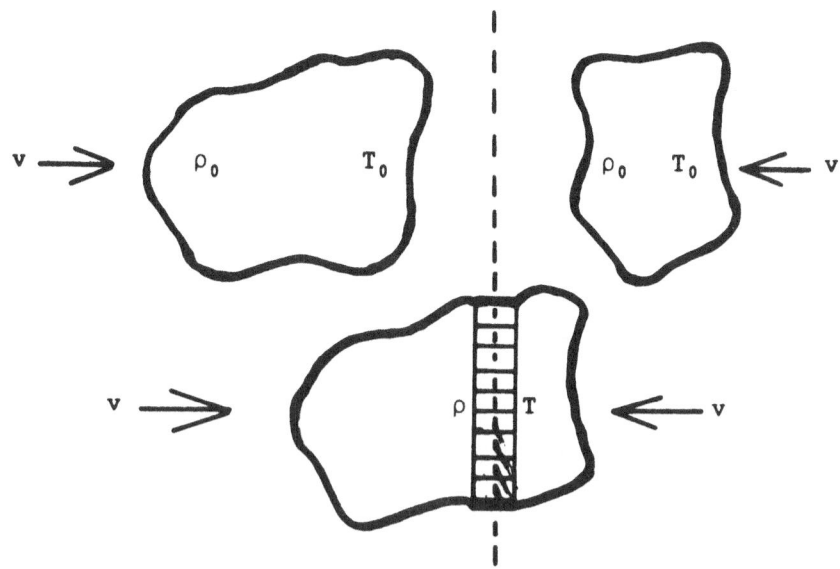

Figure 2. Two clouds in face-on supersonic collision

        Consider two clouds that do move towards each other so as to under-
go a face-on collision: choose a frame of reference in which their
velocities are ± v as indicated in Figure 2. We write

$$v^2 = \alpha^2 RT/\mu, \tag{4.1}$$

where T has the same significance as in (3.5), and we suppose we are
dealing with a case where

$$\alpha \gg 1, \tag{4.2}$$

so that by virtue of (3.4), (3.5) v is supersonic for the inflowing
material. The material that meets at the interface is halted there. The
inflowing material then simply piles up to form a layer of density ρ,
say, because v being supersonic, the material has no warning of its fate
before it is stopped at the layer boundary. Directly after stopping, the
material will be much hotter than T, but according to the stated property
of T it will rapidly cool to this temperature. So long as it exists as
such, the layer will therefore be effectively at temperature T throughout. Also neglecting any change of v during the time concerned, conditions at the layer boundary remain the same so long as there is still any
inflowing material, and so we may treat ρ as uniform throughout the layer.
The equality of pressure across the boundary is expressed by

$$\rho_0 \, v^2 \approx RT\rho/\mu, \qquad\qquad\qquad (4.3)$$

the right-hand side being the gas pressure inside and the left-hand side
the dynamical pressure of the inflow. In consequence of (4.2) we neglect the effect of the gas pressure of the uncompressed material – that
is why (4.3) is not written as a strict equality. Using (4.1), (4.3)
becomes

$$\rho \approx \alpha^2 \rho_0 \qquad\qquad\qquad\qquad (4.4)$$

If, before all the material on either side of the layer is used up,
its thickness λ reaches about or somewhat more than the value λ_J,
according to our stated understanding about gravitational instability, the shocked material then breaks up to produce *primary
condensations*. For in that state, and not before, it can break up into
roughly spherical portions of diameter about λ_J. And we have remarked
that any such portion will readily pass into a state of gravitational
collapse.

When such condensations are being produced, if there is still incoming material on both sides of the layer, some may be added to these
condensations, some may become gravitationally bound to the set of condensations as a whole, or if the amount is comparable with that already
used, the continuation of the process would form a second set of condensations. If there is still incoming material on only one side, some or
most of it may move away from the set of condensations. All these considerations indicate that when an aggregate of primary condensations is
formed the amount of uncondensed material left with it would generally
be less than that of condensed material.

On this picture any primary condensation is formed from the encounter of two columns of cloud material only. But if, as we are about to
accept, the relative motion arises essentially from the gravitational
attraction of the two clouds involved, then the value of v depends on
all the matter in the clouds. It follows from such considerations that
it will not make much difference if the clouds meet in irregular faces.
Indeed any resulting irregularity in the layer of shocked material should
actually facilitate its break-up into condensing fragments.

It is to be noted that the present way of forming condensations is
an excellent illustration of the expected processes of *accumulation*
and of *energy-loss:-* accumulation in the pile-up of the layer, energy-loss
in keeping the temperature of the shocked material always no more than
about 3000 K.

4.2   Speeds and times

The descriptive account of a cloud-collision just given should be
valid if the clouds have simply been somehow endowed with the motion
specified by the speed v, and if no forces act upon their matter except
in the material collision. This would imply, in particular, that the

motion of the material is unaffected by the self-gravitation or mutual
gravitation of the clouds. But we are concerned with the case where the
motion is actually produced effectively entirely by this gravitation.
To deal adequately with the motion in such circumstances would be exceed-
ingly difficult. Here we shall assume that the descriptive account, in
which the motion is specified by the single parameter v, will provide a
meaningful approximation provided we assign a suitable value to it.

If particles of masses m_1, m_2 fall under their mutual gravitation
from large separation to separation r, their relative speed is 2v then,
where

$$v^2 = G(m_1 + m_2)/2r \qquad\qquad (4.5)$$

In the case of a collision of clouds of masses M_1, M_2 we shall adopt for
the parameter v the value given by

$$v^2 = G(M_1 + M_2)/2r_0, \qquad\qquad (4.6)$$

where r_0 =   distance between mass-centres at first contact
              of the bodies.   (4.7)

This is suggested merely by the crudest analogy with (4.5). In so far
as employment of the parameter v has significance, the formula (4.6)
should give the correct order of magnitude, and also a useful comparison
between various cases.

Another quantity required in the discussion is the well-known 'free-
fall time' of material of negligible pressure and initial density ρ_0,
which is

$$t_f(\rho_0) = (32\ G\rho_0/3\pi)^{-\frac{1}{2}} \qquad\qquad (4.8)$$

4.3   Particular examples of collision between two clouds

We present the simplest case and then briefly cases that illustrate
departures from it in regard to shape of cloud, symmetry and nature of
encounter - one feature at a time.

Figure 3.Simplest
case of cloud
collision

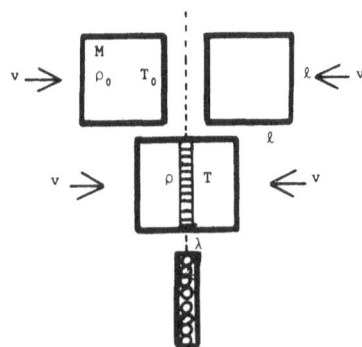

1.   Simplest case.   Clouds having the form of equal cubes, each of
dimensions $\ell \times \ell \times \ell$, mass $M = \rho_0 \ell^3$
Formulae (4.6),(4.3) give now, where for the moment we treat them as
equalities,
$$v^2 = GM/\ell = G\rho_0 \ell^2 ,$$
$$\rho_0 v^2 = RT\rho/\mu ,$$

and if all the material is in the shocked layer with thickness λ

$$2\ell \rho_0 = \lambda \rho.$$

These equations lead to

$$\lambda = 2(RT/G\mu\rho)^{\frac{1}{2}} \approx 1.1 \; \lambda_J \qquad\qquad\qquad (4.9)$$

so that

$$m \equiv \rho \lambda^3 \approx 1.4 m_J. \qquad\qquad\qquad (4.10)$$

Thus when all the material has been used up it forms a layer such that,
if this is divided into cubes, each of these has mass m somewhat in
excess of the Jeans mass for its material. They may naturally and conveniently be treated as the *primary condensations* in this case.

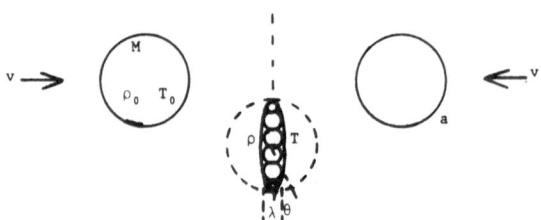

Figure 4. Collision between clouds in the form of equal
circular cylinders

2.   Clouds in the form of equal right circular cylinders, each of
radius a, height 2a, mass $M = 2\pi a^3 \rho_0$
In this case (4.6) gives
$$v^2 = GM/2a.$$
If all the material goes into a shocked layer, this has an elliptical
section as shown. Let λ be the thickness corresponding to eccentric
angle θ as shown in Figure 4. Then a calculation similar to that in
example 1 shows that corresponding to (4.9)

$$\lambda = (4/\pi)\lambda_J \sin \theta \approx 1.3 \; \lambda_J \sin \theta$$

It follows that $\lambda \gtrsim \lambda_J$ over more than 60% of the width of the layer
which includes over   70% of the mass.

     This case is thus not greatly different from the simplest case;
primary condensations could start forming near the centre before all the
material has been used up, but it seems more natural to infer that once
again nearly all the material will go into primary condensations, those
towards the centre being somewhat larger than those further out.

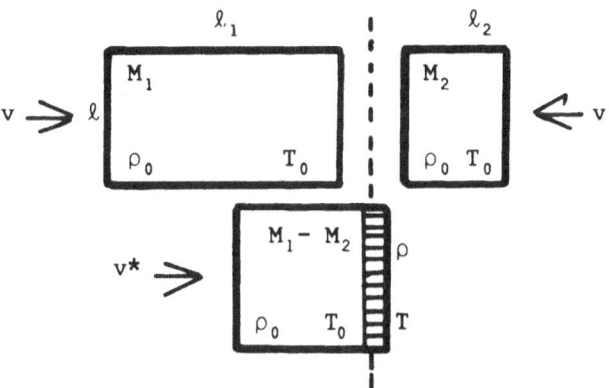

Figure 5.   Unequal clouds with $\ell_2 < \ell$, $\ell_1 + \ell_2 > 2\ell$.

3.   Unequal but comparable clouds having dimensions $\ell_1 \times \ell \times \ell$, $\ell_2 \times \ell \times \ell$, masses $M_1 = \rho_0 \ell \ell_1^2$, $M_2 = \rho_0 \ell \ell_2^2$. Thus we have the simplest case except that the clouds have different lengths in the direction of motion. Here (4.6) gives

$$v^2 = G(M_1 + M_2)/(\ell_1 + \ell_2) = G\rho_0 \ell^2.$$

which happens to be the same as in example 1, for all values of ℓ_1, ℓ_2. This makes comparison with example 1 quite simple. In particular, the value of the density ρ of the shocked material is the same as it is in that example, so long as there is still incoming material on both sides.

Bearing in mind the result of example 1, we see that if ℓ_1, ℓ_2 are both much less than ℓ, when all the material is used up the thickness of the shocked layer is less than λ_J; so we infer that it does not forthwith yield primary condensations.   If ℓ_1, ℓ_2 are both much more than ℓ, then we have the sort of case previously mentioned where there are primary condensations and a residue of diffuse material, with maybe a second production of condensations.

The novel case is that having $\ell_2 < \ell$, $\ell_1 + \ell_2 > \ell$. Then when all the smaller cloud has been used up we have the situation sketched in Figure 5, where at this stage the shocked layer is not yet thick enough to form condensations. We could re-start the calculation at this stage; in accordance with (4.6), we should see that the remaining unshocked material will move into the interface with relative speed v^*, say, greater than the originally calculated v, for the value of $M_1 + M_2$ is unchanged while the new value of r_0 would be smaller than before (when the estimated relative speed of the clouds was $2v$). In spite of the free boundary on one side, there is thus no doubt that the accumulation of shocked material will continue. So the net result should be not greatly different from that of the simplest case, apart from the fact that, in the original frame of reference, conservation of momentum

requires the shocked material ultimately to take up the whole of the
original (non-zero)resultant momentum. As regards the formation of
primary condensations, a moderate lack of symmetry between the colliding
clouds consequently makes little difference.

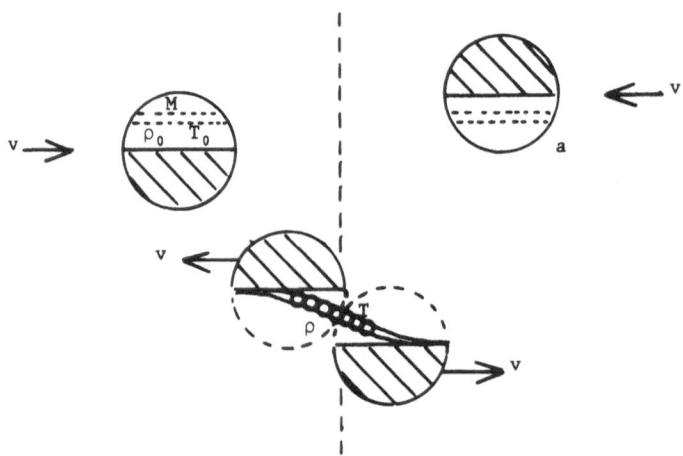

Figure 6   Cylindrical clouds in shearing collision.

4.   Equal cylindrical clouds in shearing collision. We take the same
two circular cylinders as in example 2; if we still suppose the effective value of v to depend only upon the separation of the mass-centres
as in (4.6) then the value is the same as in that example. In order to
illustrate this type of encounter, we suppose the axis of each cylinder
to move in a plane tangential to the other . Then in a very simple picture only the parts of the clouds between the planes are concerned in the
collision; the parts not concerned are shaded in Figure 6; they do, of
course, contribute to the gravitational attraction. On this picture the
collision produces a shocked layer shown schematically in the sketch. An
element of the layer is made from the material originally in the columns
along two chords such as those indicated. It is seen that over most of
the regions concerned the lengths of the two clouds are comparable, so the
material involved will behave generally as it would do in example 3.
Also if the sum of the lengths of the two chords is La, then the thickness of the shocked layer in the direction of motion, calculated in the
same way as in example 2, is

$$\lambda = (L/\pi)\lambda_J$$

Numerical working then shows that $\lambda \gtrsim \lambda_J$ over about 80% of the distance
between the two planes enclosing the material concerned. Thus it appears
that most of this material will go to form primary condensations.

     The new feature is that here there is non-zero *angular momentum* to
be conserved. Elementary calculation shows that the material in the two
half-cylinders concerned in the collision as depicted in Figure 6 has
angular momentum approximately 0.08Mva about an axis through its mass-
centre. The whole of the material of the two cylinders has angular
momentum Mva about the same axis.

The simplified picture neglects tidal interaction which may result in more of the material being affected by the collision than is indicated in Figure 6. So all we can assert at this stage about the system produced by the encounter, i.e. the material involved in the collision and any of the rest that remains gravitationally bound to it, is that its angular momentum lies in the range

$$0.08 \text{ Mva} \lesssim \text{angular momentum} \lesssim \text{Mva} \qquad (4.11)$$

where $v^2 \approx GM/2a$. But it has to be repeated that this illustrative example is chosen quite arbitrarily.

The outcome of these simple examples is to indicate that so far as the production of a set of primary condensations - forming a single gravitationally bound system - is concerned, any supersonic encounter between comparable clouds of raw material brought about by their mutual gravitational attraction produces in general about the same result as would the same material were it the material in our simplest case. It appears that the main difference between one case and another is likely to concern only the amount of angular momentum left in the system.

5. NUMERICAL VALUES

5.1. Simplest case: properties

For the reasons stated at the end of §4 it will suffice as regards numerical values - apart from considerations of angular momentum - to use only the simplest case (example 1). We keep the same notation and in particular we use

$$\rho_0 = n_0 H \text{ gram cm}^{-3}$$

and it is convenient to require that
$$M \text{ is measured in units of } 10^{44} \text{ gram.}$$
As in §3 we use the value
$$T/\mu = 3000 \text{ K}$$
Working mostly to two significant figures in numerical factors the various formulae in example 1 lead to the results in Table 1.

Table 1. Results of cloud-collision in the simplest case

$$\lambda \approx 37/n_0^{\frac{2}{3}} M^{\frac{1}{3}} \qquad \text{parsecs}$$
$$m \approx 8.6 \times 10^5/n_0^{\frac{2}{3}} M^{\frac{1}{3}} \qquad \text{solar masses}$$
$$\alpha \approx 26 n_0^{\frac{1}{6}} M^{\frac{1}{3}}$$
$$v \approx 130 n_0^{\frac{1}{6}} M^{\frac{1}{3}} \qquad \text{km/s}$$
$$\ell \approx 13 (M/n_0)^{\frac{1}{3}} \qquad \text{kiloparsecs}$$
$$t_f(n_0) \approx 5 \times 10^7/n_0^{\frac{1}{2}} \qquad \text{years}$$
$$t \approx 9.5 \times 10^7/n_0^{\frac{1}{2}} \qquad \text{years}$$

where $t = \ell/v = 1/(G\rho_0)^{\frac{1}{2}}$ is a measure of the duration of the collision.

In the example the only free parameters are n_0, M . They are on
quite different footings. In our general model, n_0 is the same for all
clouds. If the model is significant it must be so for a value of n_0
not enormously different from unity, for the reasons appearing in the
dicussion in §3. So there is actually not much freedom about this parameter. On the other hand we can have any value of M we like and the
choice of a particular value imposes no restriction on the model. For,
according to the specifications of the model, clouds of all sizes occur
and if a cloud of a certain size produces a desirable result, that size
is available. We should be suspicious of the model, were a very
particular value of M required to produce a result of interest. But
the opposite is the case. In fact, one reason for giving these formulae
for the various features is to point out how they are insensitive to the
value of M, which occurs to no power higher than one-third. Of course
we must also check that very different values of M produce no undesirable
consequences. It is interesting also to note that the properties are
also fairly insensitive to the value of n_0.

5.2   Application

The last remarks in §4 showed that it is adequate for the moment
to discuss only our simplest case. The remarks in 5.1 now show further
that it suffices then in the first instance to discuss only one numerical
example of this case.

Table 2   Numerical example compared with the Galaxy

Model	Galaxy
Postulated	
M=1 so that	
$2M=2\times10^{44}$gram $= 10^{11}$ solar masses	Mass 1.4×10^{11} solar masses
$n_0=1$ hydrogen atom/cm^3	
Inferred: whole system	
$\ell \approx 13$ kiloparsec	Diameter $\approx$ 25 kiloparsec
$\alpha \approx 26$	
$v \approx 130$ km/s	Rotation speed at Sun's distance
$t_f(n_0) \approx 5 \quad 10^7$ years $\approx$ crossing time	$\approx$220 km/s
$t \approx 9.5 \times 10^7$ years	
Inferred, primary condensations	**Globular clusters**
$\rho \approx 680$ hydrogen atom/cm^3	
$t_f(\rho) \approx 2 \times 10^6$ years $\lesssim$ collapse time	
$\lambda \approx 37$ parsec	Diameter $\approx$ 5 - 32 parsec
$m \approx 8.6 \times 10^5$ solar masses	Mass $\approx 10^5$ -10^7 solar masses
$2M/m \approx 1.2 \times 10^5$	

Gravitational energy of uniform sphere 10^6 solar masses, radius 10 pc
$\approx 5 \times 10^{51}$ergs
Energy released in supernova outburst (type II) $\approx 10^{52}$ ergs

---

We consider in fact the example
$$n_0 = 1, \quad M = 1.$$

In Table 2 some of the results in Table 1 are re-written for this case
and a few derived quantities are also given. These are compared with
certain features of the Milky Way Galaxy (mostly from Allen, 1973). The
value of M is chosen to give about the same total mass so that this
comparison may be relevant.

Referring to Table 1 we notice that, with n_0 unchanged, a
thousand-fold change in M would change $\lambda, m, \alpha, v, \ell$ by only one order of
magnitude, and t, $t_f(n_0)$ not at all; with M unchanged, a hundred-fold
change in n_0 would change λ, m by a factor about 20, and other quantities
by one order of magnitude or less. Thus when we continue to discuss
the one numerical example in Table 2 we keep within something like an
order of magnitude in the corresponding results for an enormous range
in overall features as represented by the parameters n_0 and M.

6.   EVOLUTION OF RESULTING SYSTEM

6.1 Globular cluster formation.

In the simplest case, the set of primary condensations being formed
at relative rest, they certainly form a single gravitationally bound
system as a whole. So the members of this system will proceed to move
in the gravitational field of the whole system, maintaining about the
same overall linear extent i.e. of order of magnitude 10 kpc in the
example, with crossing time of the order of 10^8 years. Each primary
condensation is by derivation a gravitationally bound system. It
may be disrupted by the tidal action of the rest of the system; presumably
this would require on average a good many 'crossings' so that the mean
life of a condensation against disruption from causes outside itself
is considerably more than 10^8 years.

The collapse time of an individual condensation, being about 2×10^6
years in the example, is very small in comparison. So the collapse
and any other early evolution of the condensation will proceed with
negligible outside influence. It is also to be noted that the manner of
production of the condensations ensures that each has small angular
momentum (zero in the simple model); so the collapse need not be hindered
by rotational effects. The collapse is therefore expected first to
produce one or more massive stars near the centre of the condensation.
After about 10^6 years, at anyrate some of these must explode like
massive supernovae. This inference seems to be valid even though at this
stage the material is simply the hydrogen-helium mixture specified in
(3.1). Any such explosions will, however, *synthesize heavier elements.*
The material of the explosion will encounter any still uncondensed raw
material, forming a mixture of shocked material in which normal stars
are expected to come into existence. Such a sequence has frequently
been discussed in various contexts. While there is no adequate theory
as to how it occurs, we may to some extent by-pass any theory by remarking

that, after the explosion of massive stars, conditions in the condensation are rendered generally similar to those under which stars are known
to be formed in the present Galaxy. Even without understanding how it all
all happens, it therefore seems safe to infer that (normal) stars are
in fact produced.

There is thus at anyrate the possibility of converting most of the
material of a primary condensation of mass about 10^6 solar masses into
stars. According to this account, these would be formed after the
gravitational collapse of the condensation is quite well-advanced. So
these stars would occupy a volume smaller than that of the original
condensation which in the example of Table 2 had diameter nearly 40 pc.
Thus we should have a non-rotating cluster, of about 10^6 stars, of
diameter less than 40 pc - this we are bound to recognize as a *globular
cluster*. It is gratifying that the model predicts so well the dimensions
of actual globular clusters in the Galaxy which are noted in Table 2.

It is interesting that the free-fall time for the condensation, which
is a lower bound to the collapse time, happens to be so near to the
1-2 million years which is the usual estimate of the lower bound to the
lifetime of a very massive star. This may be a critical feature in the
production of a globular cluster.

Another critical feature may be the fact that the gravitational
energy of a condensation might be no more than the energy expected to be
released in a supernova explosion - see last two lines of Table 2. Were
most of the energy of the explosion taken up by the material it might
be blown apart. From the standpoint of forming a galaxy, this need be
no disadvantage so long as stars are made in the process. But it does
place another hazard in the way of forming an actual stellar cluster.
If the stars are made sufficiently quickly, on the other hand, then the
energy of the explosion would not be able to blow the structure apart.
Thus it may be that the two critical features are in fact closely related.

As to the amount of material that would have to be processed through
the explosion in order to make possible the formation of the required
normal stars, this is not a large fraction of the whole. One has heard
estimates that as little as 1 part in 10^4 in heavy elements would
suffice to produce the oldest stars.

6.2   Elliptical galaxy formation

It is generally inferred, as mentioned in §2, that existing globular clusters are survivors of a much larger number in the past, and
that the rest were torn apart by tidal forces. The suggestion of the
present work is that most of the material in a galaxy started in
primary condensations that formed either long-lived or transient
globular clusters. The past life of a present galaxy, supposed built
on this model, is many crossing-times; so there has been ample time
for disruption to have occurred. Thus this simple model where, taken
literally, the system as a whole has zero angular momentum would yield
a present *elliptical galaxy*. All the stars in the main body would be

old stars, like Population II stars in our Galaxy.

As we have seen, a certain amount of the original raw material
may not have been incorporated into the primary condensations; also
some of the material in these could have been blown out by the explosions
of massive stars. So the galaxy as first formed must in general contain
interstellar material. In the simplest model, this material, like the
rest of the system, would have zero angular momentum; according to
accepted views it will fall towards the centre of the system. There it
will give rise to a galactic *nucleus* and its associated activity.

It seems important to note that the inference about the diffuse
material falling towards the centre is not obvious or trivial. One
might very well expect the opposite. For if we regard the stars in the
galaxy as behaving generally like the 'molecules' of a gas, then if much
lighter particles were to attain anything like equipartition with them,
the particles would acquire enormous kinetic 'temperature'. Consequently
the particles would spread *outwards*, not fall inwards. In the case
of ordinary interstellar gas and dust any tendency to attain such a
high temperature is defeated by the efficient cooling mechanisms of
such matter. This is why it is in fact correct to infer that it falls
inwards. Nevertheless a gas of 'molecules' in the form of, say, asteroids
would possess no such efficient 'cooling' mechanism. So were it possible
for the model to produce an appreciable mass of such bodies they would
build up an outer envelope – a galactic 'Oort cloud'. This is mentioned
here because, on current ideas about the whole extent of a galaxy, the
model may seem rather restricted. But the suggestion would be that the
model applies to the main body of a galaxy, composed of its luminous
stars, and that any extension beyond that would be some such appendage
as indicated.

6.3   Spiral galaxy formation

In a case where the system has sufficiently large angular momentum
resulting from a shearing collision the system of primary condensations
once formed would as before move in the gravitational field of the whole.
But now– through the same interactions as those that ultimately disrupt
the globular clusters – the rotation would cause the system as a whole
to flatten towards a central plane. If the flattening is sufficiently
great, then the interstellar material will fall predominantly towards
that plane, instead of towards the centre of the system. There in due
course it will give rise to the disk population of stars. In this way
we should have the makings of a *spiral galaxy*.

6.4   Dwarf galaxy formation

Let us return to the simplest case and keep to examples having
$n_0 = 1$. In order for the process to produce results such as we have been
discussing, in the first place we require $\alpha > 1$ so as to give the

supersonic motion needed to cause the accumulation of material; it is
found that this requires $M \gtrsim 6 \times 10^{-5}$ where M is still in the unit of
10^{44} gram. Also we require $\ell > \lambda$, otherwise the collision achieves
nothing; this is found to require $M > 1.5 \times 10^{-4}$. Finally, the model
supposes each cloud to hold together under its own gravitation i.e.
$M > m_J(\rho_0)$; this is found to require $M > 4 \times 10^{-4}$, or
$$M \gtrsim 2 \times 10^7 \text{ solar masses.}$$
Thus a cloud of the raw material of about 2×10^7 solar masses could
collapse upon itself, and it would gain little from participating in a
collision. So we can claim that we should then have a *dwarf galaxy
which is its own single globular cluster*.

As before the existence of stars in such a galaxy would indicate
the occurrence of exploding massive stars at an early stage. Apart
from material that might be lost to the system as a result, this would
give the *lower bound to the mass of a galaxy*. There seems to be no
generally accepted empirical estimate of this quantity in the actual
universe; Unsöld, 1977, quotes as an order-of-magnitude value of
the mass of an 'extreme' dwarf elliptical galaxy 10^6 solar masses. So
the prediction is about as plausible as could be expected. In fact it
is one of the main *predictions* of the model. Another is that from
Table 1 the typical mass of a globular cluster varies inversely as the
cube root of the mass of the galaxy with which it is associated; again
there seems to be available at present no good empirical check.

6.5 Other considerations

On the picture here presented the material of any initial fragment of raw material of mass less than $m_J(\rho_0)$ will disperse; it
will not go directly to form galaxies. In the actual universe, if the
picture is valid in a general way, presumably such material might have
another chance later if it becomes incorporated into larger clouds.

The picture gives no simple *upper bound* to the mass of a galaxy.
But a very large cloud would be more likely to encounter several smaller clouds about the same time, than to encounter another comparable
to itself and no others. So the chance of making an enormous galaxy
may be negligible.

The encounter of a large cloud with several smaller ones would
be of interest because it should give rise to several galaxies which
could be gravitationally bound to each other. In this way, the model
would provide naturally for the occurrence of multiple galaxies,
galaxies with satellites, and groups of galaxies (some more stable
than others). The formation of a large cluster of galaxies would be
a different matter.

It is to be noted that a collision between two galaxies each
containing a considerable amount of interstellar matter could cause this
matter to produce stellar clusters by the process pictured here.
Since this matter would already contain heavy elements these clusters

would be different from those formed from the original raw material.
The Large Magellanic Cloud does possess some clusters that are apparently much younger than its 'normal' globular clusters; that is one way
of accounting for their existence.

6.6 Temperature

     Most of the results are sensitive to the value of T in §§4,5. But
T is not a disposable constant of the theory. It is determined by the
atomic properties of the material. However, the best evaluation may
not have been chosen; but if the model is successful this could be
claimed as good confirmation of the correctness of the value of T
adopted.

     The value $n_0 \simeq 1$ would be reasonable for the density of a big-bang
universe at redshift $z \simeq 100$ and background temperature about 300K. This
would be later than the decoupling epoch, as required. So the general
requirements of the theory are plausible, but if it is to be pursued
further much more comprehensive numerical testing would be necessary.

7. COMMENTS

     If the model be judged significant, it indicates that collisions
of portions of raw material provide the clue to galaxy formation. The
only way to get these collisions is to have the material of the universe
broken in pieces - or something essentially equivalent like supersonic
turbulence on a very large scale. As has been said, standard big-bang
cosmology cannot provide this feature; it would be important to invent
a cosmological model possessing such a feature.

     An even more preliminary account of these ideas was published in
McCrea 1979. Over several years I have given seminars on the subject
and I am grateful for valuable comments from my hearers; I am specially
grateful to Dr. S.M. Fall for discussion and references to the literature.

REFERENCES

Allen, C.W.: 1973, Astrophysical quantities, 3rd ed., Athlone Press.
Ebert, R.: 1955, Zeits. f. Astrophysik, 37 pp. 217-232.
Fall, S.M. and Rees, M.J.: 1977, Mon. Not. R. astr. Soc. 181 pp. 37P-42P.
Frenk, C.S. and White, S.D.M.: 1980, Mon. Not. R. astr. Soc. 193,
     pp. 295-311.
Kinman, T.D.: 1959, Mon. Not. R. astr. Soc. 119, pp. 559-575.
McCrea, W.H.: 1957, Mon. Not. R. astr. Soc. 117, pp. 562-578.
McCrea, W.H.: 1979, Irish astr. Jl, 14, pp. 41-49.
Unsöld, A.: 1977, The new cosmos, 2nd ed., Springer   p 312.
Woolley, Sir Richard : 1978, Mon. Not. R. astr. Soc. 184, pp. 311-317.

GALAXY FORMATION, CLUSTERING AND THE "HIDDEN MASS"

M.J. Rees and A. Kashlinsky
Institute of Astronomy
Madingley Road, Cambridge

ABSTRACT: Various topics relevant to galaxy formation are discussed,
especially the nature of the "hidden mass", and the properties of
"population III" pregalactic objects which are expected in models for
cosmic evolution involving primordial entropy perturbations.

1. SOME GENERAL REMARKS ON GALAXY FORMATION

     Progress in understanding the structure, dynamics, and morphology
of individual galaxies depends on quantifying the processes that
gradually convert gas into stars over the lifetime of a galaxy. Our
poor understanding of star formation is the main stumbling-block
preventing us from quantifying this. We do not know what determines
the initial mass function (IMF); still less do we know how the IMF
depends on physical conditions. Nor do we know much about the
efficiency of star formation: this determines how much gas can be
turned into stars on the free-fall timescale, and how efficiently
enriched gas can be recycled into new stars. Some aspects of galactic
evolution, where good progress is nevertheless being made, are discussed by Fall in his paper.

     There has been less progress in understanding how galaxies first
formed. Galaxy formation straddles the interface between cosmology
and astrophysics. It occurred at a cosmic epoch very different from
the present, and thus falls in the cosmologist's province. On the
other hand, once galaxies have formed, the phenomena within them that
interest astrophysicists proceed more or less regardless of the
broader cosmic context. We do not know why the most conspicuous
luminous entities in the universe are aggregates of 10^{11} stars, with
dimensions $\sim 10^4$ parsecs. Even worse, we do not know whether the
explanation we are seeking lies within the province of the astrophysicist
or the cosmologist. Are these characteristic dimensions a direct consequence of initial conditions? Or is there a physical reason for this
preferred scale, just as we now know there is a physical reason why all
stars have a mass within an order of magnitude of the Chandrasekar mass?

259

A. W. Wolfendale (ed.), Progress in Cosmology, 259–273.

There are some straightforward physical ideas, based on the cooling,
collapse and fragmentation of massive gas clouds (Rees and Ostriker
1977; Silk 1977), that single out a characteristic mass and radius
which seem relevant to galaxy formation. If these ideas indeed have
something in them, there is no more need to relate the masses of
galaxies to a preferred scale of initial cosmic irregularities than to
invoke a preferred fluctuation scale in the interstellar medium to
explain the masses of stars. (But the characteristic scales of the
biggest galaxies must to some extent be a function of cosmic epoch.
If we came back in 10^{11} years, we should find that our galaxy and
Andromeda would have merged into a single elliptical system, and that
the entire content of a cluster like Coma would be an amorphous cD
galaxy.)

It is thus not clear to what extent galaxies are permanent
structures manifesting some 'magic mass' for which we should seek a
physical explanation (maybe this is so for spirals and disc systems,
even if not for ellipticals). Even if so, it is unclear whether this
mass scale, and the corresponding length scale, stems from local
physics, or is a consequence of selective growth or damping mechanisms
in the early Universe.

2.   INITIAL SPECTRUM OF FLUCTUATIONS

Scenarios for galaxy formation all postulate that the early
universe was not completely smooth (see section 7 for some further
comments). Some initial irregularities must have given rise to bound
systems, which then either themselves turned into galaxies, or triggered
the formation of galaxies by an indirect route.

Two classes of perturbation can be envisaged: isothermal (or
entropy) perturbations, in which the radiation pressure is unaltered;
or adiabatic (isentropic) perturbations in which the photon/baryon
ratio is unperturbed. Isothermal perturbations are essentially 'frozen
in' before recombination; adiabatic perturbations oscillate before recombination, and all scales below $\sim 10^{15}$ $M_\odot$ are attenuated. A general
perturbation whose oscillatory component is damped by viscosity can
leave behind an isothermal component. The oscillatory behaviour
and damping of these various modes prior to recombination has been
extensively discussed in the literature. After recombination, when
the gravitational instability of the matter is opposed only by gradients
in the gas pressure (less than radiation pressure by a factor $\sim 10^8$
$\Omega^{-1}{}_{baryon}$), all scales of $\gtrsim 10^6$ $M_\odot$ grow, at least until bound systems
form and generate enough energy to heat and reionise the gas. To obtain
a bound system by the present time, the necessary amplitude at recombination must still be at least 10^{-3}. The growth of perturbations on
the mass scales of galaxies is inhibited before recombination by the
effects of ionisation pressure and viscosity; this means that amplitudes
of $\gtrsim 10^{-3}$ may be necessary even at the (earlier) epoch when such scales
are first encompassed within the particle horizon. (In models involving

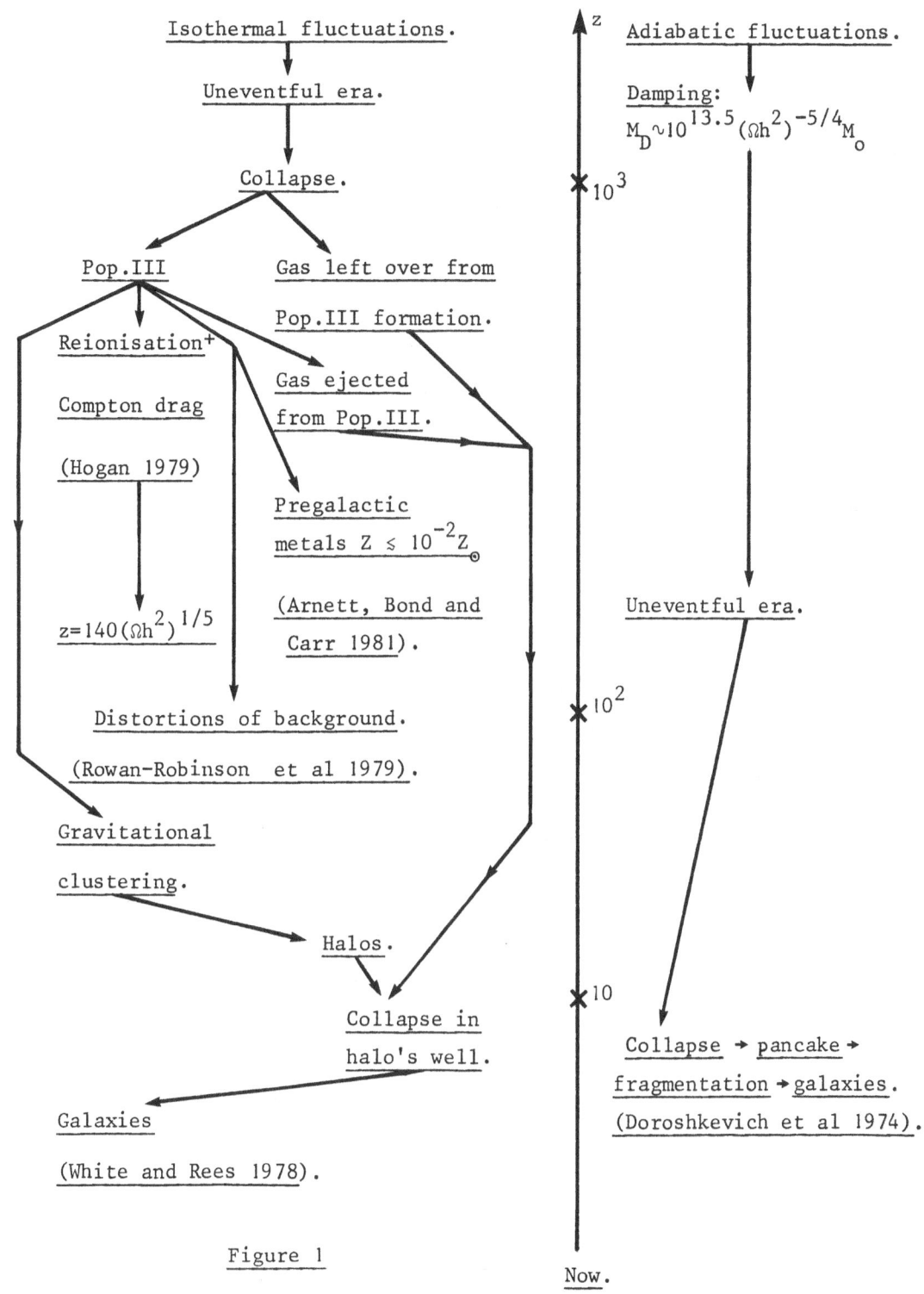

Figure 1

massive neutrinos, when growth can occur before recombination, one can
get by with perturbations whose amplitude on entering the horizon is
somewhat smaller.)

Key aspects of the two basic "scenarios" for galaxy (and cluster)
formation are summarised in figure 1. The two schemes contrast most
starkly if we compare what the universe would have been like at, say,
$Z = 20$. In the hierarchical picture (requiring entropy perturbations),
bound systems would already exist. Indeed, nearly all the initial
mass could have been incorporated in pregalactic stars and Population III
objects, which could have already augmented or distorted the microwave
background spectrum. (These pregalactic objects may amplify perturbations
on large scales, giving rise to galaxies by secondary effects.) On the
other hand, in the adiabatic picture essentially nothing of interest
has happened by $Z = 20$. Density contrasts are still less than order
unity, and the universe would still be in the form of expanding
neutral gas.

3. SUPERCLUSTERING(?)

The details of the large scale distribution of galaxies, and the
issue of whether there is a preferred scale rather than a simple power-
law covariance function, may offer some clues to which picture is the
right one. The universe undoubtedly looks smoother on larger scales:
no-one is gainsaying this general trend. But recent redshift surveys
(particularly those by Davies and collaborators at Harvard, and by
Ellis and his associates in the UK) suggest that the fluctuations on
scales of 10 - 100 Mpc are larger than previously thought. This point
had already been convincingly made for a number of years by Einasto
and his colleagues. The key question is whether the data are consistent with gravitational clustering of individual (galactic) masses,
or whether there is evidence for gas-dynamical dissipative processes,
leading to giant sheet-like or filamentary structures and to a favoured
mass for "superclusters" (Doroshkevich, Sunyaev and Zeldovich 1974).
Everyone would agree that on scales of galaxies and below, dissipation
has been crucial, but on sufficiently large scales linear structures
have evolved directly by gravitation clustering. But what is the
transition scale between these regimes? Is it individual galaxies,
clusters, or still larger?

The most direct evidence on superclustering is provided by studies
of the spatial and velocity distribution of galaxies. The question of
whether there is a large "preferred scale" in the galaxy distribution,
crucial though this is for discriminating between different galaxy
formation scenarios, remains controversial. Three indirect lines of
attack on this question are perhaps worth mentioning.

(i) Studies of the spatial (and redshift) distribution of quasars
may be a useful probe of clustering on scales $\gtrsim$ 10 Mpc at $Z \simeq 2$. An
interesting recent analysis by Osmer (1981) yields results that are not

quite sensitive enough to be interesting. However, if surveys using
objective prisms (or related methods) could be extended even just one
magnitude fainter (corresponding to an increase of $\sim$ 5 in the number
per square degree) the method would be at a level of sensitivity where
the *absence* of clustering would be a significant constraint. Of course,
we do not know to what extent quasars serve as an adequate probe of the
galaxy distribution (still less of the overall mass distribution) - the
quasar phenomenon may depend on the galactic environment; indeed present
data suggests that quasars "avoid" the centres of clusters.

(ii) Quasar absorption lines probe the distribution of clouds along
the line of sight. So far, however, there is no evidence for any large-
scale clustering of the clouds responsible for the Lyman α systems
(Sargent *et al*. 1980, Young *et al*. 1981b).

(iii) In some extreme versions of the "pancake" picture for galaxy
formation, X-rays from Z $\gtrsim$ 3 would already have been detected. The
evolution with redshift of the X-ray emission from clusters, and the
possible "graininess" of the diffuse background on angular scales of
$\lesssim$ 1°, will be one of the many to be settled by the next (post-"Einstein")
generation of X-ray telescopes.

One must of course be cautious about inferring the distribution of
all mass from observations of luminous objects which may comprise only
10% of the total amount of gravitating material — the more so because
within individual galaxies the dark material is less centrally concentrated than the visible stars and gas. However, while it is easy to
envisage how such segregation could have occurred in the course of an
individual galaxy's formation and evolution, it is less easy to envisage
how the two kinds of matter could be segregated on scales exceeding a
few megaparsecs. Thus on large scales it is more justifiable to regard
galaxies as valid tracers for the mass distribution. If large "voids"
really exist, which *cannot* be explained by gravitational clustering
(because, for instance, the associated peculiar velocities are too low),
one's mind should be open to more radical possibilities:

(a) Galaxy formation could have been inhibited or modified in
some large regions, so that the effective M/L for those regions is very
high even though ρ is more or less the same as elsewhere. The morphology
of galaxies is sensitive to their environment; in particular, the
presence of extensive luminous discs may require continuing infall for
a Hubble time (cf. Gunn 1981). One could imagine a more drastic environ-
mental change whereby galaxy formation is completely prevented in large
"voids" — for instance, if pregalactic gas could be heated to $\gtrsim$ 10^{6}K,
the Jeans mass would exceed a galactic mass and the condensation of gas
clouds into galaxies would be prevented. It would require a large
localised energy input to heat all the gas within a void to this
temperature. However, the energy per atom ($\sim$ 100 ev) is still much
more modest that the $\sim$ 10^{5}ev per particle that would have been needed to
push the material over $\sim$ 100 Mpc distances and physically evacuate a
void region.

(b) More radically still, the early universe could have split into "domains" with different baryon/photon ratios.

4. HIDDEN MASS

It needs no great ingenuity to invent possible forms of hidden mass-"massonium". Among them are these:

(1) <u>Neutrinos of rest mass $\sim$ 10 ev.</u> If neutrino rest masses are non-zero, then it is a straightforward consequence of the hot big bang model that they contribute a density parameter $\Omega_\nu \simeq 0.04$ $(H_0/50$ km/sec/ Mpc$)^{-2}(m_\nu)_{ev}$. This is because their number density is comparable to that of the photons in the microwave background; i.e. they outnumber the baryons by a factor $\sim 10^8 \, \Omega^{-1}_{baryon}$. These primordial neutrinos would have cooled adiabatically with the expanding universe, and would become sufficiently slow-moving that they could participate in gravitational clustering. We shall hear more from other speakers about the idea that all the baryons studied by astronomers - in stars, galaxies and gas - are just traces of "sediment" in a neutrino-dominated cosmos.

(2) <u>Low-mass stars.</u> Unevolved low-mass stars (or 'Jupiters') with individual masses $\leqslant 0.08$ $M_\odot$ would have a high enough M/L to escape detection. The present limits could perhaps be tightened by better background measurements in the near infra-red, by searching for high-proper-motion faint red objects in our Galaxy, or by improving the limits to the surface brightness of halos in edge-on spirals.

(3) <u>Remnants of massive stars.</u> Black holes could in principle have formed as the endpoint of the evolution of now-defunct massive stars which formed either in the early history of galaxies or at a possible pregalactic era of activity $(10 \leqslant z \leqslant 10^3)$. There are then some other constraints. Limits on the extragalactic night sky bright-ness imply that any stars with masses (say) 5 - 100 $M_\odot$ must have formed and completed their bright evolutionary phases at $z \gtrsim 10$ (unless their optical and ultraviolet emission is absorbed by dust and re-emitted in the infra-red). Such considerations do not, however, constrain so severely a 'Population III' of massive stars that formed at $z \simeq 100$ (see the accompanying discussion by Rowan-Robinson and Tarbot).

(4) <u>Remnants of supermassive stars.</u> Here the background light constraint is more ambiguous, because supermassive stars may collapse before having gone through a prolonged phase of hydrogen burning. However there are other ways of detecting them. For instance, there would be conspicuous accretion effects in our galaxy if individual masses exceeded $\sim 10^6$ $M_\odot$. Also, dynamical friction would be important for $\gtrsim 10^7$ $M_\odot$, (This would cause the dark objects to concentrate towards the centres of galaxies, contrary to the aim of the game which is that they should be predominantly <u>outside</u> the luminous material.)

(5) <u>Primordial black holes</u>. If black holes form via conventional "astrophysical" processes, their masses would be $\gtrsim 1$ $M_\odot$. But there is no reason why *primordial* holes should not extend down to $\ll 1$ $M_\odot$. There are however very stringent limits on the *very* low mass range: 'miniholes' of $\lesssim 10^{15}$ g would explode conspicuously by the Hawking (1974) process, and they cannot contribute more than $\Omega_{hole} \simeq 10^{-8}$ (Carr 1978).

5. GRAVITATIONAL LENS IMAGING

If the hidden mass is in the form of compact discrete objects (i.e. any of the options (2) - (5) listed in the last section), the detection of gravitation lensing would allow us to estimate their characteristic masses.

The optics of a gravitational lens are quite straightforward when the lens is a point mass. For a source at redshift z_S and a lens redshift z_ℓ the angular separation of the images is $\theta = f(z_S, z_\ell)$ 10^{-6} $(M_\ell/M_\odot)^{\frac{1}{2}}$ arc sec where M_ℓ is the lens mass and f is of order unity for $z_S \simeq 1$ and $z_\ell \simeq z_S/2$. (Refsdal 1966). This means that galactic-mass lenses can yield images separated by a few arc seconds. Interest in this subject has been stimulated by the discovery of two quasars that each show multiple imaging indicative of the lens effect (Walsh *et al.* 1979; Weymann *et al.* 1980). The effects are not of any profound interest for gravitation theory — the deflection is about the same amount (an arc second or so) as the gravitational light-bending within the Solar System, and is therefore not a manifestation of strong-field gravity. The study of gravitational lens images will, nevertheless, when more instances of the phenomenon are found, be of great potential importance in probing the distribution and nature of the mass in galactic halos and clusters.

The optics are more complicated when the lens is an extended transparent object such as a galaxy. The number of images can be 3, 5 or even more; the magnification and shape of each image depends on the detailed density profile within the lensing galaxy. The general principles were discussed by Bourassa and Kantowski (1975). (One notes, as a curiosity, that a disc of uniform surface density acts as a proper lens with a well-defined focal length independent of how far the ray path is from the axis.) Young *et al.* (1981a) have carried out detailed "ray-tracing", to see whether the triple quasar Q1115-080 (Weymann *et al.* 1980) can be due to lensing by a disc galaxy (with halo). It is plainly not yet possible to derive a unique model, but this work shows how much more one could learn if the images could be resolved and their shapes determined. The size of each image will be similar (apart from the magnification factor) to the intrinsic angular size of the quasar.

The above comments have been based on the supposition that the mass in the lens is distributed smoothly. However, if this mass is in stars or other discrete objects of characteristic mass m_* there will

be fine structure in each image due to this granularity with angular
scale $\theta_f \simeq 10^{-6}(m_*/M_\odot)^{\frac{1}{2}}$. This will affect the magnification if the
intrinsic size of the source is $\lesssim \theta_f$, and lead to non-intrinsic
variations as the individual stellar-mass objects move across the
quasar (Chang and Refsdal 1979, Gott 1981, Young 1981). If $m_* \simeq 1\ M_\odot$,
micro-arc-second resolution would be required to resolve the images;
but then the non-intrinsic variations are rapid enough to be detectable.
The line-emitting region would probably be too large to be affected by
this small-scale "graininess"; the central continuum may will be $\lesssim 10^{-7}$"
and would be affected even by Jupiter-sized masses (Gott 1981); only a
halo of massive neutrinos (individual masses $\sim 10^{-32}$ gm) is smooth
enough for this complication to be ignored. If the "hidden mass" was
in the form of $10^6 M_\odot$ black holes, the fine-structure in the lens images,
with a milli-arc-second scale, would affect the broad line region and
VLBI radio structure as well as the optical continuum.

An object at redshift $\gtrsim 1$ has a probability of order Ω_G of having
its image significantly distorted by lensing along the line of sight
(Ω_G being the fraction of the critical density contributed by objects
compact enough to act as efficient lenses). Thus, any attempt to use
high-z objects for classical cosmology must take account of lensing:
if one infers a high Ω (as, for instance, did Baldwin *et al.* (1978)
using quasars out to $z \simeq 2$ with "calibrated" luminosity) the possibility
must be considered that one's sample is biased towards objects that
are anomalously magnified. This uncertainty will remain until individual
QSOs could be imaged with adequate resolution.

[Note that the same problem arises in any attempt to use angular
diameter measurements applied to some feature of distant galaxies or
quasars that might serve as a standard "rigid rod". Here the distortion
of images (which is even more drastic than the change in brightness)
would be a major complication. However, if the objects could be
resolved well enough to identify those where the lensing is important,
not only could this difficulty be evaded, but the structure of the
images would offer clues to the nature of the matter contributing to
Ω].

6. POPULATION III AND PREGALACTIC STARS

If the inhomogeneities in the early universe involve entropy
fluctuations, rather than being purely adiabatic, then they will not be
destroyed by radiative damping before recombination. For a smooth
spectrum with the amplitude decreasing towards larger scales (as is
indicated by the data on galaxy clustering) the amplitude at recombination may be $\gtrsim 1$ on scales up to $10^6 - 10^9\ M_\odot$. This number is
uncertain because of the unknown slope of the fluctuation spectrum.
Since the Jeans mass M_J just after recombination is $\sim 10^6 \Omega^{-\frac{1}{2}}\ M_\odot$, objects
of mass $> 10^6\ M_\odot$ would therefore be expected to start contracting
gravitationally promptly after recombination, i.e. at $z \simeq 1000$. Before
one can formulate a real theory of galaxy formation based on initial

entropy fluctuations, it is essential to have some idea of what happens to these post-recombination condensations of sub-galactic mass. Do they turn into supermassive black holes; or do they fragment down to stellar masses? Does the energy input from these objects affect the residual gas - by heating it up, generating 'secondary' fluctuations via bulk motions, or injecting heavy elements? What fraction of the initial gas gets trapped in these 'Population III' objects or their remnants, and can these objects, after they have undergone hierarchical clustering, provide the 'hidden mass'?

Two interesting papers written in the 1960s addressed this question, but reached quite different conclusions. Peebles and Dicke (1968) considered the fate of clouds of just above the Jeans mass; they argued that the clouds would fragment into stars, yielding systems resembling globular clusters. In contrast, Doroshkevich et $al.$ (1967) envisaged that each cloud formed a single supermassive object; the energy thereby released would heat the intergalactic medium, exert a negative feed-back on the formation of further supermassive objects, and would generate 'secondary' inhomogeneities on larger mass-scales from which galaxies would later condense. (Ostriker and Cowie (1981) offer the interesting suggestion that the feed back from the first-generation objects may be positive - rather than negative, as Doroshkevich et $al.$ had supposed. These objects then trigger a runaway process generating random velocities, shock fronts, etc. on progressively larger scales; galaxies and clusters could then be a secondary consequence of a few small-scale "seed" fluctuations.)

We have recently considered the evolution of clouds substantially above the post-recombination Jeans mass - i.e. with M in the range $10^7 - 10^8$ $M_\odot$. This is the mass-scale on which $(\delta\rho/\rho)_{rec}$ would be ~ 1 if one takes a "white noise" spectrum of isothermal fluctuations whose amplitude is matched to the data on galaxy clustering (cf. White and Rees 1978). We have noted some physical effects which could be important:

(i)   Each cloud would have acquired some angular momentum via tidal interaction with its neighbours before it started to collapse. The amount of angular momentum can be conveniently parameterised in terms of binding energy E and angular momentum I by the quantity $\lambda = I|E|^{\frac{1}{2}}/GM^{5/2} = V_{rot}/V_{free\ fall}$. Numerical simulations (Aarseth and Fall, 1980; Efstathiou and Jones, 1979) show that clouds should acquire a range of angular momenta, the quartile points for λ being 0.04 and 0.10. The cloud would thus become rotationally supported after collapsing by a factor $\sim \lambda^2$, if it had not already fragmented before that stage.

(ii)   If we are concerned with collapse soon after recombination, the background temperature will be 1000°K. This means (obviously) that no part of the cloud can get cooler than this; also, even if only a small fraction of the material gets re-ionized, Compton drag may become important (Hogan 1979). If there is some reionization, caused by shock heating due to irregularities in the (supersonic) collapse, then drag would slow down the overall collapse provided that $\tau_C(V_{\text{free fall}}/c)aT_{\text{rad}}^4$ exceeded $\rho V^2_{\text{free fall}}$. This condition is fulfilled when $r/r_o \gtrsim 10^{-2} X^{-2/3}$, r_o being the clouds initial radius and X the fractional ionization. X itself can be estimated from the rate of release of binding energy in free fall collapse as $X = 10^{-2}(M/10^7 M_\odot)^{1/3}$ $(\rho/10^{-20} \text{ gm cm}^{-3})^{-1/12}$.

A recent discussion by Tohline (1980) suggests that fragmentation may not occur until the cloud has collapsed by a large factor. The possibility of Compton drag could make the collapse slower than free-fall, and would inhibit fragmentation still more. For this reason, we consider it plausible that a cloud of mass $\sim M_J$ may increase its density by a factor $\lambda^{-6} - \lambda^{-8}$ before fragments have separated out. (The factor λ^{-6} is reached when the central part of the cloud becomes affected by rotation; the full factor λ^{-8} would be attained if the entire cloud developed into a thin rotation-supported disc.) Formation of molecular hydrogen may be favoured if the material experiences strong shocks during the infall (cf. Hollenbach and McKee 1979) and the free electrons cool below $\sim 10^4$K by the Compton effect before recombining. If enough H_2 forms to permit cooling below 10^4K, these densities are enhanced further by a factor $(T_{\text{gas}}/10^4\text{K})^{-1}$.

The naively-estimated Jeans mass at the end of such a collapse would be $\sim 20(\lambda/0.07)^4(T_{\text{gas}}/10^4\text{K})^2(M_{JO}/M)^{1/3} M_\odot$. Obviously we cannot estimate it precisely because of the steep dependence on T_{gas}; but this rough argument suggests that, in principle, the gas could   all end up in low-mass stars. On the other hand, most of the material in the original cloud *might* end up in a single supermassive object. This could happen either for the general reasons outlined by Larson (1978), or because, when the outer part of the cloud attempts to settle into a disc, viscous effects (including those arising from the gravitational instabilities themselves) heat and thicken the disc, causing the gas to drain onto a central object before individual stellar-mass objects fragmented out.

It would of course be unrealistic to *expect* to predict the masses of Population III objects with any precision - after all, we are still unclear about what determines the IMF in the sites of present-day star formation in our Galaxy, and must be even further from understanding the fragmentation process at early epochs where physical conditions were very different. However, our considerations have made us even more sceptical and open minded than earlier workers have been. Fragmentation depends critically on very uncertain parameters: the degree of inhomogeneity in the primordial clouds on mass scales below $10^6 M_\odot$, the amount of viscosity in massive discs, etc. Until these parameters can be quantified, it seems worthwhile to explore the consequences for galaxy formation of two extreme alternative hypotheses: (i) the hidden mass is in a population of low-mass stars; or (ii) there was a pregalactic population of supermassive objects or "VMOs" (Arnett, Bond and Carr 1982) which could have generated significant energy and heavy elements. We note also that the fate of a $10^6 - 10^8 M_\odot$ cloud may be sensitive to the value of its angular momentum parameter λ.

The fate of $10^7 - 10^8$ $M_\odot$ clouds is sensitive to the angular momentum — high powers of λ enter all the relevant expressions. The numerical simulations of tidal torques show that there is a spread of ~ 3 in the value of λ. Thus the fate of neighbouring clouds, even collapsing contemporaneously, may be very different. A cloud with low λ could, for instance, form such a compact disc that, if it goes globally unstable (to bar-modes, etc.) the power released would approach the Eddington limit, and radiation pressure would re-inflate it.

If the clouds collapse at slightly different times (because they have different initial values of $(\delta\rho/\rho)$), then the collapse of later ones can be affected by heat input from its neighbours that have already collapsed. Partial reionization of the later clouds makes them subject to Compton drag and erases their angular momentum; their subsequent collapse will then be characterised by *very* low λ.

If there is efficient fragmentation into low-mass stars, our systems would superficially resemble those which Peebles and Dicke (1968) identified with globular clusters. However we are envisaging systems with larger total mass, which would not have retained their individual identity. Another difference is that, in our systems, star formation is delayed until the escape velocity is ~ 100 km/sec. This means that photoionized gas is retained rather than being expelled, and the conversion into stars is more efficient.

If it turns out that no kind of neutrino (or other light particle) has a suitable mass to provide the "hidden mass" in intergalactic space, then the most likely possibility would, we believe, involve the idea that most of the mass condensed at $Z \gtrsim 100$, into objects whose remnants now provide the hidden mass.

Such processes as pregalactic nucleosynthesis and energy generation (Rowan-Robinson, Negreponte and Silk 1979; Rowan-Robinson and Tarbet, these proceedings) are obviously very sensitive to the mass function for Population III. The possibility (alluded to above) that the post-recombination evolution of the $\sim 10^7$ $M_\odot$ clouds may be "bimodal" — some fragmenting into stellar-mass objects, but some forming one, or a small number of, supermassive objects — is a further complication. However, a "bimodal" scheme permits the interesting possibility that a Population III (of supermassive objects) and (some) Population II stars may be coeval - in other words, that some of the presently-observed Population II stars may be "pregalactic", in the sense that they formed in a $\sim 10^7$ $M_\odot$ cluster which only later became incorporated (via hierarchical clustering) in a galaxy. The intergalactic medium could then, in principle, have been contaminated with heavy elements even at $Z \simeq 100$ to a level far exceeding the extreme Population II abundances. This would ease the task of those authors who have seriously considered thermalising the entire microwave background at $Z \simeq 100$. For such a "bimodal" scheme to be consistent with present-day galaxies, the relative proportions of clouds developing into supermassive and "ordinary" stars would need to be about 10:1. The "halos" of galaxies would then

consist of the remnants of supermassive objects; the clusters of
"ordinary" stars would have settled via dynamical friction towards the
centres of the "halos" before merging and losing their identity.

7. THE VERY EARLY UNIVERSE

      The initial fluctuations - necessary in any scheme for galaxy
formation - are still a mystery. There is still no fully plausible
idea which explains why our Universe contains inhomogeneities of
large enough amplitude to yield the presently-observed structure, yet
is not so "chaotic" that the overall "Robertson-Walker" geometry is
invalid. We can make direct inferences about the amplitude of
fluctuations on the scale of galaxies and clusters; but it might offer
some clues to the underlying mechanism if we had evidence about the
spectrum beyond this mass range. The angular structure of the microwave background can in principle reveal Doppler shifts and gravitational
perturbations at the epoch of last scattering, which may be at a
redshift as large as 1000. In any theory these perturbations would be
linear on scales exceeding 10^8 solar masses; they may extend up to much
larger scales even than superclusters. These fluctuations could have
been imprinted as initial conditions or via exotic processes at very
early times. Any quadrupole dependence (or indeed any fluctuations on
angles exceeding a few degrees) corresponds to scales exceeding the
horizon size at last scattering. They are therefore "acausal" fluctuations, in the context of a classical Friedmann model. Microwave
background measurements are now tantalizingly close to the level of
precision where a wealth of positive anisotropy measurements may become
available.

      Granted the assumption of a big bang model subject to gravitational
instability, the observed properties of galaxies and clusters tell us —
at least in order of magnitude — the amplitude of the necessary perturbations on these particular scales. The most convenient measure of the
'lumpiness' of a Friedmann universe on a particular scale is the
'curvature fluctuation' $\epsilon(M)$, which is essentially equivalent to the
amplitude of the fluctuation on mass-scale M when that scale is first
encompassed within a particle horizon.

      If curvature fluctuations with $\epsilon \simeq 10^{-3} - 10^{-4}$ are 'given' on the
appropriate scales, one can explore scenarios whereby gravitational
clustering and dissipative processes gradually transform them into the
galaxies and clusters we now observe. It is then natural to inquire
whether the curvature fluctuations required for galaxy formation can
be part of a spectrum extending over a much broader range of scales.
There are some (albeit rather insensitive) constraints on subgalactic
scales from consideration of primordial element formation, distortion
of the microwave background, and the necessity to avoid overproducing
'miniholes'. The limits on *large* scales, from the microwave and X-ray
background, are much more stringent (see Rees 1980 for a survey of
these limits).

The simplest assumption would be that the amplitude of curvature fluctuations depends on a power law: $\varepsilon \propto \ell^x$. If $x \neq 0$, the amplitude 'blows up' leading to a non-Friedmannian universe, on scales which are large or small according as $x \lessgtr 0$. The absence of effects due to mini-holes suggests that the universe is Friedmannian (i.e. $\varepsilon \lesssim 1$) on scales down to 10^9g, implying that $x > -0.1$ (Carr 1976). We cannot use the data to set such a strict limit on positive values of x because there is a shorter 'level arm' between cluster scales and ℓ_H than there is on the small-scale side.

The 'constant curvature fluctuation' spectrum ($x = 0$) is obviously especially attractive. It was advocated particularly by Sunyaev and Zeldovich (1970). This spectrum is scale-independent: there is no characteristic mass where ε becomes the order unity. This spectrum is entirely consistent with the small-scale constraints. However it is a delicate matter to decide whether ε can be big enough to yield galaxies and clusters without violating the limits on large scales set by the microwave background. Detailed discussion by Press and Vishniac (1980) and Silk and Wilson (1980) suggest that the constant curvature spectrum may be *in*consistent with those scenarios for galaxy formation where the initial fluctuations are purely adiabatic.

The present microwave (Smoot *et al.* 1977; Cheng *et al.* 1979) and X-ray background measurements (Warwick *et al.* 1980) are tantalisingly close to the sensitivity level at which they can usefully delineate the fluctuations on scales from 100 Mpc to ℓ_H. In the next few years we can expect improved techniques (keeping pace, one hopes, with theoretical progress aimed at predicting what the spectrum should be).

The spectrum of fluctuations could extend continuously upward to the Hubble radius ℓ_H. High precision observations of the microwave background should then reveal quadrupole, octupole and all higher harmonics. On the other hand, if the 24h microwave anisotropy (cf. Smoot 1980) is induced 'locally', and the amplitudes are undetectably small on scales $\gtrsim \ell_H$, there is no reason for expecting any associated quadrupole effect.

Curvature fluctuations can in principle, in an open universe, have a spectrum which extends to wavelengths *much larger than* our present particle horizon. Such fluctuations can be detected because they yield gradients across the presently-observable part of the universe, and temperature anisotropies on the last scattering surface. Grishchuk and Zeldovich (1978) show that, subject to a random phase assumption, growing mode fluctuations would have $\delta\rho/\rho \lesssim 10^{-4}$ on all scales. The corresponding limit on the curvature fluctuations is $\varepsilon \lesssim 10^{-4} (\ell/\ell_H)^2$. (Limits can also be placed on large scale gravitational waves, on overall vorticity, etc.) Kaiser (1981) has generalised Grishchuk and Zeldovich's results from a flat to a general open Friedmann model.

If the amplitudes of protogalactic fluctuations are already $10^{-3} - 10^{-4}$ when they are first encompassed within the particle horizon, then they must have been imprinted acausally into the initial data. If we consider very tiny scales containing only (say) $10^6 - 10^8$ particles, which enter the horizon soon after the grand unification era, then some fluctuations may arise from discreteness effects. However, such effects yield a spectrum $\delta\rho/\rho \propto M^{-7/6}$ (Zeldovich 1965, Carr 1976) corresponding to x = -3/2, and fall off too rapidly with increasing scale to give significant effects on astronomical mass-scales. So again their origin is pushed back further (to the era of quantum cosmology?).

The equation of state at supernuclear densities is sufficiently uncertain that phase transitions cannot be ruled out: at the grand unification epoch, at the era of symmetry-breaking in the Weinberg-Salam theory, or related to the quark-nucleon transition density (see Klinkhamer's contribution to these proceedings). This offers the prospect that macroscopic fluctuations can suddenly appear, attaining amplitude unity in a single expansion timescale. The most optimistic assumption on the scale of these irregularities is that it corresponds to the horizon mass (or maybe the sound speed) when the transition occurs. The latest stage at which a phase transition could occur is when the density has dropped to a few times nuclear density; the horizon mass is then $\sim 1 M_\odot$. However, even this 'optimistic' case is not good enough to yield galaxies: Hogan (1980) shows that — even if the matter is cold so that gravitational clustering proceeds unimpeded by pressure gradients — the non-linear mass scale still cannot reach cluster order by the present era. This is because the $M^{-7/6}$ law applies, with the result that the non-linear scale grows only as $t^{4/7}$ and would still be $\lesssim 10^{10} M_\odot$ at $t \simeq 10^{10}$ yrs. Nevertheless, there are tenable models which can generate the observed large-scale structures without involving acausal fluctuations (Carr and Silk 1982; Carr and Rees 1982); these involve tepid cosmologies (where most of the cosmic entropy is not primordial), or else the idea that only a few rare "seeds" are required, which then inject energy to generate secondary fluctuations on larger scales.

Of special interest in this context is the reality of quadrupole anisotropies (cf. Boughn, Cheng and Wilkinson 1981, Fabbri *et al.* 1980, Smoot and Lubin 1979); or, indeed, any other fluctuations on large angular scale. These anisotropies could be reconciled with an acausal fluctuation spectrum with x > 0, but any "causal" spectrum would fall off so steeply that there would be no detectable amplitude on scales $\sim \ell_H$.

REFERENCES

Aarseth, S.J. and Fall, S.M. 1980. Astrophys.J., <u>236</u>, 43.

Arnett, W.D., Bond, R. and Carr, B.J. 1982. Proceedings of Cambridge Supernova Conference (in press)

Baldwin, J.A., Burke, W.L., Gaskell, C.M. and Wampler, E.J. 1978. Nature, <u>273</u>, 431.

Boughn, S.P., Cheng, E.S. and Wilkinson, D.T. 1981. Astrophys.J.(Lett) <u>243</u>, L113.

Bourassa, R.R. and Kantowski, R. 1975. Astrophys.J., 195, 13.

Carr, B.J. 1976. Astrophys.J., 206, 8.

Carr, B.J. 1978. Comments on Astrophys., 7, 161.

Carr, B.J. and Rees, M.J. 1982. Mon.Not.R.astr.Soc. (submitted).

Carr, B.J. and Silk, M.J. 1982. Mon.Not.R.astr.Soc. (submitted).

Chang, K. and Refsdal, S. 1979. Nature, 282, 561.

Cheng, E.S. *et al*. 1979, Astrophys.J.(Lett.), 232, L139.

Doroshkevich, A.G., Zeldovich, Y.B. and Novikov, I.D. 1967. Sov.Astron.,
    41, 233.

Doroshkevich, A.G., Sunyaev, R.A. and Zeldovich, Y.B. 1974. in
    "Confrontation of Cosmological Theories with Observational Data",
    ed. M.S. Longair (Reidel).

Efstathiou, G. and Jones, B.J.T. 1979. Mon.Not.R.astr.Soc., 189, 203.

Fabbri, R. *et al*. 1980. Phys.Rev.Lett., 44, 1563.

Gott, J.R. 1981. Astrophys.J., 243, 140.

Grishchuk, L.D. and Zeldovich, Y.B. 1978. Sov.Astron.A.J., 22, 125.

Gunn, J.E. 1981. in Proc. Vatican Conference on "Cosmology and
    Fundamental Physics" (in press).

Hawking, S.W. 1974. Nature, 248, 30.

Hogan, C.J. 1979. Mon.Not.R.astr.Soc., 188, 781.

Hogan, C.J. 1980. Nature, 286, 360.

Hollenbach, D. and McKee, C.F. 1979. Astrophys.J.(Supp.), 41, 555.

Kaiser, N. 1981. Mon.Not.R.astr.Soc. (in press).

Larson, R.B. 1978. Mon.Not.R.astr.Soc., 184, 69.

Osmer, P. 1981. Astrophys.J. (in press).

Ostriker, J.P. and Cowie, L. 1981. Astrophys.J.(Lett.), 243, L127.

Peebles, P.J.E. and Dicke, R.H. 1968. Astrophys.J., 154, 891.

Press, W.H. and Vishniac, E.L. 1980. Astrophys.J., 236, 323.

Rees, M.J. 1980. in "Quantum Gravity II" ed. R. Penrose *et al*. (O.U.P.)

Rees, M.J. and Ostriker, J.P. 1977. Mon.Not.R.astr.Soc., 179, 451.

Refsdal, S. 1966. Mon.Not.R.astr.Soc., 132, 101.

Rowan-Robinson, M., Negreponte, J. and Silk, J. 1979. Nature, 281, 635.

Sargent, W.L.W., Young, P.J., Boksenberg, A. and Tytler, D. 1980.
    Astrophys.J.(Supp.), 42, 41.

Silk, J.I. 1977. Astrophys.J., 211, 638.

Silk, J.I. and Wilson, M.L. 1980. Physica Scripta, 21, 708.

Smoot, G.F. 1980. Physica Scripta, 21, 619.

Smoot, G.F., Gorenstein, M.V. and Muller, R.A. 1977. Phys.Rev.Lett.,
    39, 898.

Smoot, G.F. and Lubin, P.M. 1979. Astrophys.J.(Lett.), 234, L83.

Sunyaev, R.A. and Zeldovich, Y.B. 1970. Astrophys.Sp.Sci., 7, 3.

Tohline, J.E. 1980. Astrophys.J., 239, 417.

Walsh, D., Carswell, R.F. and Weymann, R.J. 1979. Nature, 279, 381.

Warwick, R.S., Pye, J.P. and Fabian, A.C. 1980. Mon.Not.R.astr.Soc.,
    190, 243.

Weymann, R. *et al*. 1980. Nature, 285, 641.

White, S.D.M., and Rees, M.J. 1978. Mon.Not.R.astr.Soc., 183, 341.

Young, P.J. 1981. Astrophys.J., 244, 756.

Young, P.J. *et al*. 1981a. Astrophys.J., 244, 736.

Young, P.J. *et al*. 1981b. Preprint.

Zeldovich, Y.B. 1965. Adv.Astr.Astrophys., 3, 241.

QUASARS AT LARGE REDSHIFTS

Malcolm G. Smith
Royal Observatory, Edinburgh, Scotland

Quasi-stellar objects have been discovered with redshifts up to z = 3.53.
The object of record redshift was discovered 8 years ago. The only
other object with z > 3.5 was discovered (using purely optical techniques)
about 8 weeks ago. A critical discussion of the reality of this
apparent upper limit to QSO redshifts is given. Because numerous
quasars are now known with z > 3, such a limit could imply that quasars
first became visible within a rather short time interval.

A group of four QSOs at z = 0.37, with size of $\sim 75h_o^{-1}$ Mpc has been
discovered, but the recently reported lines of QSOs in the sky are
probably chance associations. Star counts strongly suggest the existence
of pronounced evolution of the intrinsic luminosity function with
redshift, when using standard Friedmann models. A novel technique is
described for exploring the intrinsic optical luminosity function as a
function of redshift to beyond redshift 3, while simultaneously reaching
luminosity levels similar to ordinary Seyfert galaxies.

Considerable progress has been made in the last five years towards
understanding the location of the material responsible for the absorption
lines in QSOs. It appears that most of it is not associated with the
QSOs themselves (though a number of ejection events involving outflow
velocities $\sim 0.1c$ are known). The unassociated absorbing material
appears to be chiefly (i) in the haloes of intervening galaxies and (ii)
in intergalactic clouds of neutral hydrogen.

This article is divided into the following sections:

1.   On the apparent cutoff in the redshift distribution for quasars.
    1.1   Some history - radio searches for large redshift quasars.
    1.2   More history - optical searches for large redshift quasars.
    1.3   Osmer's evidence for an apparent turn-down in the co-moving
        space density of QSOs.
    1.4   Discussion.

A. W. Wolfendale (ed.), Progress in Cosmology, 275–290.

2.   The observed distribution of QSOs.
      2.1   Are QSOs scattered in space independently at random?
      2.2   QSO counts to faint apparent magnitudes - evidence of
             luminosity evolution.
      2.3   QSO spectroscopy to faint apparent magnitudes.

3.   Absorbing material in the line of sight to QSOs.

1.   ON THE APPARENT CUTOFF IN THE REDSHIFT DISTRIBUTION FOR QUASARS

1.1   Some history - radio searches for large-redshift quasars

We shall spend some time examining the history of the QSO redshift cut-
off problem (i) to illustrate the difficulties that have been encountered
so far, (ii) to indicate the care needed in order to establish a reliable
value for the redshift at which the comoving space density turns over
and (iii) to provide some information and references for the techniques
involved.

Ten years ago, in view of the fact that quasars had not been found much
beyond redshift z = 2.5, Lynds and Wills (1970), Sandage (1972a),
Schmidt (1970) and others argued that this redshift marked the sought-
after formation epoch of the quasars (and, by assumption, the galaxies
as well).   Sandage (1972a) then wrote "Observational steps to make the
cutoff problem more secure are straightforward, though tedious.
Optical identification of radio sources in the 4C and 5C lists, and in
the deep surveys with the Bologna cross, for example, can be made
without regard to colour .... spectra of these objects will uncover the
highly redshifted quasars if they exist."   Though the presence of
selection effects was appreciated, it seems that they were not thought
likely to be important.   However, within a year, 0642 + 449 (OH471)
z = 3.40 and 1442 + 101 (OQ172) z = 3.53 had been discovered.

Unlike the QSOs found up to that time, OH471 had no ultraviolet excess,
and so most people felt that objects of higher and higher redshift would
soon yield to the approach of observing high-latitude neutral or red-
coloured star-like radio sources.   Indeed, discovery rates of such
radio-selected QSOs with z > 3 improved (e.g., Peterson et al 1978;
Jauncey et al 1978a, b;   Wright et al 1978) as it was realised that
high-frequency searches, such as the 2700 MHz Parkes survey (e.g. Wall
1977) were a better way to find QSOs.   Essentially every large telescope
in the world, using the finest available auxiliary instrumentation, was
involved in this rather mad scramble.

An illustration of the intensity of this effort is given by a story
which suggests that Wampler was invited to Steward Observatory to give a
colloquium about QSOs six weeks after the discovery, at Steward Observatory, of OH471 (Carswell and Strittmatter 1973).   It is alleged that
Wampler planned his presentation of objects in order of increasing
redshift, intending to surprise his audience as he passed on from a

confirming spectrum of OH471 (with supposed record redshift z = 3.40) to
a slide of the spectrum of OQ172 - which had just been discovered, at
Lick Observatory, to have redshift z = 3.53 (Wampler et al 1973)!   (The
plan fell through because news of the Lick discovery had leaked to
Steward a couple of days before Wampler's colloquium).

1.2 More history - optical searches for large-redshift quasars.

Amid the pressures of this scramble for time on large telescopes, a
(then) much less trendy method for finding QSOs was being quietly
developed at Kitt Peak by Arthur Hoag and Dan Schroeder (1970).   Using
photographic plates, a transmission grating and a small (91cm) telescope
they showed that emission lines from QSOs could be detected directly in
slitless spectra.   Being so limited in field coverage and telescope
aperture, they were not able to discover any unknown QSOs, but they did
show that hypersensitised Kodak IIIaJ was an ideal emulsion for this
kind of work (Hoag 1972).

Markarian (1967), in Soviet Armenia, had earlier been the first to show
how Schmidt telescopes could be used to great effect for extragalactic
spectroscopic survey work far beyond the Local Group.   Thus, when a
thin (low-dispersion) objective prism was ordered by Victor Blanco
(1974) for the 61cm Curtis Schmidt telescope, I had all the tools and
techniques necessary to make the first <u>discovery</u> of QSOs by the slitless
spectroscopy technique (Smith, 1975, 1976).   Two years later, Art Hoag
was able to get his transmission-grating/prism ("grism") combination to
the prime focus of the AURA 4-metre telescopes (Hoag et al 1977;   Hoag,
Burbidge and Smith 1977), which led to the discovery of fainter, weaker-
lined objects (see, e.g. Clowes, 1981).

Prisms and grisms were used by many people to find numerous high-redshift
quasars (see, e.g. references in the reviews by Smith 1978, 1981) and
then to demonstrate that the co-moving space density of such objects,
rather than decreasing near z $\sim$2.5, continues at least at a constant
level out to the limit of the IIIaJ search technique at z $\sim$3.3 (where
Lyα is redshifted to the long-wavelength limit of the IIIaJ emulsion
response near λ5300Å - Osmer and Smith 1977;   Lewis, MacAlpine and
Weedman 1979).

The next step was to try to find QSOs with even higher redshifts using
red-sensitive emulsions.   Hoag was again the first to make the
appropriate experiments, but his telescope was too small to make much
progress (Hoag 1976).   I also carried out some obvious experiments in
1974 using 127-02 emulsion (the pre-cursor of IIIaF), a 4° prism and the
Curtis Schmidt, but the limiting magnitude was too bright to give useful
results.   Art Hoag and I got together in late 1976, using his prime-
focus grism arrangement with Kodak IIIaF emulsion on the CTIO 4-metre
telescope (Hoag and Smith 1977);   the grism had the advantage that its
dispersion did not decrease rapidly at longer wavelengths (c.f. prisms).
Our aim was to be able to detect Lyα over the range 3400A to 6900A, (1.8
< z < 4.7).   Lyα had been the strongest line in most of the QSOs we had

discovered earlier with IIIaJ emulsion. As the comoving space density of
QSOs showed no sign of decreasing out to the limit of IIIaJ (Lyα
redshifted to 3.3), nor was such a decrease really expected any longer,we
were fairly optimistic that even with the four half nights available to
us, we would find several QSOs with z ⌄4. Indeed, one of our more
marginal candidates turned out to have a redshift of z = 3.45 (Smith et
al 1977) which, until two months ago, was the second highest redshift
known.   However, even though nearly all our 70 or so candidates were
later confirmed by Osmer (1980) as QSOs, none had redshift higher than
OQ172 (z = 3.53).   Our first naïve reaction was to believe that we had,
this time, found the redshift cutoff.   Emission lines (though not of
Lyα) were visible on our plates out to λ6900Å.   The volume of space 3.5
< z < 4.7 is greater than that bounded by 2.5 < z < 3.5.   However Bob
Carswell had seen this sort of thing before - a "cutoff" (in ultraviolet
excess objects) at z ⌄2.5 followed by a discovery (of a neutral-coloured
object) at z = 3.41 (Carswell and Strittmatter 1973).   We calculated
the effect of a steep luminosity function and a blue-blazed grism on the
numbers of detectable QSOs in the 5 deg² covered by Hoag and Smith and
found that the observed data provided no evidence at all for a cutoff
(Carswell and Smith 1978).

Koo and Kron (1980) tried next - this time with a CCD detector and IV-N
plates to help boost red response.   Again they were unsuccessful -
indeed they found no new quasars at all; however, they were able, at
least provisionally, to set tighter limits on the evolution of comoving
space density beyond z ⌄3.3 than had Carswell and Smith.

The highest redshift QSO found by optical techniques, also with z =
3.53, was discovered about eight weeks before this conference, by Hazard
(1981).   He used the UK Schmidt, its (thin) objective prism and IIIaF
emulsion and found an object (1159 + 123) with such strong Lyα that it
was detectable even though the prism dispersion had by λ5500Å fallen to
⌄5100Å/mm.   Higher contrast (and hence better sensitivity to emission
lines on the IIIaF plates) can be provided at the UK Schmidt by use of a
higher dispersion prism, but unfortunately delivery of this large piece
of optics is now scheduled to be more than two years after the date
originally expected.

Osmer (1977) and Carswell and Smith (1978) realised that, meanwhile, use
of a red-blazed grism was needed at a superb site like Cerro Tololo, in
order to make much further systematic progress.   Though a smaller total
number of QSO candidates was likely in a given area of sky, compared
with the survey by Hoag and Smith, greater efficiency would be achieved
towards the red-end of the IIIaF emulsion response. Osmer (1981b) has
recently circulated an important preprint giving the results of a survey
with a red-blazed grism.

1.3 Osmer's evidence for an apparent turn-down in the co-moving space
     density of QSOs.

The main features of Osmer's (1981b) paper are as follows:

a.   He obtained 17 plates, using a red-blazed grism, Kodak IIIaF
     emulsion and a (filtered) pass band of 5700Å $< \lambda <$ 6900Å. His
     survey, like the earlier work with a blue-blazed grism by Hoag and
     Smith, covered 5 $\deg^2$. He states that he is sensitive to Lyα
     at redshifts $3.7 < z < 4.7$ (mid range $z = 4.2$) and CIVλ 1549 at
     redshifts $2.69 < z < 3.47$.

b.   He found 40 candidate objects which he examined individually with
     the CTIO vidicon spectrometer system on the 4-metre telescope.   He
     confirmed 15 of these candidates as emission-line QSOs.

c.   Seven objects had $z > 1.0$.   Five were discovered by means of their
     CIV emission, yielding redshifts in the range $2.77 < z < 3.36$.
     Two were found from their MgII emission, with redshifts near
     $z = 1.1$.

d.   Eight objects had $z < 1.0$, and were discovered from their Hβ
     and [OIII] emission lines, in the range $0.03 < z < 0.31$.
     Most of these low-redshift objects that show Hβ in the slit spectra
     appear to be similar to the distant 'HII galaxies' discovered on UK
     Schmidt plates by Terlevich and others (see, eg, Terlevich and
     Melnick 1981).   They may not differ greatly (except in absolute
     luminosity) from the so-called 'Tololo' galaxies listed by Smith,
     Aguirre and Zemelman (1976) and MacAlpine and co-workers (see
     reference list).   However, be they quasars or not, these low-
     redshift objects do not enter critically into the discussion,
     other than to reinforce Osmer's point that his survey is sensitive
     to weak emission features.

e.   Osmer gives various arguments which show that his survey reached
     emission-line flux levels $\sim$3 times (1.2 mag) weaker than the
     earlier Hoag-Smith survey.

f.   The Hoag-Smith Survey yielded in 5 $\deg^2$, 7 QSOs with $2.5 < z < 3.5$
     (mid range $z = 3.0$).

g.   The cosmological dimming from $z = 3$ to $z = 4.2$ [for $q_o = 0$, it
     varies as $z^{-1}(1+z)^{-2}$] is also 1.2 mag .

h.   The volume ratio of shells V $(3.7 < z < 4.7)$/V$(2.5 < z < 3.5)$
     $=1.37$ (for $q_o = 0$) - see e.g., the explicit expressions for V(z) in
     Carswell and Smith (1978).

i.   For a uniform comoving space density, Osmer expected $7 \times 1.37 =$
     9.6 QSOs with $3.7 < z < 4.7$ in the new survey with the red-blazed
     grism ($q_o = 0$).   In his preprint, Osmer gives no discussion of the
     effects of uncertainties in the number of objects (7) found by
     Hoag and Smith (1977) in the redshift interval $2.5 < z < 3.5$.

j.   Osmer found <u>no</u> QSOs with $3.7 < z < 4.7$.

k.. Osmer's calibration of sensitivity is sufficient to show that the
     discrepancy between predictions and observations is not likely to
     result from insufficient sensitivity.

l.   The probability of getting zero in a Poisson distribution with
     expected value 9.6 is $\sim 10^{-4}$.

m.   The 95% upper limit in surveys like this is 3 in 5 $\deg^2$, for if the
     expected number is 3, then the observed value of zero will occur
     with 5% probability.

n.   Thus, according to Osmer, the comoving space density of these
     quasars turns down.   For $q_o = 0$, ρ $(3.7 < z < 4.7)$ is a factor ~ 3.2

less than ρ (2.5 < z < 3.5).   For q_o=1, the factor is $\sim$4.2.

1.4 Discussion.

There is a flaw in Osmer's argument as presented in his preprint.   He
computes the probability of finding zero in independent samples from a
distribution with expected value 9.6.   This assumes that the Hoag-Smith
point has no uncertainty.   As we have no 'a priori' reason for a cutoff
at z $\sim$3.5, one should instead ask what is the probability of finding 9.6
and zero in independent samples from a distribution with expected value
possibly somewhere in between - for example, 5.   Then, even by these
very crude arguments we see that the case as presented by Osmer is not
entirely convincing; the data he has used in his discussion may yield
little more than a 2σ effect.

We need some kind of check as to whether the single Hoag-Smith value of
seven QSOs with 2.5 < z < 3.5 in 5 deg^2 is likely to be representative;
could it instead be an unlucky, high, fluctuation?   Fortunately, Osmer's
(1981b) own QSOs, discovered from their CIV lines, can be used to
provide a first-order check.   To keep the discussion as straightforward
as possible, we shall consider q_o=0 and predict only the effects of a
uniform co-moving space density extending to z = 4.7.   Osmer (1981b)
demonstrated that his filter permits discovery of quasars in the range
2.69 < z < 3.47 from their CIV lines, with essentially the same detectivity that Hoag and Smith had for Lyα in the range 2.5 < z < 3.5.
Counting Q0000-398 as an "independent" discovery, Osmer found five QSOs
with 2.69 < z < 3.47 in 5 deg^2.   The ratio of volumes in the shells
V(2.5 < z < 3.5)/V(2.69 > z > 3.47) from, for example, the specific
formulae quoted in Carswell and Smith (1978), is 1.25.   For uniform
comoving space density ρ(z) therefore, one predicts from Osmer's own
data 5 x 1.25 x 1.37 = 8.6 for the expected number of QSOs with 3.7 < z
< 4.7 in Osmer's (1981b) survey.   (I have done the volume shell cor-
rections in two parts to facilitate comparison with Osmer's argument).
This new prediction of 8.6 from Osmer's own CIV data is sufficiently
close to the value of 9.6 from the Hoag-Smith data to give more confidence (i) that the expected value is, after all, close to that adopted
by Osmer and (ii) that his conclusions therefore hold, at least
qualitatively.   Furthermore, I have looked at the spatial distribution
of QSOs in the surveys and find no obvious clumping which would be
likely to make these values unrepresentative.

Taking the mean predicted value (from two largely independent samples)
of 9.1 QSOs in the range 3.7 < z < 4.7 for Osmer's (1981b) survey, the
value of zero actually obtained then almost certainly does become a
significant indication of a turn down in the co-moving space density
near z $\sim$3.5.   I believe more observations are needed to set limits on
the expected value of ρ(z) at redshifts above and below z $\sim$3.5 before a
reliable quantitative value for the rate of turn down can be given.

Having finally established a statistically significant likelihood of a
turn-down in the comoving space density of quasars near z $\sim$ 3.5, let us

examine its cosmological significance.   Sandage, Schmidt, Osmer and
others have speculated that when observing the QSOs of highest known
redshift, we may be seeing very close to the edge of the world, beyond
which galaxies will not been seen.   On the other hand, theoretical
estimates of the epoch of galaxy formation are extremely uncertain.
What makes further progress difficult is uncertainty as to whether
initial collapse of density fluctuations in the universe was adiabatic
(producing 'pancakes' which broke up into cluster-sized units from which
galaxies later formed - with the supercluster formation occurring
perhaps at z $\sim$5, and the galaxy formation occurring even more recently)
or isothermal (producing galaxies first, perhaps at redshifts z $\sim$10 or
more from which clusters and superclusters of galaxies formed later in
an heirarchical sequence.)   Isothermal (dissipational) collapse is most
often discussed probably because adiabatic collapse (or for that matter,
neutrino-dominated universes) are more difficult to model (see, eg, Gott
1973, 1975; Rees and Ostriker 1977; Binney, 1977; Doroskevich, Shandarin
and Saar 1978; Rees 1978, 1982; Peebles 1980; Rowan-Robinson 1982;
McCrea 1982; Fall 1982, and the discussion of galaxy clustering analyses
in Shanks 1982 and references therein.)

The work on active galaxy nuclei (eg, Sandage 1972b, 1973; Weedman
1976; Wilson and Penston 1979; Phillips 1981) and the direct imaging of
low-redshift quasars (see, e.g., discussion and references in Hutchings
et al 1981, and in Wyckoff, Wehinger and Gehren 1981) suggests that many
galaxies may have had at least some degree of quasar-like activity in
their nuclei.   In this sense, the discovery of large numbers of high-
redshift quasars sets a lower limit to the age of the associated galaxies.
Thus it is possible that useful constraints may emerge from observations
of the highest redshift quasars and their clustering properties.

2.   THE OBSERVED DISTRIBUTION OF QSOS.

2.1 Are QSOs scattered in space independently at random?

Webster (1976) has examined a number of radio-source catalogues, using
power-spectrum analysis techniques, and finds no evidence for large-
scale clustering.   Oort, Arp and de Ruiter (1981) find 12 pairs of
neighbouring quasars from a number of different samples which, they
state, have a high probability of being physically associated.   Optical
surveys yield the largest surface densities of QSOs on the sky, so in
spite of their many disadvantages (see, eg, the reviews by Smith 1978,
1981), increasing efforts are being made to search for clustering in
such surveys.   Osmer (1981a), using a variety of techniques, finds no
evidence for clustering or other deviations from randomness on scales of
order 50h^{-1} to 1500h^{-1} Mpc (in present epoch coordinates) in the Hoag-
Smith (4-metre telescope plus grism) and Osmer-Smith (1980)(Curtis-
Schmidt objective-prism) slitless surveys at Cerro Tololo.   Note that h
is the Hubble constant expressed in units of 100 km s^{-1} Mpc^{-1}.   Webster
(1982) has independently analysed the Curtis Schmidt data, and, even
though he attempts to enhance the contrast of any potential clustering

through searches of various subsamples, he finds no evidence for
clustering, apart from one group of four QSOs near redshift z = 0.37.
Both Osmer and Webster conclude that there is not enough data yet to
evaluate the significance of any clustering on scales less than $50h^{-1}$
Mpc (At z=2, 1 degree on the sky corresponds to $70h^{-1}$ Mpc in present
epoch coordinates, and is thus equivalent to   z = 0.07).   In order to
improve the signal-to-noise ratio, Osmer (personal communication) has
now obtained "a good set of new grating prism fields taken on a strip of
sky in the Northern Galactic Hemisphere that was chosen at random and
has been unstudied so far......The fields are connected and in fact,
slightly overlap", so Osmer hopes to be able to calibrate the plate-to-
plate variations independently of the quasar results.   Webster (1982)
points out that deeper surveys are likely to show up quasar clusters,
even at high redshift;   furthermore, the lack of definitive observational
evidence for quasar clustering may not be taken as evidence against the
occurrence of quasars in superclusters.

Arp and Hazard (1980) have recently reported the existence of several
triplets of QSOs, so well aligned that the line joining the outermost
QSOs passes through the seeing disc of the central object.   If these
apparent associations were found to be real, then, in view of their
disparate redshifts, there would have to be a large, non-cosmological
component to at least some of the redshifts.   Edmunds and George (1981)
have carried out tests with a series of randomly generated points using
a mean surface density of six QSO deg^{-2} and conclude that triple alignments of the kind found by Arp and Hazard, are not unexpected.   Trew et
al (1982) have looked at real data, as well as making simulations, in
order to assess the significance of these alignments.   They have
discovered a number of new alignments and include a photograph of a
region of sky near the South Galactic pole, containing four alignments.
They find that [Number (N) of alignments detected in the real quasar
sample - Mean Number (M) of alignments produced by the simulated control
fields of random points] reaches ∿3.6 times the standard deviation of a
random field in some cases.   However, any clustering present in the
data will increase the number of alignments that can be detected.   They
"do not suggest that the observed two point correlation functions
necessarily imply real clustering in space of the quasars;   it is
equally if not more probable that the observed clustering is caused, for
whatever reason, by an uneven detection rate".   [This illustrates one
of the basic difficulties one faces with optical data;   another example
in a different context is given in the paper by Shanks (1982) where, in
order to get around the effects of galactic absorption, one has to
perform the co-variance analysis to different apparent magnitude limits].
Trew et al (1982) conclude that there is no strong evidence for an
excess of aligned quasar triplets over that expected in a random, or
reasonably clustered distribution of points, but that this statement
depends crucially upon the degree of clustering present in the data.

2.2   QSO counts to faint apparent magnitudes: evidence of luminosity
      evolution.

Koo and Kron (1982) in a preprint entitled "QSO counts : a complete
survey of stellar objects to B=23" report on a sample of stellar-appearing
objects at high galactic latitude which lie outside the (small) volume
of U-J, J-F, F-N space (photographic, not infrared, notation) occupied
by common stars. They find that these candidate extra-galactic objects
have colours similar to brighter QSOs, and there is no evidence in their
data for a class of very distant, compact (primeval) galaxies which
radiate predominantly by starlight.   Their Figure 7 compares A(m), the
number of QSO candidates per deg^2 per one magnitude interval, against
the luminosity evolution model of Braccesi et at (1980);   the fit of
model to data is good, although the lack of redshift information prohibits
a complete test of the model.   The "QSO" counts flatten for J > 21 as
indeed had been expected from less definitive star counts coupled with
models of the Galactic halo (see also Peterson et al 1979;   Tyson and
Jarvis 1979;   Bohuski and Weedman 1979;   Shanks, Phillips and Fong 1980;
Vaucher and Weedman 1980;   Bahcall and Soniera 1980;   Gilmore 1981;
Tyson, 1981;   Tritton and Morton 1982).   Kembhavi and Fabian (1981)
have represented the integral Braccesi-Bònoli counts (see, e.g., Bònoli
et al 1980) by log N($<m_B$) =0.86 (m_B - 18.33) for $m_B \leq$ 19 and by log
N ($<m_B$) = 0.36 (m_B - 17.40) for m_B > 19.   From these, and similar
counts, Cheney and Rowan-Robinson (1981) have concluded that some degree
of luminosity evolution seems inescapable in view of the small magnitude
interval in which the number counts flatten out.   Segal, Loncaric and
Segal (1980) show that while this may be true for standard Friedmann
models, the chronometric cosmology (Segal 1976) has the potential for
providing a non-evolutionary solution to this problem;   however, they
predict an infinite value of $\partial log N/\partial m$ at finite values of m, beyond
which N is constant.   If Koo and Kron's (1982) new data is considered a
true reflection of the behaviour of QSOs at faint magnitudes, then
$\partial log N/\partial m$ does not seem to become infinite.   Tyson (1981) offers an
alternative speculation - involving gravitational lensing of Seyfert
galaxies and changes in the K-correction for Seyferts at faint magnitudes
- in order to explain the flattening of the counts.

Because positive identifications and redshifts of QSOs have not been
published fainter than B $\sim$21.8 (Richer and Olson 1980), better observational data is needed to resolve some of the above problems in a
reliable manner (e.g., by confirming the number and nature of faint
candidate objects and by measuring their redshifts).   What is needed is
some spectroscopic method for observing objects an order of magnitude
fainter.

2.3   QSO spectroscopy to faint apparent magnitudes.

Telescope time has been assigned on the Anglo-Australian Telescope (AAT)
to attempt to detect QSOs and Seyfert galaxies to apparent magnitudes
$\sim$24½ and to measure their redshifts.   The proposal, by Atherton,
Smith, Taylor, Axon and Reay is initially to use the scanning/imaging
Fabry-Pérot system TAURUS (Taylor and Atherton 1980; Atherton et al
1981), a grating-prism (grism) and a focal-plane mask to search for and
confirm emission-line objects in a 400Å wavelength range down to 2-3

magnitudes below the sky background over two fields each of 9 arc
minutes diameter.

The focal reducer of TAURUS has a 9 arc-minute field (Taylor and
Atherton 1980) and converts the f/8 cassegrain field of the AAT into an
f/2 output to the Boksenberg Image Photon Counting System (IPCS,
Boksenberg 1972).   Using 512 x 512 40-micron pixels will give one-
arc-second resolution across the entire field.   The tunable filter
(Atherton and Reay 1981) will be used in the ninth order of interference,
giving a passband which varies from 40Å at 4500Å to 20Å at 4900Å. The
other orders will be blocked using a 400Å band-pass four-period filter.
Thus, the device gives a series of digital images whose wavelengths
correspond to the voltage settings of the piezo-electrically scanned
tunable filter.   Emission-line objects will stand out (in two or three
succcessive frames) at least for broad-lined objects such as QSOs and
Seyfert galaxies, as the wavelength corresponding to the strong emission
line (expected to be Lyα in most cases) is transmitted.

The sources detected from a single emission feature with the imaging
Fabry-Pérot will be further investigated using TAURUS and the Boksenberg
IPCS with a grism and a focal-plane mask to identify their nature and
redshift.   The mask will be constructed so as to let light from the
emission line sources pass through appropriately positioned two-arc-
second diameter holes, while suppressing the bulk of the night sky
radiation that would otherwise underlie the grism spectra;   final sky
subtraction will be performed from holes immediately above or below the
object holes integrating on the sky.   If this technique proves successful
(the practical problems are considerable, and the chance of zero detections
non-neglible), then one has a method for direct and systematic exploration
of the variation of the intrinsic luminosity function for emission-line
QSOs as a function of redshift;   one would then follow up with proposals
to cover wider wavelength (and redshift) ranges.   The advantages of the
technique lie in the high quantum efficiency of the IPCS, its ease of
calibration and the narrow passband (allowing better sky rejection and
deeper limiting magnitudes in practicable exposure times).   The main
disadvantage lies in the poor field coverage (0.017 deg^2 per field),
which will make it difficult to ensure that one has representative
samples.

The ability to go nearly three magnitudes fainter than any spectroscopy
so far attempted on QSOs means that (even for $q_o = 0$ and $H_o = 50 \text{km s}^{-1}$
Mpc^{-1}) an apparent magnitude of $24\frac{1}{2}$ at redshift 2, corresponds to an
absolute magnitude of $-21\frac{1}{2}$, typical of many unexceptional Seyfert type 1
galaxies. We shall thus gain the first information concerning the
existence and nature of objects of such luminosities at these distant
epochs.

3.   ABSORBING MATERIAL IN THE LINE OF SIGHT TO QSOS.

Although still a subject of controversy, it seems that absorption features

in the spectra of quasi-stellar objects are mainly produced by material
not actually associated with the objects themselves (see, e.g., the
review by Weymann, Carswell and Smith 1981).   There are exceptions.
The very broad absorption features often extending over several hundred
angstroms of the observed spectrum and recently observed to possess
considerable substructure are generally thought to be associated with
events occurring in and around the QSOs themselves.   Unfortunately we
do not know, even to within several orders of magnitude, what is the
distance between the central source of ionising radiation and the broad-
lined absorbing material.   We know that every case so far exhibits a
high level of ionisation (manifested in features attributable to O^{5+},
N^{4+} and C^{3+}). Ejection velocities well in excess of 0.1c can be involved.
Somewhere between 1% and 10% of all high redshift QSOs seem likely to
show these broad absorption features.

Most of the absorption lines in QSOs are, however, very sharp.   They
are often unresolved even at resolutions of ~ 0.25Å.   It is now thought
that nearly all these sharp lines arise at cosmological distances from
the QSOs in whose spectra they are seen.   There is some controversy
concerning the frequency of occurrence of sharp-lined absorption from
clusters of galaxies in which the QSOs might be embedded (Weymann,
Carswell and Smith 1981;   Young, Sargent and Boksenberg 1982) so I shall
not consider these further.

We are thus left with two main remaining classes of absorbing material.
The first contributes to the very obvious forest of absorption lines
which sets in at wavelengths just shortwards of Lyα emission.   Key
facts about these are:
i)   Cross-correlation analyses show strong peaks at the separation of
      Lyα/Lyβ (Young et al 1979;   Chen et al 1981), thus supporting the
      original suggestion by Lynds (1971) that most of the lines in the
      forest are from neutral hydrogen.
(ii) No clumping in the distribution of these lines has been detected on
      velocity scales in the range 300-30,000 km s^{-1}, thus supporting the
      idea that the absorbers are unlikely to be concentrated in clusters
      of galaxies, i.e., they exist in intergalactic space (Sargent et al
      1980).
(iii) From profile fitting to data with ~ 0.25A resolution, HI column
      densities generally $<10^{14}$ cm^{-2} and Doppler b- parameter in the range
      20-40 km s^{-1} are found for these intergalactic clouds of hydrogen.
(iv) No obvious bunching is found around redshift values $z_{abs} \sim z_{em}$.
(v)   The data is consistent with a uniform or slightly evolving co-
      moving space density of such absorbers (Sargent et al 1980;   Phillips
      and Ellis 1981;   Carswell et al 1982).   A uniform co-moving
      distribution of such absorbers yields for q_o =0, a number density of
      lines per unit redshift, N(z) = (17.7±1.3) (1 + z) for a rest
      equivalent width above 0.32Å.

The other main class of sharp-lined absorbing material produces, over a
wide range of redshift, "metal" lines (e.g., of C, Si, N, O, Mg, etc) in
addition to the hydrogen lines.   The observed frequency of occurrence

of such systems (Burbidge et al 1977; Sargent et al 1979; Weymann
et al 1979) suggests effective cross-sectional radii of $\gtrsim$50-100 Kpc.
The similarity between these spectra and those of our Galactic halo
(e.g., Savage and Jeske 1981; Songaila, Cowie, and York 1981), such as
mixed ion absorption, an absence of excited fine-structure levels,
similar column densities and similar multi-component line profiles, are
all consistent with the absorption being produced in large, low-density
haloes of normal galaxies at great (i.e., cosmological) distances from
the QSO. Similar absorption in haloes of galaxies other than our own has
been found directly in three published cases - see Boksenberg and Sargent
1978; Boksenberg et al 1980; Blades, Hunstead and Murdoch 1981). The
facts about these "metal-lined" systems are:

(a) The physical conditions are not well defined, as spectral resolution
     has generally been insufficient to sort out blends. Nevertheless,
     the many analyses which have been published point towards 'metal'
     column densities in the range 10^{13} to 10^{16} cm^{-2} (i.e. total hydrogen
     column densities $\sim 10^{17}$ to 10^{21} cm^{-2}) and velocity dispersions in the
     range 20-100km s^{-1}.

(b) The absence of the excited fine structure line CII*1335.7 and the
     strong presence of CII 1334.5 allow tight upper limits $\leq$1cm^{-3} to
     be set on the electron density.

(c) The ionisation equilibrium in the absorbing material seems more
     suggestive of photoionisation rather than collisions (see, e.g.
     Arons 1972; Röser 1975; Sargent et al 1979).

(d) The metal-lined systems, unlike the hydrogen-only systems, appear
     to be clumped. Sargent et al find a peak in the 2-point correlation
     function at $\sim$150km s^{-1}, though observations reveal correlations on
     scales ranging from 40km s^{-1} to thousands of km s^{-1}. Young et al
     (1982) suggest that "the fine slittings frequently observed in CIV
     doublets at high spectral resolution are produced by galaxy clustering
     rather than by multiple clouds in single galaxies".

Thus, most of the material producing sharp absorption lines in QSO
spectra is located at very large distances from the QSO. The metal-lined
systems are most likely produced in extensive haloes of galaxies while
the "hydrogen only" systems are produced in inter-galactic clouds of low
column density. In an uncertain number of cases, (Weymann, Carswell
and Smith 1981; Young, Sargent and Boksenberg 1982) metal-lined systems
may also be produced in galaxies physically associated with the QSO
while some of the ejected broad-absorption-line systems do seem to break
up into sharp-lined components.

The fact that intervening material does not completely blot out the
spectrum shortwards of the Lyα emission, (even in OQ172 and 1159 + 123,
the new object discovered by Hazard) means that hydrogen in the universe
had been re-ionised (following recombination at much earlier epochs)
prior to the epoch corresponding to z=3.5. As discussed by Sciama (1981,
1982), it remains to be seen whether or not this ionisation was caused
by the quasars themselves or whether some other source, such as heavy-
neutrino decay, could have been responsible.

REFERENCES

Arons, J.: 1972, Ap.J., 172, p.553.
Arp, H.C., and Hazard, C: 1980. Ap.J., 240, p.726.
Atherton, P.D., and Reay, N.K.: 1981, preprint "A Narrow-Gap
    Servo-Controlled Tunable Fabry-Pérot Filter for Astronomy".
Atherton, P.D., Taylor, K., Pike, C.D., and Hook, R.N.: 1981,
    preprint - "Taurus: A Wide-Field Imaging Fabry-Perot Spectrometer
    for Astronomy".
Bahcall, J.N., and Soniera, R.M.: 1980, Ap.J. (Suppl.), 44, p.73.
Binney, J.: 1977, Ap.J., 215, p.483.
Blades, J.C., Hunstead, R.W., and Murdoch, H.S.: 1981, M.N.R.A.S.,
    194, p.669.
Blanco, V.M.: 1974, P.A.S.P., 86, p.41.
Bohuski, T.J., and Weedman, D.W.: 1979, Ap.J., 231, p.653.
Boksenberg, A.: 1972, Proc. ESO/CERN Conf., Auxiliary Instrumentation
    for Large Telescopes, Geneva, p.295.
Boksenberg, A., Danziger, I.J., Fosbury, R.A.E., and Goss, W.M.: 1980,
    Ap.J. (Letters), 242, L145.
Boksenberg, A., and Sargent, W.L.W.: 1978, Ap.J., 220, p.42.
Bònoli, F., Braccesi, A., Marano, B., Merighi, R., and Zitelli, V.:
    1980, Astron. and Ap., 90, L10.
Braccesi, A., Zitelli, V., Bònoli, F., and Formiggini, L.: 1980,
    Astron. Astrophys., 85, p.80.
Burbidge, G., O'Dell, S.L., Roberts, D.H., and Smith, H.E.: 1977,
    Ap.J., 218, p.33.
Carswell, R.F., and Smith, M.G.: 1978, M.N.R.A.S., 185, p.381.
Carswell, R.F., and Strittmatter, P.A.: 1973, Nature, 242, p.394.
Carswell, R.F., Whelan, J.A.J., Smith, M.G., Boksenberg, A., and
    Tytler, D.R.: 1981, M.N.R.A.S., in press, "Observations of the
    Spectra of Q0122-380 and Q1101-264".
Chen, J.-S., Morton, D.C., Peterson, B.A., Wright, A.E., and Jauncey,
    D.L.: 1981, M.N.R.A.S., in press.
Cheney, J.E., and Rowan-Robinson, M.: 1981, M.N.R.A.S., 195, p.497.
Clowes, R.G.: 1981, M.N.R.A.S., in press.
Doroskevich, A.G., Shandarin, S.F., and Saar, E.: 1978, M.N.R.A.S.,
    184, p.643.
Edmunds, M. and George, G.H.: 1981, Nature, 290, p.481.
Fall, S.M.: 1982, paper presented at this conference.
Gilmore, G.: 1981, M.N.R.A.S., 195, p.183.
Gott, J.R.: 1973, Ap.J., 186, p.481.
Gott, J.R.: 1975, Ap.J., 201, p.296.
Hazard, C. 1981, "Survey and Rare-Object Searches", paper presented at
    the Workshop on Automated Photographic Analysis, Cambridge, U.K.
Hoag, A.A.: 1972, A.A.S. Photo. Bull. No. 2 (Issue 6), p.12.
Hoag, A.A.: 1976, P.A.S.P., 88, p.844.
Hoag, A.A., Burbidge, E.M. and Smith, H.E.: 1977, "L'Evolution des
    Galaxies et ses Implications Cosmologiques", Colloque IAU No. 37,
    Colloque CNRS 263, Paris, Sept. 1976, p.521.
Hoag, A.A., and Schroeder, D.J.: 1970, P.A.S.P., 82, p.1141.
Hoag, A.A. and Smith, M.G.: 1977, Ap.J., 217, p.362.

Hoag, A.A., Smith, M.G., Burbidge, E.M., Smith, H.E,. and Sandage, A.R.:
    1977, Bull. A.A.S., 9, p.308.
Hutchings, J.B., Crampton, D., Campbell, B., and Pritchet, C.:   1981,
    Ap.J., 247, p.743.
Jauncey, D.L., Wright, A.E., Peterson, B.A., and Condon, J.J.:   1978a,
    Ap.J. (Letters), 219, L1.
Jauncey, D.L., Wright, A.E., Peterson, B.A., and Condon, J.J.:   1978b,
    Ap.J. (Letters), 223, L1.
Kembhavi, A.K., and Fabian, A.C.:   1981, preprint, "X-ray Quasars
    and the X-ray background".
Koo, D.C., and Kron, R.G.:   1980, P.A.S.P., 92, p.537.
Koo, D.C., and Kron, R.G.:   1982, preprint "QSO counts: A Complete
    Survey of Stellar Objects to B = 23 mag".
Lewis, D.W., MacAlpine, G.M., and Weedman, D.W.:   1979, Ap.J., 233, p.787.
Lynds, C.R.:   1971, Ap.J. (Letters), 164, L73.
Lynds, C.R., and Wills, D.:   1970, Nature, 226, p.532.
MacAlpine, G.M., and Lewis, D.W.:   1978, Ap.J. (Suppl.), 36, p.587.
MacAlpine, G.M., Smith, S.B., and Lewis, D.W.:   1977, Ap.J. (Suppl.),
    35, p.197.
Markarian, B.E.:   1967, Astrofizika, 3, p.55.
McCrea, W.H.:   1982, paper presented at this conference.
Oort, J.H., Arp, H., and de Ruiter, H.:   1981, Astron. and Ap., 95, p.7.
Osmer, P.S.:   1977, Ap.J. (Letters), 218, L89.
Osmer, P.S.:   1980, Ap.J. (Suppl.), 42, p.523.
Osmer, P.S.:   1981a, Ap.J., 247, p.762.
Osmer, P.S.:   1981b, Ap.J., in press, "Evidence for a Decrease in the
    Space Density of Quasars at z ≥ 3.5".
Osmer, P.S., and Smith, M.G.:   1977, Ap.J.(Letters), 21, L73.
Osmer, P.S., and Smith, M.G.:   1980, Ap.J. (Suppl.), 42, p.333.
Peebles, P.J.E.:   1980, "The Large Scale Structure of the Universe",
    Princeton University Press, p.389.
Peterson, B.A., Ellis, R.S., Kibblewhite, E.J., Bridgeland, M.T.,
    Hooley, T., and Horne, D.:   1979, Ap.J. (Letters), 233, L109.
Peterson, B.A., Jauncey, D.L., Wright, A.E., and Condon, J.J.:   1978,
    Ap.J. (Letters), 222, L81.
Phillips, M.M.:   1981, Proc. Herstmonceux Conference "Active Galaxies",
    Observatory, in press.
Phillips, S., and Ellis, R.S.:   1981, preprint, "The Distribution of
    Absorption Lines in QSO Spectra".
Rees, M.:   1978, in "Observational Cosmology", Proc. Saas Fe School
    (Geneva Observatory Publications) p.259.
Rees, M.J.:   1982, paper presented at this conference.
Rees, M.J., and Ostriker, J.P.:   1977, M.N.R.A.S., 179, p.541.
Richer, H.B., and Olsen, B.I.:   1980, P.A.S.P., 92, p.573.
Roser, H.-J.:   1975, Astron. and Ap. 45, p.329.
Rowan-Robinson, M.:   1982, paper presented at this conference.
Sandage, A.R.:   1972a, Q.J.R.A.S., 13, p.282.
Sandage, A.R.:   1972b, Ap.J., 178, p.25.
Sandage, A.R.:   1973, Ap.J., 180, p.685.
Sargent, W.L.W., Young, P.J., Boksenberg, A., Carswell, R.F., and
    Whelan, J.A.J.:   1979, Ap.J., 230, p.49.

Sargent, W.L.W., Young, P.J., Boksenberg, A., and Tytler, D.: 1980,
    Ap.J. (Suppl.), 42, p.41.
Savage, B.D., and Jeske, N.A.: 1981, Ap.J., 244, p.768.
Schmidt, M.: 1970, Ap.J., 162, p.371.
Sciama, D.W.: 1981, "Massive Neutrino Decay and the Ionisation of the
    Intergalactic Medium" - in preparation.
Sciama, D.W.: 1982, paper presented at this conference.
Segal, I.E.: 1976 "Mathematical Cosmology and Extragalactic Astronomy"
    (New York: Academic Press).
Segal, I.E., Loncaric, J., and Segal, W.: 1980, Ap.J., 238, p.38.
Shanks, T.: 1981, Observatory, in press (Proc. R.A.S. Specialist
    Discussion -"Quasars").
Shanks, T.: 1982, Paper presented at this conference.
Shanks, T., Phillips, S., and Fong, R.: 1980, M.N.R.A.S., 191, 47p.
Smith, M.G.: 1975, Ap.J., 202, p.591
Smith, M.G.: 1976, Ap.J. (Letters), 206, L125.
Smith, M.G.: 1978, Vistas in Astronomy, 22, p.321.
Smith, M.G.: 1981, in "Investigating the Universe", ed. F.D. Kahn
    (Reidel: Holland), p.151.
Smith, M.G., Aguirre, C., and Zemelman, M.: 1976, Ap.J. (Suppl),
    32, p.217.
Smith, M.G., Boksenberg, A., Carswell, R.F., and Whelan, J.A.J.: 1977,
    M.N.R.A.S., 1981, 67P.
Songaila, A., Cowie, L.L., and York, D.G.: 1981, Ap.J., 248, p.956.
Taylor, K., and Atherton, P.D.: 1980, M.N.R.A.S., 191, p.675.
Terlevich, R., and Melnick, J.: 1981, M.N.R.A.S., 195, p.839.
Thuan, T.X., and Gott, J.R.: 1975, Ap.J., 204, p.649.
Tritton, K.P. and Morton, D.C.: 1981, Proc. 2nd IAU Asian-Pacific
    Regional Meeting, Bandung, Indonesia, in press "Galactic Structure
    from a faint-object survey in a field in Aquarius".
Trew, A.S., Clube, S.V.M., Savage, A., and Clowes, R.G.: 1982,
    M.N.R.A.S., submitted, "An Assessment of the Significance of Quasar
    Alignments".
Tyson, J.A.: 1981, Ap.J. (Letters), 248, L89.
Tyson, J.A., and Jarvis, J.F.: 1979, Ap.J. (Letters), 230, L153.
Vaucher, B.G., and Weedman, D.W.: 1980, Ap.J., 240, p.10.
Wall, J.V.: 1977, I.A.U. Symposium No. 74, "Radio Astronomy and
    Cosmology", ed. D.L. Jauncey, Reidel, Holland, p.55.
Wampler, E.J., Robinson, L.B., Baldwin, J.A. and Burbidge, E.M.: 1973,
    Nature, 243, p.336.
Webster, A.: 1976, M.N.R.A.S., 175, p.61.
Webster, A.: 1982, M.N.R.A.S., in press ("The Clustering of Quasars
    From an Objective-Prism Survey").
Weedman, D.W.: 1976, Ap.J., 208, p.30.
Weymann, R.J., Carswell, R.F., and Smith, M.G.: 1981, Ann. Rev. Astr. Ap.,
    19, p.41.
Weymann, R.J., Williams, R.E., Peterson, B.M., and Turnshek, D.A.: 1979,
    Ap.J., 234, p.33.
Wilson, A.S., and Penston, M.V.: 1979, Ap.J., 232, p.389.
Wright, A.E., Peterson, B.A., Jauncey, D.L. and Condon, J.J.: 1978,
    Ap.J. (Letters), 226, L57.

Wyckoff, S., Wehinger, P.A., and Gehren, T.:   1981, Ap.J. 247, p.750.
Young, P.J., Sargent, W.L.W., Boksenberg, A., Carswell, R.F., and
    Whelan, J.A.J.:   1979, Ap.J., 229, p.891.
Young, P., Sargent, W.L.W., and Boksenberg, A.:   1982, Ap.J. (Suppl.),
    in press "CIV Absorption in an Unbiased Sample of 33 QSOs:
    Evidence for the Intervening Galaxy Hypothesis."

THE AGE PROBLEM

A.C. Edwards
Department of Astrophysics
Oxford University
South Parks Road, Oxford OX1 3RQ

"The fairy tales of science, and the long result of Time"
                                    from Locksley Hall by A. Tennyson.

The Age Problem may be stated as follows:
a) Does the Universe have an age?
b) If so, is it possible to estimate that age?
c) How does one go about making such estimates in practice?
d) Have we any reason to believe that the various estimates made to
   date present us with any conflicts?

Like nearly all philosophical questions, question a) is plagued
by assumptions. One generally avoids answering the question entirely,
as is done here, but, instead, responds to it by taking the pragmatic
position that the answer is "yes" and to lay down ground rules under
which subsequent questions can be interpreted and discussed. At the
end of the inquiry any conflicts arising on reaching question d) – or
elsewhere on the road to d) – imply returning to the ground rules,
changing them, and iterating to consistency. However, it is important
to remark that there may be several sets of consistent ground rules.

The discussion here is to be founded on ground rule zero that the
universe is simply as we see it and may be interpreted as simply as we
care to make it; namely, as a dynamically evolving homogeneous and
isotropic totality emerging from a condensed state that it occupied
some finite time ago. The goal is to gauge the age, as best one can,
to that (last?) epoch of maximum compression. The "last" is a tacit
admission that a closed universe, a phrase endowed with meaning in the
ground rules vocabulary, may be an oscillating one of indefinitely many
cycles of imponderable total age.

A ground rule is now required that states precisely the physical
laws that govern the dynamical universe and the behaviour of its
constituent parts. Strictly speaking we only know physics 'here and
now' whereas if we are to attempt to gauge ages up to the lifetime of
the universe we have to know physics at all 'there and then'. As our

A. W. Wolfendale (ed.), Progress in Cosmology, 291–303.

physical ground rule it is taken that physics 'here and now' is the
same as physics 'there and then', that standard general relativity
gives us the means of consistently constructing a correct cosmology, of
interpreting observations correctly and of supplying us with the
necessary equations with which to construct and evolve models of the
physical objects that populate the universe. Thus I cling grimly to the
paths of orthodoxy - but I must admit in passing to being impressed by
the diligence shown by the developers of Scale Covariant Cosmology in
their exertions with that theory. Their approach seems to lead at
present to an as equally consistent a set of ground rules as does the
orthodox (Canuto et al., 1981 and references therein).

AGE IMPLIES CHANGE

      If we are to gauge an age for the universe, and give an
affirmative to question b), then we have to look at it and its
constituents for causes and effects whose relative "time constants"
are of suitable magnitude; not too short - for that only tells us of
recent and specific behaviour, nor too long - for then we have the
problem that things would have changed too little to be usefully or
accurately measurable.

      The first and most obvious thing to look at is the universe
itself - or at least at its global dynamical aspects as imprinted on
the galaxies etc. The 'change' that one is seeking so as to obtain an
age is, of course, the change in the separation between the constituents
of the universe. The basic premise of the primordial condensed state
tells us that the original separation between constituents was as near
zero as makes no difference - but what of their separation today?
This leads directly to the treacherous ground occupied by the Hubble
constant, the deceleration parameter and the cosmological constant.
The Hubble constant must be unique in the realm of science being not
only not a constant (since it depends on cosmological epoch), but also,
today, in having two 'observed' values - namely one value around
50 km/sec/Mpc and the other around 100 km/sec/Mpc - each to within
±10% internal accuracy! These correspond respectively to timescales of
19.6 billion (10^9) years and 9.8 billion years. The general attitude
taken by the community at large seems to be that these two values limit
the range of possibility for the time-scale (although there does appear
to be more observational support for the shorter timescale). Whichever
school of classical observers turns out to be nearer the truth in the
long run, astronomers in general can only benefit by learning of the
'how, when and where' of the errors that crept in. The essence of the
current observational problem is, of course, that the means by which
the Hubble constant is being obtained is through differently constructed
calibration ladders that even start off based on different assumptions.
Rather than clambering up these ladders to the distant and receding
galaxies what one needs is a method that can leap over the distances at
will.

The Hubble constant, will not of itself yield an age but only a
timescale. For the simplest cosmologies one needs the other two
dynamically significant parameters as well in order to calculate an
age. However, the current determinations of the deceleration parameter
seem insufficiently accurate to determine anything with confidence at
present. Its proper determination may well have to wait for a better
understanding of QSO's.

The cosmological constant presents equivocal difficulties. Many
particle theories seem capable of producing "cosmological constants"
at will - generally temperature (time) dependent - and then of
brushing them under the renormalization carpet. In the strict sense,
most of these cosmological constants are not cosmological constants
at all in the general relativistic context but rather extra physics,
i.e. in Einstein's equation - $G_{ik} = T_{ik}$ - the particle physicist's
"cosmological constants" are part of T_{ik} and not of G_{ik} as would be
required by the cosmologist. Nevertheless, even if it were Einstein's
"greatest mistake", the cosmological constant can no longer be
considered as a dirty word and the problem of its evaluation must be
addressed carefully - even though it may turn out to be totally
insignificant. An alternative and comforting solution would be the
discovery of a proof that the cosmologists cosmological constant has
no meaning and no existence physically, i.e. that it really is a 'great
mistake".

The verdict must surely be that whereas a study of the dynamics
of the universe will one day yield an accurate age estimate for the
universe - it is not doing so today. At best we are only getting a
very uncertain timescale.

Turning to the menagerie of the universe, one is able to discover
a few types of object and phenomena that can apparently help us in our
quest. However, as we shall see, in each case one has to introduce a
physical model to interpret the observations and it is generally this
aspect that prevents rapid progress. The phenomena can help in one of
two ways; either as specific distance indicators that can leap over
the previously mentioned extragalactic calibration ladder and enable
individual estimates of the Hubble constant and timescale to be made;
or else as specific age indicators yielding lower limits to the age
of the universe. An example of the first class is the occurrence of a
supernova (of either type) in a distant galaxy; and of the second are
the long lived radio-nuclides and the globular clusters. In each case
the requisite observations for making timescale/age estimates can be
made with adequate precision. The problems arise when one has to
accept a lot of dubious physics in the process of interpreting the
observations. I shall take each quoted example in turn but, of the
supernovae types, I will only consider type II for the sole reason that
one has a greater (misplaced?) trust in one's knowledge of the
composition (hydrogen rich) of the emitting layers.

Strictly, the principle of "age implies change" can only be applied to the radio-nuclides and the globular clusters. In the case of the radio-nuclides, when one asks what it is that changes it is, of course, the abundance ratio of two nuclides: but the question one really has to ask is - what has the ratio changed from? And, what processes have affected that change? One has to respond to these two points before one can reverse the process and from the implied change infer the age. For the globular clusters one is in a much better position. Although no one has consciously seen the birth of a globular cluster yet one would consider it a strong assumption that a uniformly populated co-eval main sequence would have existed and it is that that has been the thing to change through the process of stellar evolution - thus here one is asking how far forward does one have to evolve something simple before it is recognizable as the complex of measurement that is the globular cluster of the current epoch.

However, one has to admit that neither the radio-nuclide nor globular cluster ages 'age' anything but themselves and only place lower limits on the universe. Within the ground rules and certainly from the conservative viewpoint the lower limit is generally considered to be close enough. The final caveat is that the 'age' is only that of our own part of the universe - we know of no way of aging other parts of it to any reliable degree. Any other method of gauging ages or timescales based on sound physical principles are to be welcomed. But it is not good enough to develop matters just to the point where everything 'agrees' within the error bars; each and every method must ultimately be pushed to its limits - and if ages then disagree then so be it: that is the universe.

"There's many a slip 'twixt cup and lip"   A proverb.

The steps one takes in determining the distance of an extra-galactic SUPERNOVA (SN) and hence the distance of its parent galaxy (and an estimate of the Hubble constant by dividing the galaxies recession velocity by the distance) are as follows:-

1. Catch a distant extra-galactic SN in the act of exploding as soon as possible, i.e. early on the rise.

2. Over several weeks take a representative and sensibly spaced set of high resolution calibrated spectra and gather intermediate/broad band photometry over as wide a waveband as possible. Correct as best one can for reddening etc.

3. Using theory, interpret and fit the spectra and photometry in terms of an atmospheric model constructed from a picked chemical composition and physical parameter profiles specified throughout the atmospheric layers i.e. velocity v. radius etc.

It is step 3. at which the critical approximations arise and that

determines the current state of the art, Branch et al., 1981. At its
simplest, the interpretational approach so far followed for SNII is to
assume that the many chemically identifiable 'P-Cygni' type lines so
typical of the SNII spectrum are formed by resonant scattering in an
homologously expanding atmosphere in which the line source function is
given by a diluted black-body radiation field radiating from a
photospheric layer deeper down. The time-dependent black-body
temperature, $T(t)$, is determined by a fit to the measured and slowly
changing continuum over as broad a waveband as possible. By fitting
model line profiles to the observations one can determine the velocity
of the atmospheric layers and, through the model, the velocity of the
photospheric layer. Given sufficient time and data, especially an
accurate knowledge of the epoch of the outburst, one can then estimate
the photospheric radius $R(t)$ as the integral over time of the the the
inferred photospheric velocity, i.e. at late times the initial radius
may be ignored. The distance of the SN is then given by
$D^2 = R^2(t)\sigma T^4(t)/F_{OBS}(t)$ where $F_{OBS}(t)$ is the received bolometric flux.
Internal consistency requires that D be independent of time. If it is
not, then according to the interpretation it means that the photospheric
level has begun to migrate (inwards) through the expanding SN mass
distribution and one's modelling must be refined.

The first problem with the whole procedure is connected with
understanding the formation of the original photospheric flux
distribution which is to act as the diluted source function for the
overlying line forming atmosphere - which is the second problem. The
pre-SNII envelope and certainly the post-outburst envelopes are
attenuated and scattering and sphericity play important roles in
determining the emergent flux distribution - as also does the fact that
the envelope is rapidly and differentially expanding. Typically one
expects that the continuum colour temperature determined from the current
observational windows - an optical window is generally used in the
interpretations - are not the same as the equivalent black body
temperature. Thus the temperature which goes into the distance formula
can easily be too large or small. Similarly, seductively simple line
formation models may easily lead one astray in predicting incorrect
photospheric velocities. Other possibilities to be considered are
those of NLTE, of dust formation in the atmosphere, of atmospheric
inhomogeneity and stability and of a strong coupling between the
radiation and the hydrodynamics.

However, the current state is that the derived Hubble constants
- one for each SN once the recession velocity of the parent galaxy is
corrected for its (classically determined!!) local flow - are averaging
out at 60 ± 10 km/sec/Mpc, i.e. a timescale of 16.5 billion years. The
most important thing to note is that, despite the tedium of waiting for
suitable SNe, the observational weaponary needed to acquire the data is
available today. It is only the physical understanding of radiative
transfer and of line formation in differentially expanding NLTE
envelopes and atmospheres that is preventing a confident acceptance of
the result. If any field of astrophysics has huge potential it is this
one.

NUCLEOCOSMOCHRONOLOGY and the nucleocosmochronological timescale, to which we now turn, have one great failing in common and that is the ridiculous length of the words themselves. They shall be referred to here as NC and NCT.

The ingredients required to determine the NCT, or rather several NCTs, are as follows:-

1. The relative abundances of

$$^{232}Th/^{238}U; \quad ^{235}U/^{238}U; \quad ^{187}Re/^{186}Os; \quad ^{187}Os/^{186}Os$$

as observed today in the solar system, i.e. meteorites, earth and moon.

2. The value of the radio-active decay constants of the long-lived radio-nuclides ^{187}Re, ^{232}Th, ^{235}U, ^{238}U.

3. A physical theory to predict the production ratio of the actinides and of $^{187}Os/^{186}Os$.

4. An astrophysical understanding of the overall rate at which the elements have been made in stars (presumably) and recycled through the interstellar medium.

Once again the 'observational' aspects are sufficiently well known and, taking error bars in to account, have held up very well for roughly twenty years; Fowler and Hoyle, 1960, Symbalisty and Schramm, 1981. At worst, there may still be some minor question over the $^{232}Th/^{238}U$ ratio due to chemical fractionation effects during the solidification of the solar system. Most people would accept that, of the physical problems, it is the question of the overall production rate of the nuclides through cosmic time that prevents the derivation of a truly useful number, i.e. we have no sensible notion yet of the chemical/population evolution of the galaxy. It is only in the case of an extreme prompt initial enrichment (PIE) epoch of production of the radio-nuclide isotopes that one can derive a number from the data that may strictly be called an age. For the rest, the interpretation of the 'age' one derives from the data is galaxy model dependent and one has to admit that the radio-nuclide data only produces a timescale. The timescale one obtains from the current actinide data gives a minimum PIE age of the nucleosynthetic galaxy of 7.2 ± 2 billion years. However, it is the other aspect of the physical problem that is of more interest here since it is remarkable that the theory of the production of the radio-nuclides has gone virtually unchanged for twenty four years. $B^2FH(1957)$, accepting the necessity of two nucleosynthetic processes involving free neutrons, used the s-process to produce the valley of stability nuclides $(A > 70)$ and the r-process to produce the isolated isotopes on the neutron rich side of the valley as well as the actinides. Picking a value for the temperature of the s-processing regions in stars one can calculate the ratio of all valley s-isotopes from their (n, γ) ground state cross sections, i.e. one can calculate ratios such as $^{187}Os(s)/^{186}Os(s)$ (which is needed for the ^{187}Re NCT). However, this

last ratio needs careful consideration since ^{187}Os has well populated
low lying states under s-process conditions and the ground state (n, γ)
cross section is not necessarily the major contributor in the stellar
environment. For the r-isotopes one has to perform several acts of
imagination to calculate their ratios; Norman and Schramm, 1979,
Klapdor et al., 1981. It is still true today that the region(s) in
stars, even which stars, in which the r-isotopes are forged are not
known, the masses of very neutron rich isotopes (or in general any of
their physical attributes) are not accurately known, (n, γ) cross
sections are not known, β-decay rates are unknown - all except insofar
as they can be calculated from highly uncertain theories, e.g. Klapdor
and Oda, 1980. (However, one need only know that the r-process occur
to be on fairly safe ground with ^{187}Re.) It is, of course, with the
production ratios of the actinides that one is concerned since it is
these that determine nearly all their NCT error. The actinide ratios
that are used, and have been used since the days of B^{2}FH, assume that
the r-process path crosses the neutron-rich fission line that lies well
beyond the common actinide region. On freeze out - the terminating
stages of any process - the fully populated actinide r-path -decays
back and shortly afterwards produces the classical ratios a la B^{2}FH.
However, the only semi-serious calculations of processes with r-like
conditions in something approaching a realistic model can run out of
neutrons before hitting the fission line; Hillebrandt et al., 1976,
Blake et al., 1981. One must also consider the realistic problem of
having not only a stellar zone where the classical r-process may be
running true to form but also neighbouring zones where r-process
conditions are not optimal and the neutrons are exhausted before
processing has proceeded far up the r-path. On averaging over zones
one may still obtain a good reproduction of valley r-isotopes,
especially the r peaks, yet may not attain the standard r-ratios for
the actinides. It is problems like these and of the physical processes
that can modify the actinide ratios in the r-freeze out (and, possibly,
in stars in which they later get incorporated and mildly re-astrated)
that make the actinides not quite as useful as one would like nor their
NCTs particularly trustworthy.

More hopefully, an NC based on ^{187}Re (Clayton, 1964) has distinct
advantages and has now been developed to an interesting stage. The
^{187}Re NC is based on the assumption that the observed ^{187}Os abundance is,
to a very good approximation, the sum of s-process ^{187}Os(s) and radiogenic
^{187}Os(r) resulting from the β-decay of ^{187}Re which is produced solely
in the r-process. The theoretical s-process production ratio of
^{187}Os(s)/^{186}Os (^{186}Os is currently believed to be essentially pure 's')
is based on the (n, γ) cross section ratio corrected for stellar
environment effects and has now been calculated - although it is
stressed by Woosley and Fowler (1979), the calculators, that the result
should be treated with caution. Subtracting the calculated ratio from
the accurately known solar system ratio one obtains the radiogenic
fraction that must have been produced, under the NC interpretation,
from the decay of ^{187}Re. Geochemically Re and Os are minimally
fractionated and so one is able to obtain very easily the ^{187}Os(r)/^{187}Re

ratio and, thence, the PIE age of 10.7±2 billion years. Assuming some galactic evolution model one can stretch that age to heart's content. To reduce the ^{187}Re age one is going to have to increase the calculated value of ^{187}Os(s)/^{186}Os, i.e. decrease the radiogenic fraction or else find models in which ^{187}Re can be destroyed (possibly along with other r-process elements) without destroying the radiogenic ^{187}Os(r), i.e. one requires a differential effect that effectively decreases the ^{187}Re decay lifetime. In the past it has been suggested that the certain decrease of the lifetime of ^{187}Re in the stellar environment would be effective in decreasing (Talbot, 1973) the ^{187}Re NCT. Perhaps a fresh look should be taken (Takahishi and Yokoi, 1981) since it is quite obvious that the ^{187}Re NC is extremely important. It is very probable that the ^{187}Re PIE age will decrease.

A general point in closing is that the nearer the age of the universe (obtained from another source say) encroaches upon the minimum PIE age then the more biased must the galactic evolution model be to initial enrichment and the more similar the NCTs derived from the radionuclides.

The GLOBULAR CLUSTERS (GCs) are the only objects for which one can calculate cosmologically useful ages as opposed to timescales and, when all is done, the statement will be made that the universe is older than the oldest cluster. As stated earlier, the age that is being gauged for a GC is the time it takes for a supposed co-eval main sequence to change into the complex array of stars in a colour magnitude diagram (CMD) that we observe today. Hand in hand with the age determination goes the specification of the primordial composition of the GC stars, i.e. the composition they had on the main sequence. That in itself is cosmologically interesting. Should it turn out that age and composition are strongly correlated then the GCs become triply important in that they may also map out the initial stages of the chemical evolution of the galaxy. Therefore they are worth the investment of a lot of effort – more so than most astronomical objects.

Demarque (1980) has given a very useful synthesis of the classical GC age data as it stands today. The basic result, that the oldest and most metal poor GCs have ages circa 15±2 billion years, rests on catalogues of computed stellar evolutionary models that rely on the precepts of the '60s and early '70s. However, the point that will be made here is that with existing knowledge of the 'updates' that have to be incorporated into the next generation of models, one can have confidence that the age estimates are bound to decrease. In that same paper, Demarque described the procedures one has to go through to obtain the necessary data – photometric and spectroscopic – so that one can estimate an age for a GC and, rather than repeat the message, we refer the reader to that paper.

The first step one takes in interpreting the data on a given GC is to determine a set of abundances. The GC metal abundance is deduced by

using the spectra – most probably only of bright cluster giants – and/or
metallicity colour indices (which are themselves only useful if such
indices have been calibrated against reduced spectra of other stars, i.e.
nearby field subdwarfs etc. This raises the question of whether field
pop II≡cluster pop II). To show how uncertain all this can be, the
metallicity scale has undergone a major revision (downwards) in recent
years (Cohen, 1980, Pilachowski et al., 1980) – but only at the metal
rich end of the scale. One senses some confidence for the metal poor
end of the scale but shifts of factors of two would probably not cause
disquiet. What has become clear though (Kraft, 1979) is that the
metallicity scale can only be used for the heavier elements, i.e. iron
peak. It used to be taken as a rule of thumb that C, N and O (and all
intermediate mass elements) followed Fe in proportion; however, this
is no longer the case and is the basis of the observational turmoil of
the '70s. The mounting observational evidence that star to star CN, CO,
CH, O and N variations can be spectrally detected in nearly all parts
of the CMDs of GCs that have been adequately surveyed leaves one gasping
with admiration at how sneaky GCs really are. One is left with the
(receeding) hope that C+N+O is constant in a given cluster and that
mixing is variable from star to star. However, even if C+N+O should be
constant in a given GC, the observation that $\left[O/Fe\right]$ is apparently of
order +1 for the most metal poor field stars (Sneden et al., 1979,
Clegg et al., 1981) forces one to acknowledge that the CNO abundances
in GC stars have not been correctly represented in the construction of
models. It may well be worth thinking about repeating the multi-colour
photometry – for those GCs for which it is reasonable to expect the
better age estimates – in a band system whose extra colour indices are
sensitive to CNO abundances.

The second point concerning the metallicity revolution is that the
intermediate mass nuclei are also enriched with respect to the iron peak;
Khokhlova, 1977, Luck and Bond, 1981. (Perhaps it is time to introduce
the old Cosmic Ray terminology of Light, Medium and Heavy into GC
research?) This all means that the opacity tables etc. used in the
construction of the model catalogues have underestimated the opacity –
possibly grossly – due to the previously unknown complexity of the
abundance patterns. The effect is that a real cluster in the region of
its evolved main sequence (EMS), turn off point (TO) and sub-giant
branch (SGB) will, therefore, have been matched to a model isochrone
that, for the isochrone's age, has not been red enough. Correcting for
the metallicity in the models of the next generation will yield a
decreased fitted age for the real cluster.

Finally, to determine the helium abundance (Y) that applies to a
cluster's stars, one uses number counts of stars (hard to accurately
determine) to estimate the (helium sensitive) ratio of the number of
stars on the Upper Giant Branch (UGB) to the sum of those on the
Horizontal Branch (HB) and the Asymptotic Giant Branch (AGB). This
does not lead to an observational estimate of Y because it is the models
themselves that have predicted the dependence of the ratio on Y, i.e.
there is no confirmational calibration. For some clusters one can use

the systematics of the horizontal branch variables to determine another
Y (again inferred from a match to physical models) and, in these cases,
the pulsation and ratio Y's tend to agree - with the ratio Y being, if
anything, lower at a value of approximately 20% by mass. However this
value depends very much on how one goes about treating semi-convection
in HB stars - also one is not at all sure what influence rotation would
have if it were to be included as a necessity in the models. The recent
2-D Helium Flash calculations of Cole and Deupree, 1981, Deupree, 1981,
are also most interesting in that they suggest that the core shell flash
is not quite the orderly affair people have assumed it to be but more of
a riot. The core gets through the flash, probably without mixing through
the hydrogen burning shell, but is left with appreciably less He fuel to
burn on the HB; thus shortening its life there and requiring a higher
He abundance for the original star. The current agreement on the low Y
value may, therefore, only be a case of theory clapping theory on the
back.

    In addition, there are still many features of the CMD that have not,
as yet, received adequate attention or explanation that, apart from
revealing continuing ignorance of stellar structure and evolution, may
contain abundance or structure information. Gaps on the UGB of several
clusters with well populated GBs have been known for a long time but
many feel that they may just be statistical glitches. A gap on the GB
at approximately the luminosity level of the HB is put down to the stage
where the hydrogen burning shell hits an interior mild composition
discontinuity left behind by a prior deep penetration of the convective
envelope. This gap would be abundance sensitive and, probably, rotation
sensitive. One grossly overlooked gap that I believe exists in all GC
CMDs is to be found at the base of the lower giant branch (LGB)
(somewhere between late on the SGB and 1^m up the LGB) where, in every
adequately acquired CMD, see for examples Philip et al., 1976, Alcaino
and Liller, 1980 a,b,c, the GB appears to be ruptured. (I have split the
GB into the LGB and UGB at the HB luminosity.)   A gap phenomena that
far down the LGB has no accepted theoretical explanation at present but
could be associated with H-shell flashes, Bolton and Eggleton, 1973,
or to a vigorous disturbance if energetic ^{3}He (which is too often
neglected in Pop II studies) is mixed into deep and hot layers, i.e.
during the envelope penetration which occurs then; or, most probably,
to rotation effects; Demarque et al., 1972, Sweigart and Mengel, 1979.
It may be at this stage that the CNO chaos first appears and, possibly,
Y enrichment above and beyond that currently calculated for the deep
mixing phase.

    Nevertheless, the abundance parameters as learned from the bright
stars are accepted as those of the original main sequence stars and it
is these stars in the mass range now arriving at TO that essentially
determine the shape of the isochrone and the GC age. The accuracy of
the calculated age depends not only on the abundances assumed but also
on the input physics used in the stellar evolution programs. Of the
gravest potential danger with the theory is the fact that the solar
neutrino problem is still with us - and a doubt raised on our understanding

of stellar physics. Rather than belabour the opacity and convection
theory problems - both of which are steadily improving; Merts, 1981,
Deupree and Varner, 1980 - it is clear that many observers would like
rotation to figure more regularly in the calculations (difficult though
that is to do sensibly) as well as an abundance mix richer in CNO and
intermediate mass nuclei. One effect that most model builders have
missed and must now include is that of helium diffusion out of the
surface layers during the main sequence evolution. Noerdlinger and
Arigo state that this effect speeds the evolution and lowers the
luminosity at TO of a specific star and requires the CMDs to be
reinterpreted, the ages being lowered by 20%. (As stated earlier,
enriching in CNO etc will also lower the age). It is the fitting of
the shape of the theoretical isochrone - especially in the neighbourhood
of the SGB, TO and EMS - that, transformed into the observed CMD, fixes
the age and its error in modern work. If the transformation (e.g.
L, Te$\leftrightarrow$V, B-V) is incorrect then so is the age and there have been
uncertain suggestions that this has happened. Correcting for the
transformation errors would also lower current ages.

Thus there is every possibility that the current estimates of up to
15 billion years for the most metal poor clusters - radical sources
would even say 20 billion years - will decrease and could come down to
10 billion years with just the changes in the pipeline at the moment.
As a best buy in the 'futures'market I would suggest 11^{+3}_{-2} billion years
(to be mildly provocative).

Before leaving the GCs it is perhaps well to ask if stellar
evolution theorists are at fault for not having ready explanations, apart
from rotation, for the CNO chaos. The answer could well be that the
stellar evolutionists have done extremely well with their modelling and
that the CNO chaos is due to primordial abundance chaos. But what could
cause such primordial chaos without injecting more heavy elements, i.e.
Fe, that would certainly show up in GC giant branches, e.g. as in ω Cen?
A simple answer would be that prior to and during the formation of GCs
there had existed a large population of very to super-massive stars
which did little more than lose mass rapidly whilst very hot main
sequence stars. Once such a star is stripped down (practically
immediately) to and below its original convective core mass (that is the
requirement) it produces nothing but N with an excess of ^{4}He thrown in.
The C+N+O sum of even poorly remixed interstellar gas would not change
but one would except C/N/O variations with ^{4}He having a minor excess in
N rich pockets. The stellar core that remains to produce extra Fe may
have insignificant mass left for the metal input to be noticeable, or
else be too massive and collapse to a popular black hole. Thus one
might expect to find occasionally some very N rich pop II stars -
especially if [O/Fe] is +1 for the field. Of course, such very
massive stars would also push up the ^{4}He abundance appreciably and one
must expect net He variations between GCs of differing metallicities.
Another aspect of such mass losing luminaries is that the central
temperatures are very high for hydrogen burning stars so that the Ne-Na
and Mg-Al cycles (Barnes, 1981) might be operating. Hence Na and Al
variations are to be expected along with the CNO variations.

"There's no getting blood out of a turnip"   F. Marryat

Are there any <u>conflicts</u> between the timescales and ages? Our last
and longest question, question d), must obviously receive the answer –
'No'. The derived ages are simply not accurate enough and will not be
for sometime. The observations are good and extant technology adequate
for obtaining more data as and when required. The basic fault, if fault
it is, lies not with a lack of ideas or sparkling new insights but with
the application and input of intricate and tedious physical theories and
data into gradually improving computer programs. With regard to the
globular clusters, it may well be time for a revolution in stellar
evolution calculations in that one should aim for "realistic" 2-D
calculations that include rotation from the pre-main sequence onwards.

Within the framework of this volume "Progress in Cosmology", the
Age Problem as discussed here would be best precised as 'Slow Progress
in Computational Classical Astrophysics'.

REFERENCES

Alcaino, G., and Liller, W.: 1980a, A.J., 85, 680.
   "   "   1980b, "   ", 1330.
   "   "   1980c, "   ", 1592.
Barnes, C.A.: 1981, Prog.Part.Nuc.Phys., Vol.6, Ed. Wilkinson
    (Pergamon), pp.251-254.
Blake, J.B., Woosley, S.E., Weaver, T.A., and Schramm, D.N.: Ap.J.,
    248, 315.
Bolton, A., and Eggleton, P.: 1973, Astron.Astrophys., 24, 429.
Branch, D., Falk, S.W., McCall, M.L., Rybski, P., Vomoto, A.K., and
    Wills, B.J.: 1981, Ap.J., 244, 780.
Burbidge, E.M., Burbidge, G.R., Fowler, W.A., and Hoyle, F.: 1957,
    Rev.Mod.Phys., 29, 547. (B^{2}FH)
Canuto, V.M., and Hsieh, S.-H.: 1981, Ap.J., 248, 801.
Clayton, D.D.: 1964, Ap.J., 139, 637.
Clegg, R.E.S., Lambert, D.L., and Tomkin, J.: 1981, preprint.
Cohen, J.G.: 1980, I.A.U. Symp. 85, Ed. Hesser (Reidel), pp.385-400.
Cole, P.W., and Deupree, R.G.: 1981, Ap.J., 247, 607.
Demarque, P.: 1980, I.A.U. Symp. 85, Ed. Hesser (Reidel), pp.281-304.
Demarque, P., Mengel, J.G., and Sweigart, A.V.: 1972, Ap.J.L., 173, L27.
Deupree, R.G.: 1981, BAAS, 13, 527.
Deupree, R.G., and Varner, T.M.: 1980, Ap.J., 237, 558.
Fowler, W.A., and Hoyle, F.: 1960, Ann.Phys., 10, 280.
Hillbrandt, W., Takahashi, K., and Kodama, T.: 1976, Astron.Astrophys.,
    52, 63.
Khokhlova, V.L.: 1977, Sov.Astron.Letters, 3, 21.
Klapdor, H,V., Metzinger, J., Oda, T., Thielemann, F.K., and
    Hillebrandt, W.: 1981, Proc.4th Int.Conf. on Nuclei far from
    Stability, CERN, pp.34-350.
Klapdor, H.V., and Oda, T.: 1980, Ap.J.L., 242, L49.
Kraft, R.P.: 1979, Ann.Rev.Astron.Astrophys., 17, 309.
Luck, R.E., and Bond, H.E.: 1981, Ap.J., 244, 919.
Merts, A.L.: 1981, BAPS, 26, 815.
Noerdlinger, P.D., and Arigo, R.J.: 1980, Ap.J.L., 237, L15.
Norman, E.B., and Schramm, D.N.: 1979, Ap.J., 228, 881.
Philip, A.G.D., Cullen, M.F., and White, R.E.: 1976, Dudley Obs. Reports
    No.11.
Pilachowski, C.A., Sneden, C., and Canterna, R.: 1980, I.A.U. Symp 85,
    Ed. Hesser (Reidel), pp.467-470.
Sneden, C., Lambert, D.L., and Whitaker, R.W.: 1979, Ap.J., 234, 964.
Sweigart, A.V., and Mengel, J.G.: 1979, Ap.J., 229, 624.
Symbalisty, E.M.D., and Schramm, D.N.: 1981, Rep.Prog.Phys., 44, 293.
Takahashi, K., and Yokoi, K.: 1981, Proc. 4th Int.Conf. on Nuclei far
    from Stability, CERN, pp.341-350.
Talbot, R.J.: 1973, Ap.Sp.Sci., 20, 241.
Woosley, S.E., and Fowler, W.A.: 1979, Ap.J., 233, 411.

NEW DEVELOPMENTS IN THE THEORY OF SPIRAL GALAXIES

K.O. Thielheim
University of Kiel

About 30% of all galaxies exhibit spiral forms, 60% are elliptical and 10% irregular. It is the objective of galactic dynamics to explain these structural features. A first generation of self-consistent N-body simulations indicates that ellipticals are equilibrium configurations of gravitationally interacting multi-particle systems for which unfortunately a theory does not yet exist. Recent progress has been made on the modal analysis of Freeman disks. In a second generation of N-body simulations spiral density waves have been reproduced in disk configurations. As an alternative to the Lin-Shu conjecture based on the QSSS-hypothesis we have considered a mechanism by which spiral density waves are produced in the surrounding disk as a consequence of the slow increase of the quadrupole moment of a central oval shaped equilibrium configruation immersed in the disk.

1. STRUCTURAL FEATURES IN GALAXIES

There are essentially three types of galaxies distinguished by their morphological structure. According to the Shapley Ames Catalogue comprising one thousand of the brightest galaxies about 75% of them exhibit spiral structure, 20% are elliptical or spherical, while 5% are irregular and comprise a broad variety of forms. These numbers obviously are subject to a statistical bias since they pertain to those objects showing the largest apparent brightness. Correcting for this effect the relative numbers are somewhat different, namely about 30% spiral, 60% elliptical and 10% irregular galaxies. About two third of all bright spirals in their centre exhibit a bar type structure from the ends of which the spiral arms emerge (de Vaucouleurs 1963, 1970).

The characteristic structural features of galaxies may be demonstrated by a few examples taken from Hubble's Atlas of

A. W. Wolfendale (ed.), Progress in Cosmology, 305–321.

Galaxies (1962). Elliptical galaxies as for example NGC
3377, classified as E6, contain little gas and dust. So
what we actually see is the light coming from the stars.
This galaxy appears to have a cigar-shaped form, but there
is good reason to believe, that it actually is more like a
flat possibly triaxial ellipsoidal object, the first-sight
appearance being an effect of its obliqueness with respect
to the line of sight. Others like NGC 4111, classified as
SO2, are seen edge-on, showing the flatness in the density
distribution of stars.

One of the most beautifully developed spiral galaxies is
NGC 5364 in Virgo, classified as Sc, showing two external
arms, beginning tangent to an internal ring. These arms extend over about one and a half rotations. Inside the ring
there is a bright central region and two more not so well
defined internal arms. In general spiral structure becomes
visible through the light emitted by associations of young
bright stars belonging to population I and by the surrounding interstellar medium stimulated by stellar radiation.
The distribution of the interstellar gas responds to the
gravitational field which in turn is governed by the distribution of stars, since the latter constitute the major part
of total galactic mass. Therefore the solution to the problem of galactic spiral structure has to be searched for in
the dynamics of the stellar component rather than the gas.
It should be noted that the stars present in the central
region of galactic spirals belong to population II and thus
are similar to those observed in elliptical galaxies. Otherwise interstellar gas and dust is found in the central
region of spiral galaxies and not so in elliptical ones.
Some spirals like NGC 4565 in Coma Berenices are seen edge-
on, showing the flatness of the disk. It is therefore plausible to assume that the dynamical processes leading to the
formation of spiral arms essentially evolve in two spatial
coordinates defined in the galactic plane. NGC 1300 in
Eridanus is an example for a barred spiral galaxy, classified as SBb(s). The two arms emerge almost rectangularly
from the ends of the bar, each extending through a range of
almost one half rotation.

In what follows I will discuss a few aspects of the theoretical understanding of how these structural features come
into existence.

2. EQUILIBRIUM DISTRIBUTIONS OF GRAVITATIONALLY INTERACTING
   MULTI-PARTICLE SYSTEMS

What are the equilibrium configurations to which systems
comprising typically one hundred billion gravitationally
interacting particles may eventually settle down? Computer

experiments reveal some of what has been expected from a
still non-existing theory of such equilibrium configurations.
In self-consistent N-body simulations the equations of
motion of typically one hundred thousand gravitationally
interacting particles are integrated numerically (Miller and
Prendergast 1968; Hohl and Hockney 1969; James 1977 and
subsequent work). A sufficient amount of random velocity, or
as we may say, a minimum temperature, has to be given to the
initial distribution of stars in order to prevent local Jeans
instabilities (Toomre 1964). A typical example for a time
sequence of plane density distributions obtained in the
first generation of simulations that have been performed
till the end of the seventies is the one of Hohl (1971). He
started from an initially axial symmetric uniformly rotating
distribution with a superimposed velocity dispersion according to Toomre's criterion. The characteristic unit of time
in this case is the initial period of rotation. After no
more than about four units of time, during which typically
S-shaped short living transient phenomena occur, which
certainly have nothing to do with long living galactic
spirals, this distribution develops into an oval object,
rotating at a constant angular pattern velocity without
essentially changing its shape. The averaged surface mass
density as a function of radial distance is found to be
represented by the superposition of two exponential functions.
It has become a habit to call these oval objects bars
although they are certainly not identical with the bars seen
in barred spirals, the latter representing, as has been
said, the distributions of gas and young stars rather than
of stellar matter. Self-consistent N-body simulations thus
demonstrate that in a first stage on a comparatively short
time scale an equilibrium configuration is formed resembling
a prolate, more or less flat triaxial ellipsoid rotating
around its smallest axis. If in this process essentially all
the mass available is successfully incorporated we are left
with an elliptical galaxy.

A first step in the development of a theory of equilibrium
configurations is the finding of self-consistent mass distributions, satisfying both, the collisionless Boltzmann
equation and Poisson's equation, which combine to a non-
linear integro-differential equation. A second step is the
modal analysis of these distributions leading to dispersion
relations which deliver real and complex eigenvalues corresponding to oscillating or rotating modes as well as instabilities. Still no procedure based on fundamental physical
principles has been found to distinguish among the many self-
consistent mass distributions possible the one which is
realized by nature. Consequently the radial dependence of the
averaged surface mass density is not yet understood. Some
self-consistent axial symmetric configurations have already

been found (Lynden-Bell 1962; Kalnajs 1976). But even the
more realistic of those have so far, due to their complexity, withstood a complete modal analysis.

Therefore the problem has been approached in a more approximative way by hydrodynamical models. Instead of using the
full Boltzmann equation only the first two of its moments,
i.e. the equation of continuity and Euler's equation are
taken into account closing the higher moments by an equation
of state. It is for this reason that hydrodynamic equilibrium
configurations based on the equation of state for incom-
pressible fluids, such as associated with the names of
McLaurin, Jacobi, Dedekind, Riemann and Darvin have attracted recent attention (Chandrasekhar 1969), especially so,
since modal analysis is comparatively easy to perform for
such models (Hunter 1970, Takahara 1976; Iye 1978; Schmidt-
Kaler and Wiegand 1980). It has also been tried to substi-
tute the equation of state by conditions imposed on stellar
motion (Hunter 1970, 1979; Marochnik 1964, 1967; Kato 1968,
1970; Berman and Mark 1977, 1979). A complete modal analysis
is available for uniformly rotating stellar McLaurin disks
(Hunter 1963; Kalnajs 1971). In this context it should be
noted that the phase transitions from McLaurin spheroids to
Jacobi ellipsoids exhibit some similarity to bar instabilities in N-body simulations in so far as they are observed
at similar temperatures (Kalnajs and Athanassoula-Georgala
1974). Freeman disks which in the co-rotating frame aside
from the centrifugal potential are characterized by a two-
dimensional cut-off harmonic oscillator potential, can be
considered as the first terms of a two-dimensional Taylor
series of a more realistic potential. Therefore, and in view
of the greater richness in internal dynamics they can in
spite of their unrealistic density distribution be considered as reasonable models of stellar bars. Their modal
analysis hopefully will throw some light on the dynamics of
stellar equilibrium configurations (Schäfer and Thielheim
1981).

It has been widely discussed in literature that the formation of oval objects can be suppressed by the presence of
superimposed spherical mass distributions either in form of
a halo into which the whole system is embedded or in form of
a central bulge and that it can also be suppressed by
sufficiently high values of velocity dispersion. Ostriker
and Peebles (1973) have extracted a criterion from numerical
experiments by which the formation of the oval object can be
completely prevented. But non of these conditions seem to
hold in real galaxies so that one can always expect an oval
object to develop. Instead of considering this to be an unrealistic aspect of numerical experiments I wish to emphasize that the development of an oval object in each spiral

galaxy may be an important feature of its dynamics directly
related to the formation of spiral density patterns.

If in the first stage of development only part of the total
mass is involved in the formation of the equilibrium configuration, the latter appears to be immersed in a flat low
density stellar disk differentially rotating. This embed-
ment allows the oval object to evolve adiabatically in a
second stage on a comparatively large time scale through
interaction with the disk involving the transfer of mass,
energy and angular momentum.

3. WHICH IS THE BASIC PHYSICAL MECHANISM PRODUCING A PERSISTENT SPIRAL DENSITY PATTERN?

After the discovery of the differential rotation of our own
galaxy (Oort 1927) and the measurement of rotation curves
of other spiral galaxis (Mayall and Aller 1942) the theory
was confronted with the dilemma to explain the persistence
of spiral structure in spite of the fact that at different
distances from the galactic centre stars are rotating at
different angular velocities. The basic idea to the solution of this problem was offered by B. Lindblad (1941, 1942,
1950) suggesting that the spiral pattern in these galaxies
is essentially due to a spiral shaped density wave. The
understanding of this notion is that particles move into
and out of a spiral shaped pattern of enhanced density while
the latter is rotating stationarily a a constant angular
velocity. This fundamental conjecture has been confirmed in
a second generation of computer experiments (Berman and
Mark 1979; Sellwood 1980, 1981). In these self-consistent
N-body simulations the density distribution within the
surrounding disk has been investigated by means of Fourier
analysis with the result that a persistent trailing spiral
density wave was found rotating at a constant angular velocity corresponding to the one of the central oval object.
Still we are not yet quite satisfied with these results
since principally computer experiments do not give a direct
insight in the physical mechanisms at work when galactic
spirals are produced.

A seemingly plausible approach which has become familiar
under the name of the QSSS-hypothesis is based on the idea
that the existence of the spiral density wave is primarily
due to the mutual gravitational attraction among stars in
the galactic disk (Lin and Shu 1964). Again one starts from
Liouville's and Poisson's equation with the purpose to
search for spiral modes. This idea has given rise to a
highly sophisticated formalism which has dominated theoretical work on spiral galaxies for one and a half decade. I
will not go into the detail, instead refer to the series of

excellent review articles (Wielen 1974; Marochnik and
Suchov 1974; Schmidt-Kaler 1975; Toomre 1977). One aspect
is the necessity to have an explicit excitation mechanism
as for example by a central rotating oval perturbation of
the gravitational field (Lindblad 1941; Lin 1969; Toomre
1969; Lynden-Bell and Kalnajs 1972). Yet we have to concede
that up to now the QSSS-hypothesis has not been able to
establish a self-consistent, closed formal description of
spiral density waves observed in real galaxies as well as in
recent N-body simulations.

In view of the situation it appears to be reasonable to consider approaches alternative to the QSSS-hypothesis. In
recent work we have adopted as a working hypothesis that
the slow increase with time of the quadrupole moment of the
central rotating oval equilibrium distribution may produce
through a response mechanism a spiral shaped trailing density wave in the surrounding disk and that this mechanism
may be the primary physical cause for the generation of
spiral patterns modified but not governed by the self-
gravitation within the disk (Thielheim 1980a, b, 1981a).
The equations of motion of non-interacting test particles

$$\ddot{r} = r(\dot{\psi} + \Omega p)^2 - \partial\phi/\partial r \tag{1}$$

$$\ddot{\psi} = -2r(\dot{\psi} + \Omega p)^2/r - r^{-2}\partial\phi/\partial\psi \tag{2}$$

in polar coordinates (r,ψ) under the influence of a
gravitational potential ϕ composed of an axial symmetric
background and an oval bi-symmetric perturbation rotating
at an angular velocity Ωp, such that

$$\phi(r, -\psi) = \phi(r,\psi) \tag{3}$$

are invariant under the symmetry transformation

$$r \rightarrow r, \quad \psi \rightarrow -\psi, \quad t \rightarrow -t \tag{4}$$

In this case therefore each orbit is associated with another one mirror symmetric to the first but with the same
sense of rotation so that the density response in any case
is mirror symmetric. But this symmetry is broken if the
perturbation is non-stationary due to the increase with
time of the quadrupole moment of the oval perturbation.

We have performed N-body simulations of non-interacting
test particles using Toomre's model 1 (1963)

$$\phi_0 \propto (1 + r^2)^{-\frac{1}{2}} \tag{5}$$

as an axial symmetric background and the ansatz

$$\phi_1 \propto - \varepsilon_0 r^2 (1 + r^2)^{-5/2} g(t) \cos\{2(\psi - \Omega pt)\} \tag{6}$$

as an oval perturbation (Thielheim and Wolff 1981a, b). The
increase with time of the perturbation was described by

$$g(t) = \begin{cases} 0 & , \ t<0 \\ 1-\beta\exp(-\alpha t)/(\beta-\alpha) + \alpha\exp(-\beta t)/(\beta-\alpha), & t>0 \end{cases} \tag{7}$$

Typical parameter values are $\varepsilon_0 = 0.4$, $\alpha = 0.01$, $\beta = 0.2$,
$\Omega p = 0.135$. The pattern frequency Ωp has been chosen such
that the two inner Lindblad resonances coalesce. The other
parameters have been selected with the intention to produce
a density contrast in the response pattern which is visible
almost to the naked eye. A Gaussian random velocity dispersion
corresponding to Toomre's criterion is superimposed on the
initial distribution. A typical configuration which has developed after three rotations of the central oval perturbation when the latter has grown to approximately three
quarters of its final strength is shown in fig. 1.

A weighting factor corresponding to the mass of the test
particles is indicated by the area of the dots representing
them. The response pattern becomes obvious in a diagramme
showing the contour lines of the relative response density
excess as shown in fig. 2.

These features are not unlike those found in barred spirals
as for example NGC 1300. The density response pattern has
been further investigated by means of Fourier analysis. A
typical result is shown in fig. 3. The latter has been ob-
trained for an exponential time dependence $g(t) = \exp(-\tau/t)$
and parameter values $\varepsilon_0 = 0.2$, $\tau = 50$, $\Omega p = 0.135$.

Open trailing spirals are found emerging from the ends of
the bar near corotation and extending beyond the outer
Lindblad resonance. The spiral structure was found to persist by at least 15 characteristic orbital periods corres-

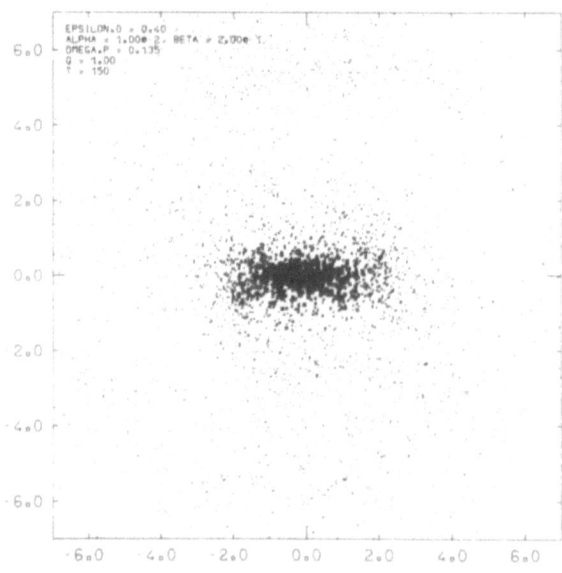

Fig. 1 Typical stellar density distribution
        simulated in an axial symmetric back-
        ground potential with slowly increasing
        oval perturbation rotating in the
        mathematically positive sense. No mutual
        gravitational attraction between the
        stars in the disk is present in this case.

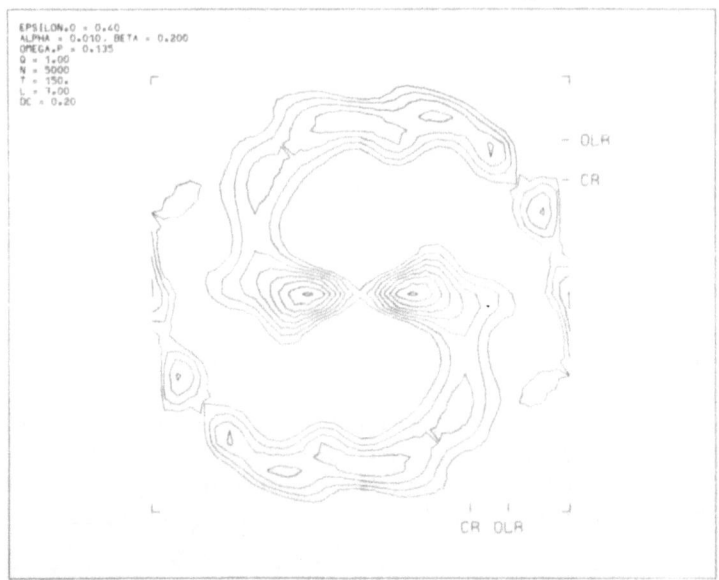

Fig. 2 Relative density excess of the
        distribution shown in fig. 1.

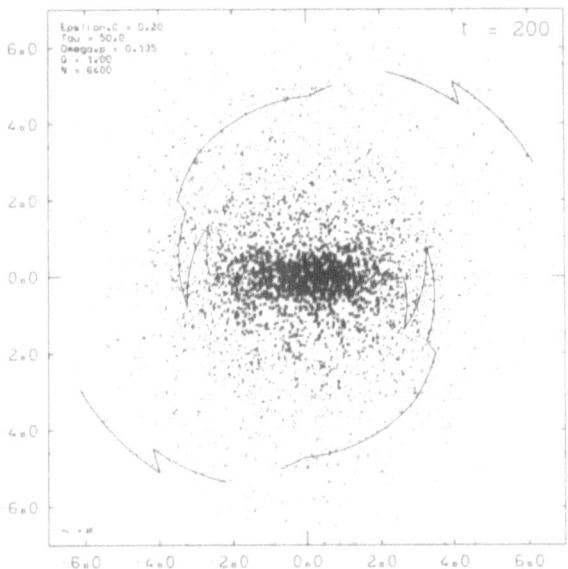

Fig. 3 Shape of maximum density response in a
        stellar density distribution similar
        to the one of fig. 1 evaluated by
        Fourier analysis.

ponding to a time interval of 1.5 x 10^9 years. These computer simulations therefore confirm the working hypothesis
mentioned before. The latter is also in agreement with recent self-consistent N-body simulations (Sellwood 1981) in
which spirals were observed whenever the oval configuration
appeared to grow.

In order to get a better understanding of this mechanism one
may use first order epicyclic approximation (Chandrasekhar
1942; Lindblad 1959) to solve the equations of motion of
non-interacting test particles. In this approach the stellar
orbits come in groups, each associated with a guiding orbit
around which the other members of the group are oscillating.
The guiding orbits may be looked upon as representatives for
the rest of orbits. In an axial symmetric gravitational
field the guiding orbits are just circles on which reference
stars move at constant angular velocity. In the presence of
a rotating stationary oval perturbation of the gravitational
field the guiding orbits are still closed in the corotating
frame but otherwise deviate somewhat from circles and the
stars moving on them perform harmonic oscillations in
epicyclic coordinates (ξ,η) defined as in fig. 4 around
the reference stars mentioned before.

The breaking of mirror symmetry in the response density

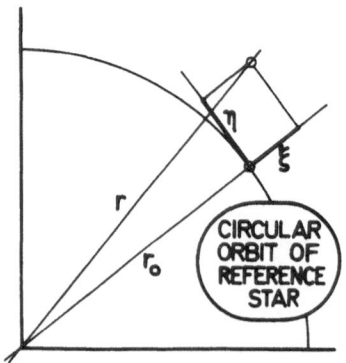

Fig. 4   Definition of epicyclic coordinates

pattern of stellar disks may be illustrated by an analogous
phenomenon in the dynamics of a harmonic oscillator

$$x + \chi^2 x = f \tag{8}$$

with a driving force

$$f = f_o \exp(-2\Gamma t) \cos(2\omega t). \tag{9}$$

The solution of its equation of motion is

$$x = A_o \exp(-2\Gamma t(\cos(2\omega t + \delta) + x_{hom} \tag{10}$$

involving a phase angle δ determined by

$$\sin \delta = 8\Gamma\omega / | \sqrt{(\chi^2 + 4\Gamma^2 - 4\omega^2)^2 + (8\Gamma\omega)^2}| \tag{11}$$

and an amplitude

$$A_o = f_o / | \sqrt{(\chi^2 + 4\Gamma^2 - 4\omega^2)^2 + (8\Gamma\omega)^2}| \tag{12}$$

For "adiabatic" changes the solution is approximately given by

$$x \propto A_o \cos (2\omega t + \delta) + x_{hom} \qquad (13)$$

This solution may be solved for t and inserted into the expression for the external force f such that the effective external force is expressed as a function of x. The latter then corresponds to the effective external potential

$$V_{eff}(x) \propto \pm \left[(\frac{x}{A_o})^2 \cos\delta \pm \{ (\frac{x}{A_o}) \sqrt{1 - (\frac{x}{A_o})^2} \pm arc \sin (\frac{x}{A_o}) \} \sin\delta \right] \qquad (14)$$

where the minus sign refers to the inner range

$$(2\omega)^2 < \chi^2 + 2\Gamma^2 \qquad (15)$$

while the plus sign refers to the outer range

$$(2\omega)^2 > \chi^2 + 2\Gamma^2 \qquad (16)$$

of the frequency of the driving force. One may now consider the symmetry of position probability response

$$p(x) = 1/2\pi \sqrt{A_o^2 - x^2} \qquad (17)$$

in comparison with the effective external potential $V_{eff}(x)$ shown in fig. 5. The latter exhibits the same symmetry if the external force is harmonic, i.e. $\Gamma = 0$. Otherwise this symmetry is broken. The phase jumps occuring at the boundary between the inner and outer range of external frequency and the resonances associated with them correspond to the Lindblad resonances observed in stellar disks. These resonances are damped in the non-stationary case. The analogue also gives a hint as to how self-gravitation may effect the response pattern. The latter is obviously constructive, i.e. position probability is large, when the potential is low inside the Lindblad region, while the reverse is true outside that region. The same behaviour is observed for guiding orbits in a stationary external spiral potential (Frahm and Thielheim 1978; Fuchs and Thielheim 1979 a, b).

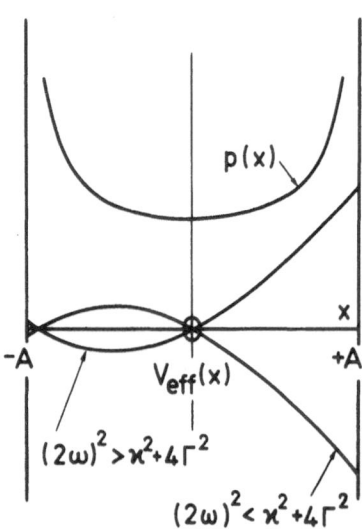

Fig. 5   Symmetry breaking in the driven linear
         harmonic oscillator. Compare position
         probability distribution with external
         effective potential $V_{eff}(x)$.

But otherwise the situation is more complicated in stellar
disks, where the crowding of guiding orbits turns out to be
essential for the generation of response patterns.

Kalnajs (1973) has remarked that what he called kinematic
spirals may thereby be generated artificially just by rotating the guiding orbits at appropriate angles dependent
on the distance from the galactic centre. This is actually
what we find in the afore mentioned approximation for
guiding orbits under the influence of a slowly increasing
quadrupole moment where the guiding orbits are no longer
closed in a strict but only approximative sense. In order to
simplify algebraic manipulations for the demonstration of
this effect the axial symmetric background potential

$$\phi_o \propto \exp\,(-\alpha r) \tag{18}$$

as well as the oval perturbation

$$\phi, \propto \exp(-\beta r)\,\exp\,(-2\Gamma r)\,\cos\,(2(\psi - \Omega pt)) \tag{19}$$

have been chosen to follow exponential laws in their radial
dependence. A typical example is shown in figure 6, where
each guiding orbit is marked by points indicating equal
intervals of time. The trailing spiral response pattern thus
becomes nicely visible.

One may exploit first order epicyclic approximation further
to estimate the shape of maximum density response. For this
purpose four stars are considered in the initial still axial
symmetric density distribution, thereby defining a surface
element. At a later time, when the oval perturbation has
grown to a certain extent, the position of these stars as
calculated in first order epicyclic approximation again
defines another surface element at a different position.
Assuming that the number of stars inside the surface element
does not change one has thus provided for a mapping of the
initial stellar density distribution to the one at a later
time from which the shape of maximum density response may be
easily obtained either numerically or in first order epicyc-

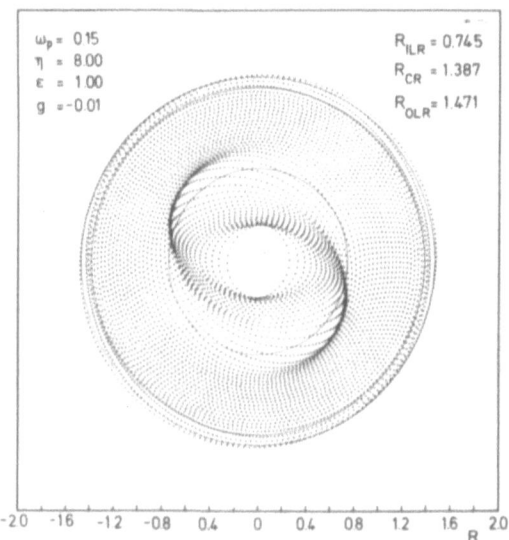

Fig. 6   Illustration of the generation of
         spiral density response patterns
         by the crowding of guiding orbits.
         The radial dependence of the axial
         background as well as the oval
         perturbation of the potential in
         this case are exponential.

lic approximation where the resulting surface mass density

is (Thielheim and Wolff 1981b)

$$
\mu(r,\psi) = \mu_o(r) \left[1 - \frac{\xi(r,\psi)}{\mu_o(r)} \frac{d\mu_o(r)}{dr} - \frac{\xi(r,\psi)}{r} \right.
$$

$$
\left. - \frac{\partial\xi(r,\psi)}{\partial r} - \frac{1}{r} \frac{\partial\eta(r,\psi)}{\partial\psi} \right]
$$

(20)

The first term describes the effect that when the axial
symmetric background mass density $\mu_o(r)$ decreases with increasing radial distance matter is transported outward in
direction of the main axis of dispersion rings, which in the
non-stationary case have to be considered instead of periodic guiding orbits when their elongation increases with
time. The second term describes the decrease of surface mass
density in the apocentres of dispersion rings as a comparison of the increased length of arc. The third term describes
the effect of crowding of orbits. The fourth term describes
the increase of mass density along the dispersion rings. The
example shown in fig. 7 has been calculated numerically for
a more realistic model, in which the dependence of surface
mass density from the radial distance is described by exponential functions (Polzin and Thielheim 1981). Again a
trailing spiral pattern is obtained.

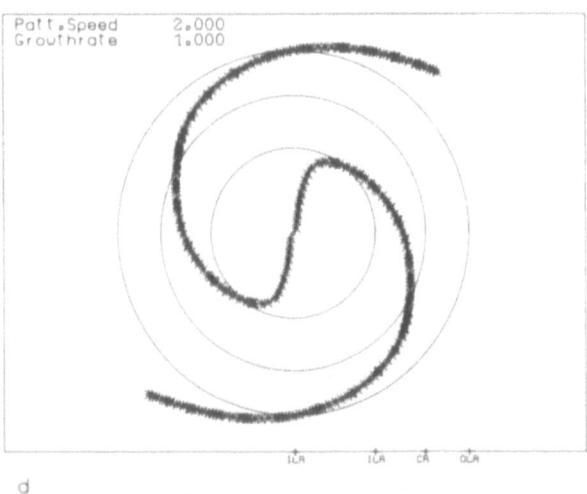

Fig. 7 Maximum density response calculated
              analytically using first order epicyclic
approximation. The radial dependence of the axial symmetric
background potential as well as the oval distortion of surface mass distribution in this case is exponential.

Remarkably first order epicyclic approximation is much better
in the time-dependent case than in the stationary case,
since the time dependence turns out to be somewhat equivalent to damping as far as the smoothing of resonances is
concerned. Still, within certain parameter regions higher
order contributions to epicyclic approximation may become
relevant (Polzin and Thielheim 1981). Additional families of
orbits may appear even for stationary perturbations (Contopoulos et al. 1977, 1980). Transitions between the population of these families occur for non-stationary perturbations (Spreckels and Thielheim 1981). For these reasons
first order epicyclic approximation cannot give more than a
semi-quantitative confirmation of the response pattern found
in N-body simulations.

4.   A POSSIBLE CONJECTURE FOR THE DYNAMICS OF SPIRAL
GALAXIES

Present results confirm so far that open barred spirals
similar to NGC 1300 can be produced through the response
mechanism invoked by an adiabatic increase of the central
quadrupole moment. Self-gravitation in the disk and interaction with the interstellar gas will certainly modify the
form as well as the density contrast of the response pattern.
Still we are left with the question by which mechanisms the
increase with time of the central quadrupole moment is induced. Mass may be transferred to the equilibrium configuration by azimuthal accretion through the trapping of orbits. Angular momentum may be extracted from the equilibrium
configuration through the torque exerted by the spiral density wave (Lynden-Bell 1979). Both mechanisms involve transfer of energy from the equilibrium configuration to the rest
of the disk. Simple models involving Freeman disks based
on this conjecture show that the adiabatic extraction of
angular momentum in the parameter region here induces indeed
an increasing of the quadrupole moment and very small changes of pattern frequency. The scheme shown in fig. 8
appears to be a possible alternative to the QSSS-hypothesis
for the interpretation of galactic spiral structure.

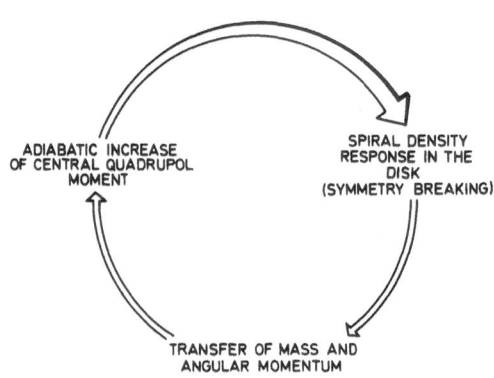

Fig. 8   Possible scheme of a mechanism
         generating a trailing spiral response
         pattern in disk galaxies.

REFERENCES

Berman,R.H., and Mark,J.W.-K., 1977, ApJ 216, 257
Berman,R.H., and Mark,J.W.-K., 1979, ApJ 231, 388
Contopoulos,G. and Mertzanides,C. 1977, Astr.Ap. 61, 477
Contopoulos,G. and Papayannopoulos,T. 1980, Astr.Ap. 92, 33
Chandrasekhar,S. 1942, Principles of Stellar Dynamics,
          Chicago: University of Chicago Press
Chandrasekhar,S. 1969, Ellipsoidal Figures of Equilibrium
          (New Haven: Yale University Press)
Frahm,R. and Thielheim,K.O. 1978, ApJ (Letters) 220, L43
Frahm,R., Fuchs,B. and Thielheim,K.O. 1979, Ap.Space Sci.
          63, 185
Frahm,R., Fuchs,B. and Thielheim,K.O. 1979a, Astr. Ap.
          72, 263
Hohl,F. and Hockney,R.W. 1969, J. Comput. Phys. 4, 306
Hohl,F. 1971, ApJ 168, 343
Hunter,C. 1963, MNRAS, 126, 299
Hunter,C. 1970, ApJ 162, 445
Hunter,C. 1979, ApJ 227, 73
Iye,M. 1978, Pub. Astr. Soc. Japan, 30, 223
James,R.A. 1977, J. Comput. Phys. 25, 71
Kalnajs,A.J. 1976, ApJ 205, 751
Kalnajs,A.J. 1971, ApJ 160, 292
Kalnajs,A.J. 1973, Proc. Astr. Soc. Austr. 2, 174

Kalnajs,A.J. and Athanassoula-Georgala,E. 1974, MNRAS 168, 287
Kato,S. 1968, Ap. Space Sci. 2, 37
Kato,S. 1970, Pub. Astr. Soc. Japan 22, 285
Lin,C.C., Yuan,C., Shu,F.H. 1969, ApJ 155, 721
Lin,C.C. and Shu,F.H. 1964, ApJ 140, 646
Lindblad,B. 1941, Stock. Obs. Ann. 13, No. 10
Lindblad,B. 1962, Stock. Obs. Ann. 14, No. 1
Lindblad,B. 1950, Stock. Obs. Ann. 16, No. 1
Lindblad,B. 1959, Handbuch der Physik, ed. S. Flügge (Berlin: Springer-Verlag) 53, 21
Lynden-Bell,D. 1962, MNRAS 123, 447
Lynden-Bell,D. and Kalnajs,A. 1972, MNRAS 157, 1
Lynden-Bell,D. 1979, MNRAS 187, 101
Marochnik,L.S. 1964, Soviet Astr. - AJ 8, 202
Marochnik,L.S. 1967, Soviet Astr. - AJ 10, 738
Marochnik,L.S., Suchkov,A.A. 1974, Usp. Fiz. Nauk 112, 275, Transl. 1974 Sov. Phys. Usp. 17, 85 (from Russian)
Mayall,N.U. and Aller,N.H. 1942, ApJ 95, 5
Miller,R.H. and Prendergast,K.H. 1968, ApJ 151, 699
Oort,J.H. 1927, BAN 3, 275
Ostriker,J. and Peebles,P.J.E. 1973, ApJ 186, 467
Polzin,D. and Thielheim,K.O. 1981, A&A 101, 409
Sandage,A. 1961, The Hubble Atlas of Galaxies (Washington: Carnegie Institution of Washington)
Schäfer,I. and Thielheim,K.O. 1981, in preparation
Schmidt-Kaler,Th. and Wiegandt,R. 1980, A&A 82, 238
Schmidt-Kaler,Th. 1975, Vistas in Astronomy 19, 69
Sellwood,J.A. 1980, Astr. Astrophys. 89, 296
Sellwood,J.A. 1981, Astr. Astrophys. 99, 362
Spreckels,H. and Thielheim,K.O. 1981, submitted for publ.
Takahara,F. 1976, Prog. Theor. Phys. 56, 1665
Thielheim,K.O. 1980, Astron. Mitt. 50, 161
Thielheim,K.O. 1980, Astrophys. Space Sci. 73, 499
Thielheim,K.O. 1981, Astrophys. Space Sci. 76, 363
Thielheim,K.O. and Wolff,H. 1980, Astron. Mitt. 48, 168
Thielheim,K.O. and Wolff,H. 1981a, ApJ 245, 39
Thielheim,K.O. and Wolff,H. 1981b, MNRAS, in press
Toomre,A. 1963, ApJ 138, 385
Toomre,A. 1964, ApJ 139, 1217
Toomre,A. 1969, ApJ 158, 899
Toomre,A. 1977, Ann. Rev. Astron. Astrophys. 15, 437
de Vaucouleurs,G. 1963, ApJ Suppl. Ser. 8, 31
de Vaucouleurs,G. 1970, Proc. IAU Symp. 38, 18
Wielen,R. 1974, Konf. Bures Sur Yvette, 357

THE MOST DISTANT GALAXIES

M.S. Longair
Royal Observatory
Blackford Hill
Edinburgh EH9 3HJ

1.   Introduction

It is entirely reasonable to ask "why study the most distant galaxies when we can scarcely claim to have an adequate understanding of nearby systems?"   The answer is that we can now observe certain classes of galaxy at such large distances that we are looking back to cosmological epochs significantly earlier than the present.   We are therefore in a position to observe directly how their properties have changed with cosmic epoch.

The way in which we have approached this problem is through the study of distant radio galaxies.   It is not difficult to find very faint galaxies nowadays with the advent of CCD cameras but it is not trivial to measure their distances.   This is one of the great advantages of studying radio galaxies.   We know from studies of bright radio galaxies that they are among the most luminous galaxies known, being as bright as the brightest galaxies in clusters, and that for powerful sources, there is only a small dispersion in their absolute magnitudes. Therefore, of the faintest galaxies we can observe, they are likely to be intrinsically the most luminous and consequently, the most distant. Indeed, for a few of them redshifts of one and greater have been measured. They also have the advantage that we know how to find them.   Accurate radio positions of bright radio sources enable the faintest radio galaxies to be found.   In addition, these galaxies can be studied in a symmetric manner so that objects of the same intrinsic type can be selected.

These distant radio galaxies are the most distant systems known whose light is dominated by starlight and it is these properties which enable us to study how the stellar component of galaxies changes with cosmic epoch.   In addition, these objects are important in pursuing the more traditional problems of the cosmic evolution of the radio source population and of the astrophysics of those systems which can become strong radio sources.   In this short report, I will concentrate upon the questions of how reliably we know that our radio source identifications

323

A. W. Wolfendale (ed.), Progress in Cosmology, 323–333.

of these distant galaxies are correct, evidence that their radio properties
change rapidly with cosmic epoch and some recent infrared observations
of them to look for changes in the stellar populations with cosmic
epoch.

2.   Identifications of Distant Radio Galaxies

      Identification programmes are underway at a wide range of flux
densities and frequencies but I will concentrate on the identifications
of 3CR radio sources.   3CR radio sources comprise the brightest sources
in the sky and for statistical purposes my colleagues and I have concentrat
upon all those sources in directions away from the Galactic plane which
could be studied with high angular resolution with the Cambridge 5 km
telescope.   The basic sample consists of a sample of about 166 sources
which have flux densities S_{178} ⩾ 10 Jy, $|b|$ > 10° and δ ⩾ 10°.

      Identifications for these sources have been sought on deep plates
taken with the Palomar 5 metre telescope but the real breakthrough
occurred when the fields of the unidentified sources were surveyed with
the JPL CCD camera (see Gunn et al, 1981).   Faint identifications were
claimed for virtually all sources in the sample, all the faint candidates
being believed to be faint galaxies on the basis of their extended
images.   There has remained, however, some undertainty about the
reliability of these identifications because many of them are extended
without the presence of a central radio-core coincident with the proposed
identification.

      This question has been addressed in a recent analysis by Robert
Laing, Julia Riley and myself (1982) for the radio sources in the "166"
3CR sample.   Part of that analysis involves a discussion of the actual
completeness of the sample.   As is expected the sample misses some
sources of large angular size and confusion can result in the inclusion
of sources which in fact have flux densities less than 10 Jy.   In the
original definition of the 166 sample by Riley, Jenkins and Longair (see
Jenkins, Pooley and Riley, 1977) each source had been scrutinised using
data available in 1976.   An improved analysis is included in the above
paper resulting in a revised sample of 173 sources.

      Our analysis of the reliability of the identifications was performed
in three stages.   First, all compact sources, $\theta \lesssim 2$ arcsec, can be
identified with certainty with the optical identifications when the
optical-radio positions coincide within about 1 arcsec.   In addition,
all extended sources which possess a compact radio core coincident with
an optical object within 1 arcsec are certain identifications.   Second,
we have investigated the reliability of those identifications with radio
sources with extended radio structure.   Those of Fanaroff-Riley class I
sources are all with bright galaxies, $V \leqslant 19$ and are certain because of
the presence of radio cores or because of the presence of strong emission
lines in the optical spectrum of the galaxy.   Third, we have discussed
those identifications with extended double radio structure.   As a first
step, we have noted the location of the identification with respect to

the outer hot spots of the doubles for all those certain identifications with V < 19 in which there is a radio core.   This distribution is shown in Figure 1a in which it can be seen that generally the identification does not lie precisely at the centre of the double structure but can be significantly displaced both along and perpendicular to the axis of the source.   However, in all but three cases, the identifications lie within a circle of radius 0.2 times the angular separation θ of the outer radio peaks which is shown on the diagram.   The same diagram has been plotted for all those double sources with V < 19 which do not possess a radio core (Figure 1b) and it can be seen that all lie within the 0.2θ circle.   Finally, in Figure 1c the same criterion has been applied to all the identifications with faint objects, all of which are faint galaxies with V < 19 and which do not possess radio cores.   It can be seen that most of the faint identifications lie within the circle and in only a few cases is the identification likely to be wrong (e.g. 3C68.2, 437).   A statistical analysis of the probability of galaxies falling at random within the 0.2θ circle for all sources indicates that at most about 1 identification could be due to chance.   Of the 173 sources in the revised sample at most 3.5% are unidentified.

Thus, we have considerable confidence in even the faintest ident-ifcations which have been proposed.   The importance of this result is that the anomalies of the steep source counts which lead to the inference of strong cosmological evolution are associated with identified radio sources, about 25.4% of them with quasars and 71.1% with radio galaxies.

Using these data, we have performed a standard V/V_{max} analysis of the spatial distribution of radio galaxies and quasars in the 3CR sample. A sample of these results is shown in Table 1.   It can be seen that those radio galaxies and quasars which are powerful radio sources ($P_{178} > 10^{26}$ W Hz^{-1} sr^{-1}) exhibit strong cosmological evolution in the sense that they have values of V/V_{max} significantly greater than the mean value of 0.5 expected for a uniform distribution.   It is noteworthy that the evolution exhibited by the powerful radio galaxies is as marked as that of the quasars.   These data provide important constraints on the cosmological evolution of the radio source population.

3.   Infrared Observations of 3CR Radio Galaxies

It has long been known that normal giant elliptical galaxies observed at large redishifts cease to be optical objects but become infrared sources, the maximum of the energy distribution occurring about a wavelength $1.3(1+z)$ μm where z is redshift.   Thus, it is of considerable interest to study these galaxies at large redshifts in the infrared waveband and to compare them with nearby galaxies.

Simon Lilly and I have concentrated on radio galaxies from the 3CR catalogue and in our preliminary survey observations of 35 galaxies spanning the range of redshift 0.03 to 1 have been made with the UK Infrared Telescope (UKIRT) on Mauna Kea, Hawaii (Lilly and Longair 1982a, b).   The observations were made in the J(1.2μm), H(1.65μm) and

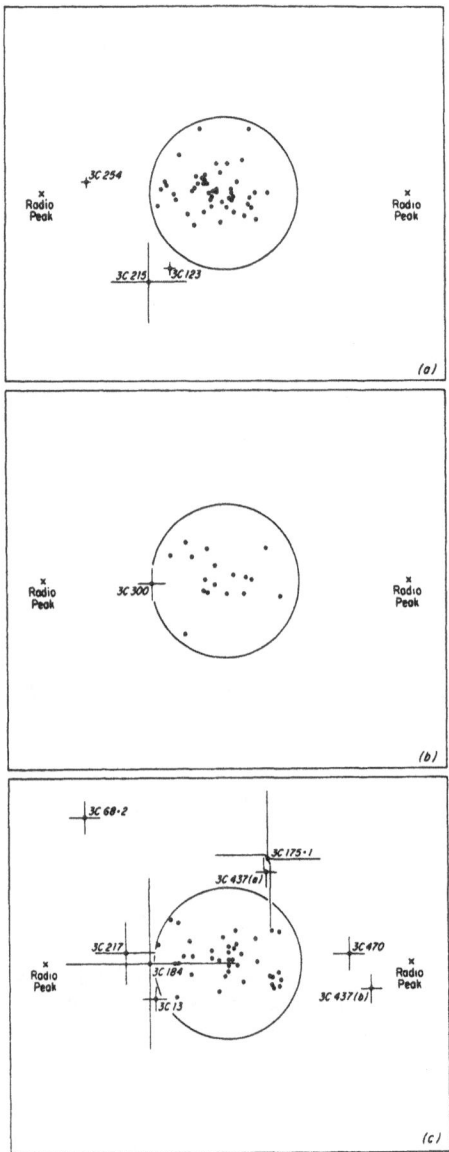

Figure 1.   Diagrams illustrating the location of the optical
identification of double radio sources relative to the hot-spots at the
leading edge of the double radio structure.   The circle shown has
radius 0.2θ where θ is the separation of the hot-spots.   Identifications
lying outside the circle are named individually with error bars showing
the uncertainty in the optical-radio position.

(a)   all certain identifications having V < 19 with radio cores.
(b)   all identification with V < 19 which do not possess radio cores.
(c)   all faint identifications, V > 19 which do not possess radio cores.

TABLE 1

Mean values of V/V_{max} for sources having $S_{178} > 10$ Jy, no optical limit.

(a) Sources with known redshifts

P_{178}(WHz^{-1}sr^{-1})	Optical type	n	$\langle V/V_{max}\rangle$	σ	D
$< 10^{25}$	Galaxy	19	0.508	0.066	0.12
$10^{25} - 10^{26}$	Galaxy	21	0.498	0.063	0.04
$10^{26} - 10^{27}$	Galaxy	32	0.599	0.051	1.9
	Quasar	4	0.682	-	-
$10^{27} - 10^{28}$	Galaxy	47	0.721	0.042	5.3
	Quasar	22	0.631	0.062	2.1
$> 10^{28}$	Galaxy	4	0.610	-	-
	Quasar	18	0.740	0.068	3.5
$< 10^{26}$	Galaxy	40	0.503	0.046	0.06
$> 10^{26}$	Galaxy	83	0.669	0.032	5.3
	Quasar	44	0.680	0.044	4.1
	All Sources	127	0.673	0.026	6.7

(b) All radio galaxies with $P_{178} > 10^{26}$ WHz^{-1}sr^{-1} with various assumed redshifts for those faint identifications for which no redshift has been measured.

Assumed redshift	$\langle V/V_{max}\rangle$	σ	D
0.2	0.656	0.031	5.03
0.5	0.669	0.031	5.45
1.0	0.683	0.031	5.90
2.0	0.696	0.031	6.32

σ is the standard error of the quoted mean values of V/V_{max}, $\sigma = (12n)^{-\frac{1}{2}}$ and D is the difference of $\langle V/V_{max}\rangle$ from the mean value of 0.5 measured in units of σ.

K(2.2μm) wavebands.   The apparent optical magnitudes of the galaxies
range from 14 to 24.   The optical morphologies of these galaxies are
similar to those of first ranked elliptical galaxies, including some cD
systems and N-galaxies, the latter being associated with broad-line
radio galaxies (BLRGs).   The current sensitivity of infrared detectors
is such that it has been possible to detect even the faintest of these
galaxies in the infrared waveband.   In addition, it has been possible
to detect some radio galaxies which have no optical counterpart in the
optical waveband.   We have confirmed the detection of 3C 68.2 reported
by Grasdalen (1980) and have obtained a marginal detection of 3C 437.

        The simplest way of presenting the results is in terms of colour-
redshift diagrams, the (H-K) and (J-K) diagrams being shown in Figures 2
and 3.

        Considering first the low redshift galaxies (z < 0.4), it is clear
that with the exception of the four galaxies classified as BLRG by
Grandi and Osterbrock (1977), and represented by crosses on the diagram,
the infrared colours of the radio galaxies occupy a well defined locus
on the colour-redshift planes.   This implies that they may all be
represented, with small cosmic scatter, by a single energy distribution.
This was derived using the observed infrared colours of the galaxies
with z < 0.4 and is shown by the solid line on the diagrams.   This mean
energy distribution for a giant elliptical galaxy is shown in Figure 4.

        The zero redshift intercepts of the colours predicted from this
energy distribution are very similar to those found in a large sample of
nearby elliptical galaxies by Frogel et al (1978),   We can therefore
conclude that the infrared energy distributions of these galaxies are
essentially the same as those of normal elliptical galaxies, and that
any additional component associated with the active nucleus must be
small.   There is no obvious relation between emission line strength and
infrared properties amongst the narrow line radio galaxies.

        At higher redshift, the galaxies broadly follow the predicted
relations computed on the basis of the infrared spectrum constructed
earlier and the energy distribution shortward of 1μm of Coleman et al
(1980).   There is no strong evidence for colour evolution in the infrared,
and this is as predicted by conventional evolutionary models of elliptical
galaxies, for example those of Bruzual (1981).

        The BGLRs are clearly red in both (J-K) and (H-K), and from the
(K,z) relation it is deduced that these are brighter than the typical
galaxies in our study, at all the wavelengths observed.   Sandage (1973a)
has shown that these systems may be successfully decomposed into a
central nuclear component situated in a normal galaxy, and we have
followed a similar procedure, for 3C 109, 234 and 382, using our
observations of the other galaxies to define the colours and magnitudes
of the underlying galaxy.   To within the uncertainties of this subtraction procedure, we find that the additional component has a power-
law spectrum, and that in the case of 3C 382 this extends to 3.5μm.   In

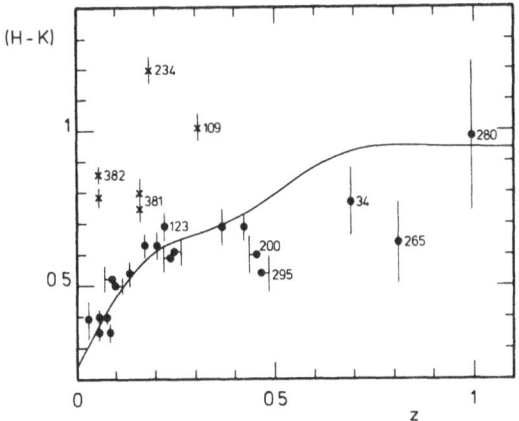

Figure 2. The (H-K) –
redshift relation for 3CR
radio galaxies. The
crosses are n-galaxies
with strong non-thermal
components.

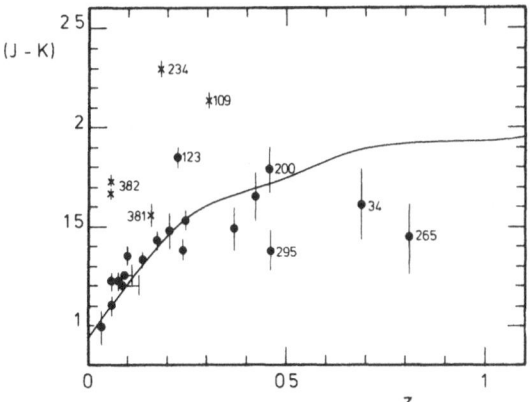

Figure 3. The (J-K) –
redshift relation for 3CR
radio galaxies. The
crosses are N-galaxies
with strong non-thermal
components.

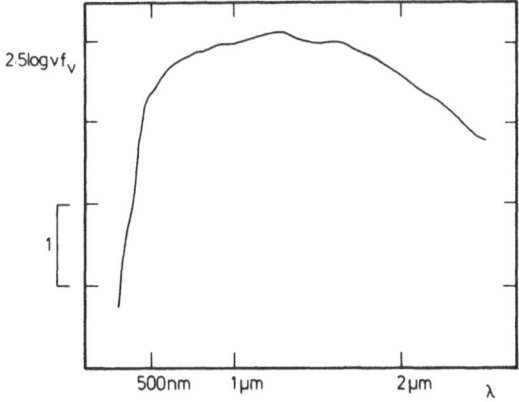

Figure 4. The standard
giant elliptical galaxy
spectrum derived from the
radio galaxies without
strong non-thermal
components with redshifts
z < 0.4 and the elliptical
galaxy spectrum of Coleman
et al (1980).

addition, the variability of this latter source which we observed over a
6 month timebase is compatible with a nuclear component of approximately
constant spectral index ($\alpha \approx 1.5$) but of varying intensity.   In the
small sub-sample for which there exists good spectrophotometry we find
that the BLRGs displaying an infrared excess have more than 10 times the
Hβ flux as compared with those other galaxies which do not.

   The optical-infrared colours may be similarly constructed from
published optical photometry (e.g. Sandage (1972b, 1973b) Kristian et al
(1978), Smith et al (1979)), although greater uncertainty will be introduce
The (R-K) and (V-K) colours as a function of redshift are plotted in
Figures 5 and 6.   At high redshift the CCD r magnitudes were transformed
to the R system using an extension of the colour equation given by Wade
et al (1979).   In the diagrams, the solid lines are the predicted
colour-redshift relations based upon the infrared energy distribution
derived earlier and the optical spectrum from Coleman et al (1980).
The other lines are various evolutionary models from Bruzual (1981).
The reddest model assumes no evolution, the others representing different
histories of the star formation rate (SFR), being either a constant
burst of duration 1 Gyr (the C model) or an exponential decay of the
SFR, with 0.7, and 0.5 respectively, of the mass of the galaxy in stars
at the end of the first 1 Gyr.   At high redshifts it is clear on both
diagrams that there are large devisions from the relations predicted by
the non-evolving models.   We have included the colours derived from the
K magnitudes of the high z 3CR galaxies observed by Lebofsky (1981), and
these are represented by open circles.   We cannot exclude the possibility
that part of the blue colours of sources such as 3C 265 may have a non-
thermal origin.   In this case, Smith et al (1979) have shown that the
optical spectrum does not possess stellar absorption features.   For the
remainder of the galaxies, however, the V-K colours can be accounted for
by the evolving galaxy models, with relatively slowly decaying SFR.   We
have also plotted on the (R-K) diagram the colours of 4 very faint
galaxies of unknown redshift, and it may be seen that these galaxies
have colours that are consistent with this conclusion.   This result has
implications for attempts to derive the redshift of distant galaxies
from their colours alone.

4.   The Infrared Hubble Diagram for Radio Galaxies

   Because essentially all the observations were made through the same
aperture, we plot on the Hubble diagram, Figure 7, the magnitude as
observed, and incorporate the K-corrections (which differ in the evolving
and non-evolving models) and aperture corrections (which are different
for different world models) into the predicted magnitude-redshift
relations for different values of q_0.   The aperture corrections were
derived from the beam profiles and from the curve of growth given by
Sandage (1972a).   The best fit model may then be determined by a Chi-
squared test, with the absolute magnitude/Hubble constant combination a
free parameter in each world model.   This method also gives the certainty
with which other models may be rejected.   Also shown on the diagram are
the sources with unknown redshifts, and the high z data from Lebofsky

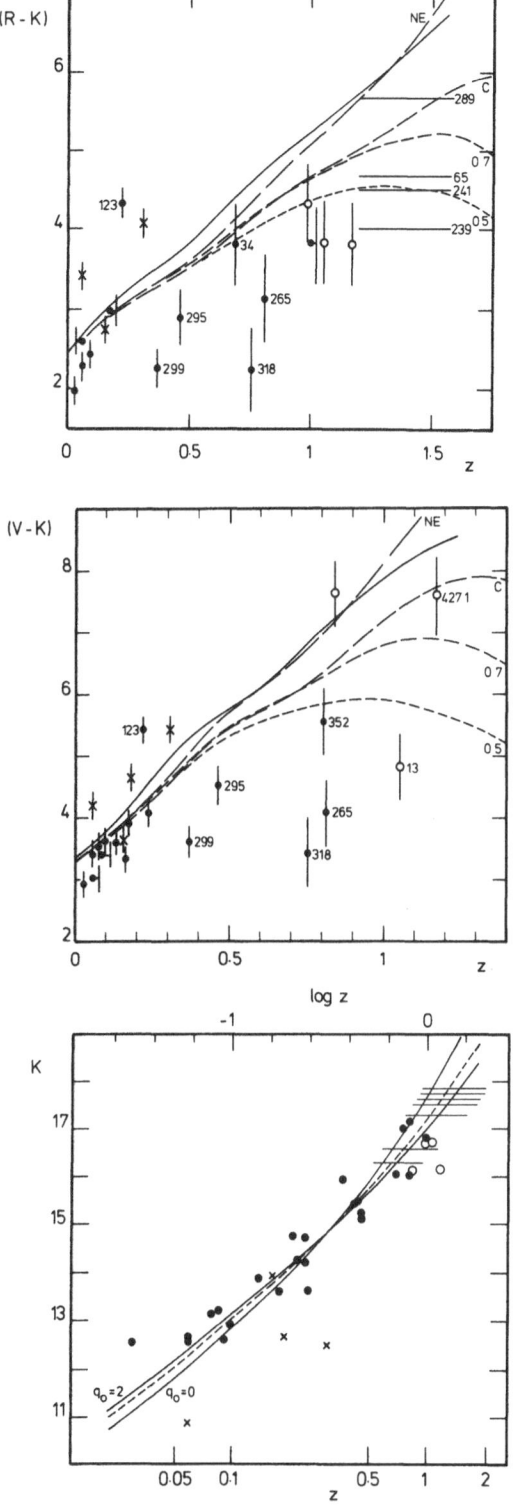

Figure 5.   The (R-K) -
redshift relation for radio
galaxies.   The straight
lines show the (R-K) values
for galaxies of unknown
redshift.   Open circles are
data from Lebofsky (1981).

Figure 6.   The (V-K) -
redshift relation for radio
galaxies.   The uppermost
lines show the predicted
relation if the galaxies
undergo no colour evolution.

Figure 7.   The redshift
- K magnitude relation for
radio galaxies.   Crosses
are N-galaxies and the open
circles are data due to
Lebofsky (1981).

(1981), as horizontal lines and open circles respectively. With the exclusion of the BLRGs, the radio galaxies form a well defined Hubble relation, the cosmic scatter about the best fit models being 0.40 mag, which is very similar to that found at optical wavelengths by Smith (1977). For the unevolving energy distribution, the apparent, uncorrected value of q_o is considerably in excess of 1. The best fit value occurs at 2.8, and the $q_o = 1$ model may be rejected with 70% confidence. The relations for $q_o = 0$ and 2 models are shown in Figure 6. However, the lack of infrared colour evolution does not imply a lack of <u>luminosity</u> evolution in the infrared. The effect of the $\mu = 0.5$ model discussed in connection with the optical-infrared colour diagrams is to increase the predicted K luminosity at a redshift of 1 by 0.8 mag. Application of the K-corrections appropriate to this model result in a best fit value of $q_o = 0.5$. Therefore, if lower values of q_o are preferred for other reasons, then the Hubble diagram provides further evidence of substantial evolution over lookback times of half the Hubble time.

5.   Conclusions

The data indicate the great potential of observations in the infrared waveband for the study of the evolution of the stellar component of giant elliptical galaxies over cosmological time-scales. The present results are preliminary but suggest that the infrared properties of galaxies are less susceptible to strong evolutionary effects than the optical properties. Combining the optical and infrared properties provides additional constraints on the evolutionary effects which influence the magnitude-redshift relation.

References

Bruzual, G., 1981.   PhD dissertation, University of California, Berkeley.
Coleman, G.D., Wu, C.C. and Weedman, D.W., 1980.   Astrophys. J. Suppl.,
    43, 393.
Frogel, J.A., Perrson, S.E., Aaronson, M. and Matthews, K., 1978.
    Astrophys. J., 220, 75.
Grandi, S.A. and Osterbrock, D.E., 1977.   Astrophys. J., 195, 255.
Grasdalen, G.L., 1980.   IAU Symposium No. 92, "Objects with Large
    Redshift" (ed. G.O. Abell and P.J.E. Peebles), 269.
Gunn, J.E., Hoessel, J.E., Westphal, J.A. Perryman, M.A.C. and Longair,
    M.S., 1981.   Mon. Not. R. astr. Soc., 194, 111.
Jenkins, C.C., Pooley, G.G. and Riley, J.M., 1977.   Mem. R. astr. Soc.,
    84, 61.
Kristian, J., Sandage, A.R. and Westphal, J.A., 1978.   Astrophys. J.,
    221, 383.
Laing, R.A., Longair, M.S. and Riley, J.M., 1982.   Mon. Not. R. astr.
    Soc., (in preparation).
Lebofsky, M., 1981.   Astrophys. J. (Letters), 245, L59.
Lilly, S.J. and Longair, M.S., 1982a.   IAU Symposium No. 97 "Extragalactic
    Radio Sources" (ed. D. Heeschen), p. 413.
Lilly, S.J. and Longair, M.S., 1982b.   Mon. Not. R. astr. Soc., (in press)
Sandage, A.R., 1972a.   Astrophys. J., 173, 485.

Sandage, A.R., 1972b.   Astrophys. J., 178, 25.
Sandage, A.R., 1973a.   Astrophys. J., 180, 687.
Sandage, A.R., 1973b.   Astrophys. J., 183, 711.
Smith, H.E., 1977.   IAU Symposium No. 74, "Radio Astronomy and
    Cosmology", (ed. D.L. Jauncey), 279.
Smith. H.E., Junkarinen, V.T. Spinrad, H., Grueff, V. and Vigotti, M.,
    1979.   Astrophys. J., 231, 307.
Wade, R.A., Hoessel, J.G., Elias, J.H. and Huchra, J.P., 1979.   Publ.
    astr. Soc. Pacific, 91, 35.

OBSERVATIONS OF GALAXY CLUSTERING

T. Shanks
Department of Physics,
University of Durham,
South Road,
Durham.

I. INTRODUCTION

Observations of galaxy clustering are important for cosmology because of the information they contain on the extent of inhomogeneity in the matter distribution of the Universe. We review here what has been learnt about galaxy clustering through statistical studies of galaxy catalogues. In statistical studies the most commonly used measure of clustering is the 2-point auto-correlation function $\xi(r)$ (Peebles, 1973). This is just a suitably normalised ratio of the count of galaxies found a distance r (Mpc) away from an average galaxy to the number expected if the galaxies were Poisson distributed. Until recently $\xi(r)$ has been obtained via its 2-dimensional angular equivalent $w(\theta)$, estimated from magnitude limited catalogues of galaxies detected by eye on Sky Survey photographs.

Intercomparison of $w(\theta)$'s at small angular scale from catalogues of different average depths allows tests to be made on the homogeneity of clustering on scales greater than $50h^{-1}$ Mpc.* We review the conclusions from this work in §II and present new results from recent machine detected galaxy catalogues on deep photographs.

The correlation analyses more directly supply information on the galaxy clustering inhomogeneities on $0.1 - 10h^{-1}$ Mpc scales. A well established result first found by Peebles (1974) is the power law $(r^{-1.8})$ behaviour of $\xi(r)$ at small scales. Not as well established is a break from $\xi(r)$'s power law behaviour found between 3 and $10h^{-1}$ Mpc. The power-law form of $\xi(r)$ is usually interpreted as evidence in favour of hierarchial model of galaxy clustering.

In §III we review the observational evidence and discuss the strength of support from statistical studies for the hierarchical model.

* h = Hubble's constant in units of 100 Km s^{-1} Mpc^{-1}

A. W. Wolfendale (ed.), Progress in Cosmology, 335–346.

Recently there has become available magnitude limited samples of galaxies with complete redshift information. These have allowed inspection of 3-dimensional distributions of galaxies. For statistical studies the lack of smoothing by projection has meant that weak clustering features (such as the $\xi(r)$ break) may be more readily estimated. We present in §IV preliminary estimates of $\xi(r)$ from one particular redshift survey - the Durham/AAT Redshift Survey. The results include estimates of galaxy correlations on scales between 10 and $50h^{-1}Mpc$.

II LARGE SCALE STRUCTURE (50 - $1000h^{-1}Mpc$)

The homogeneity of clustering at large scales can be investigated by comparing the amplitude of the 2-point galaxy correlation function in magnitude limited catalogues at a variety of depths. The correlation function is used because it is relatively easy to derive projection formulae relating $w(\theta)$'s amplitude from catalogue to catalogue. These projection formulae assume that clustering properties are the same everywhere. Thus by testing these formulae we also test the homogeneity of the Universe. Peebles and Hauser (1974) first used these "scaling" formulae and showed that from the amplitude of clustering found in the 15m limited Zwicky catalogue the scaling laws reasonably predicted the amplitudes in the 18.9m and 20m limited Lick and Jagellonian catalogues (Groth and Peebles 1977). The average depths of these catalogues ranged from 50 - 350 h^{-1} Mpc.

These results were confirmed and extended by the Durham group (Phillipps et al. 1978, Shanks et al., 1980) to depths of $650h^{-1}Mpc$. The catalogues analysed here were detected by the COSMOS Machine (Pratt et al., 1975) from U.K. Schmidt photographs over 40 sq. deg of sky to magnitude limit 21.5.

Preliminary results from correlation analyses of COSMOS detected galaxies on Anglo Australian Telescope 4 metre photographs are now available which extend the tests of the scaling laws to depth of $1000h^{-1}Mpc$ and magnitude limits B $\stackrel{<}{\sim}$ 23 (see figure 1.). The area of sky covered is small (0.3 sq. deg.) and at these depths the scaling laws become affected by models of K-corrections and luminosity evolution. Even so it is encouraging to see that models which successfully predict the observed form of deep galaxy colour magnitude diagrams (Phillipps et al. 1981) also produce scaling relations that are reasonably consistent with these observations. This agreement between catalogues of average depths between $50h^{-1}$ - $1000h^{-1}Mpc$, is a good argument for the homogeneity of the Universe in its clustering properties over these scales.

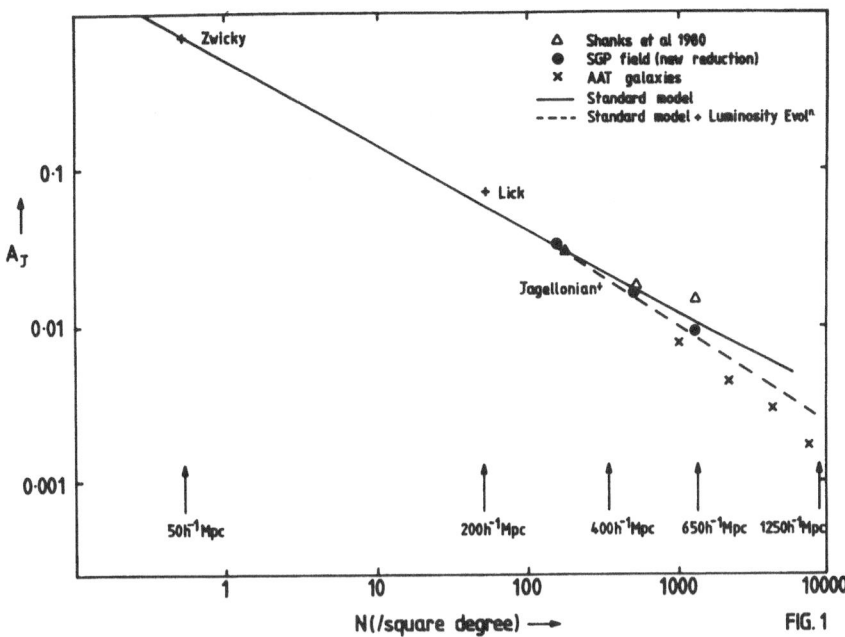

Figure 1 : The observed scaling relation for the 2-point
correlation function. A_J is the amplitude of clustering found
by estimating $w(\theta)$ in galaxy catalogues at various (J)
magnitude limits. N is the observed number density of galaxies
at each limit, which can be used to obtain the estimates of
survey depth as shown. The solid and dashed lines are the
predictions of the scaling formulae assuming clustering
homogeneity and varying amounts of galaxy luminosity
evolution.

III SMALL SCALE STRUCTURES ($0.1 - 5h^{-1}$Mpc)

What do statistical studies tell us about galaxy clustering on
small scales? The best measured statistics is $\xi(r)$ and we first review
its observed form.

At small scales from $0.1 - 3h^{-1}$Mpc it is well established that
$\xi(r)$ carries on as a power law out to at least $20h^{-1}$Mpc with no sign
of anticorrelation (holes) at any scale length. Subsequently, after
a reduction of the Lick catalogue, a break from $\xi(r)$'s power-law
behaviour was found at $r = 9h^{-1}$Mpc (Groth and Peebles 1977). The
Durham group's deeper machine measured galaxy catalogue also showed
evidence for a break in $\xi(r)$ but this time at $r = 3h^{-1}$Mpc (Shanks
et al. 1980) (see Figure 2).

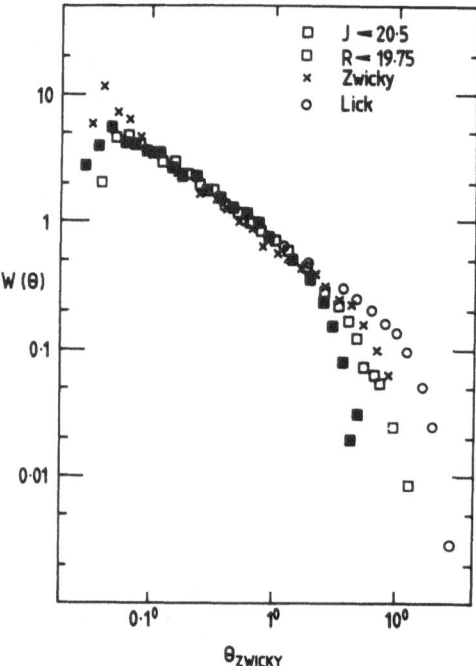

Figure 2 : Angular correlation function for the deep
COSMOS/Schmidt galaxy catalogues together with the Zwicky
and Lick catalogue results. All are scaled to the depth of the
Zwicky Catalogue. The break-away from w(θ) power law
behaviour in the deep catalogues occurs at smaller separations than in the Lick catalogue.

The scatter in the estimates of where $\xi(r)$ breaks from its
power law behaviour demonstrates the difficulty with analysing
clustering in projection. Where w(θ) is small any artificial large
scale gradient in galaxy detection will affect the results. The effect
is systematic rather than statistical and cannot be overcome by
simply analysing larger numbers of 2-dimensional projected galaxy
positions. $\xi(r)$ is expected to be near zero on scales larger than
the break length and the systematic problems also mean that there is
no believable estimate of $\xi(r)$ for $r \gtrsim 10h^{-1}$ Mpc.

We now consider the implications of the observed $\xi(r)$ for models
of small scale galaxy clustering. Although $\xi(r)$ does not contain all
the information about the galaxy distribution it has enough power to
eliminate some very simple galaxy clustering models.

For instance models where galaxies are distributed in randomly
placed clusters with gaussian density fall-offs are excluded because
they would not produce $\xi(r)$'s power law form.

A model which does produce a power law form for $\xi(r)$ and also has strong appeal because it suggests a physical process which might explain the formation of galaxy clusters is the hierarchical model. (Soneira and Peebles 1978). Here galaxies are distributed in nested clusters of clusters of galaxies. The smoothness of $\xi(r)$'s power law is thus explained as a lack of preferred scales of clustering, at least at the smallest galaxy separations.

The physical process suggested by the hierarchy was one where galaxies form early and cluster under their own gravity, building on chance occurrences of high density fluctuations in their original distribution. Subsequent gravitational clustering of the clusters provided the hierarchy's successive levels. (For detailed reviews see Fall 1979, Peebles 1980). Simple arguments based on $\xi(r)$'s - 1.77 slope suggested that the distribution of particles of matter (galaxies ?) at $Z \div 1500$ might just have been a Poisson distribution. With this model no anti-correlation of galaxies would be observed, holes produced by infall of clusters being filled (on the average) by infall of more distant clusters. A break in $\xi(r)$'s power law might correspond to the boundary between linear and non-linear regimes of galaxy clustering. (Davis et al. 1977).

Other simple phenomenological clustering models also fit the 2-point correlation function results. Many large galaxy clusters have power-law number density fall-offs away from their centres (Seldner and Peebles 1977) - the Coma cluster is one example. A model with a random distribution of clusters with a reasonable index (-2.4) of power-law fall-off can also produce a power law $\xi(r)$ with approximately the observed power law slope. The break length here would correspond to some average cluster diameter. Another model where all galaxies are distributed in straight lines or filaments can even produce an r^{-2} $\xi(r)$ fall-off, reasonably close to the observed $r^{-1.8}$.

To distinguish between these models we have to consider statistics which depend on higher moments of the clustering distribution. The next highest moments, the 2 - and 4 - point correlation functions have also been measured in the Zwicky and Lick catalogues (Peebles 1975, Groth and Peebles 1977 Fry and Peebles 1978). These provide strong support for the hierarchical model as compared to the simplest versions of the power law cluster or straight lines model. With the pleasingly simple theoretical model as a ready interpretation the case for the hierarchy has therefore gained much credibility.

However the hierarchy is still not a unique solution to the constraints implied by the 3 - and 4 - point functions and it is worthwhile to look at other aspects of the statistics of galaxy clustering to make further tests.

One problem for the hierarchy comes from a statistical analysis using Mead's Statistic (Shanks 1979), a statistic which depends on

all moments of the galaxy distribution. This statistic was
specifically chosen as being sensitive to hierarchical structure and
indeed did detect such structure in N-body simulations where
particles are allowed to cluster under their own gravity. Yet the
statistic failed to detect hierarchical structure in either the
Zwicky, Jagellonian or COSMOS/Schmidt catalogues.

The class of rich galaxy clusters known as Abell clusters (Abell,
1958) do not fit into the hierarchical picture in a variety of ways.
Firstly they are found to be much more strongly clustered amongst
themselves than galaxies. Figure 3 shows that at similar scale
lengths these are a factor of 10 more clusters around an average
cluster than galaxies around an average galaxy. If galaxies in Abell
clusters were a random selection of all galaxies then clusters and
galaxies at large scales should show the same amplitude of clustering.

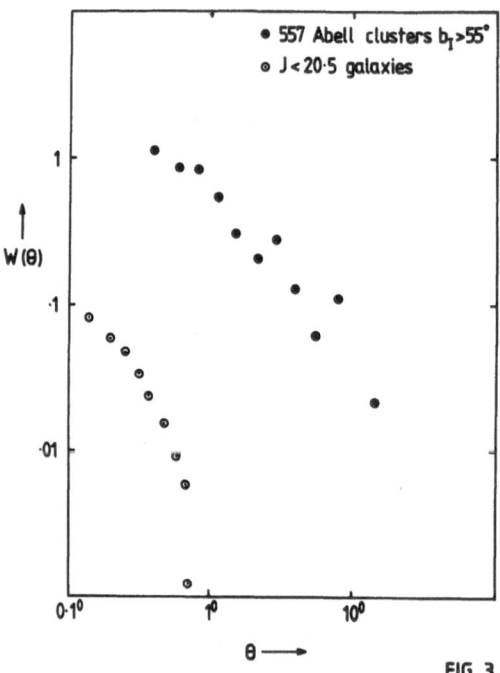

FIG. 3

Figure 3 : Angular correlation functions for the Abell
(1958) catalogue of clusters (first calculated by Hauser
and Peebles, 1974) compared with the galaxy-galaxy (J < 20.5)
correlation function which is at a similar depth. At
similar separations the clusters are more strongly clumped
than the galaxies. The clusters' w(θ) carries on out to
scales of 3^0 ($\sim 30h^{-1}$ Mpc), well past the galaxies' maximum
clustering scale-length.

Secondly the clustering of Abell clusters is detected out to $30h^{-1}$Mpc, well past any estimate of the galaxy-galaxy correlation break scale length.

Thirdly their internal structure is overtly not hierarchial. As mentioned above an average Abell cluster is well represented by a cluster with a power-law density fall-off.

Although Abell clusters may only contain 10% of all galaxies it is worrying that the most clearly defined and best studied clusters do not fit naturally into the hierarchical clustering pattern. Although the hierarchy is still probably the best and most popular statistical description of the galaxy distribution it is still an approximation whose validity must be further tested.

IV GALAXY REDSHIFT SURVEYS

The most recent opportunity to extend our knowledge of galaxy clustering on scales $5 - 50h^{-1}$ Mpc comes with the increasing availability of magnitude limited catalogues of galaxies that contain complete redshift information. The advantage of this type of sample is that the galaxy distribution can be analysed in 3-dimensions. With no smoothing by projection clustering features too weak to be detected above systematic effects before (e.g. the $\xi(r)$ break) can now be better estimated.

Redshift surveys have been carried out by various authors in the past few years (Einasto 1978, Chincarini 1978, Tifft and Gregory 1976). Claims have been made for empty holes between clusters and filamentary clusters of galaxies. However, varying degrees of incompleteness have left these samples generally unsuitable for statistical analysis. Davis et al. (1978) have produced a B < 13.0^m limited redshift survey. Although complete this sample is possibly unrepresentative at large scales through its being dominated by the Local Supercluster.

There are a number of groups at present involved in producing deeper complete redshift samples. Kirshner, Oemler and Schecter (1978) (KOS) have produced a survey of 164 galaxies J < 14.9 in 8 small fields across the sky. Kirshner et al. (1981 - KOSS) have recently extended this survey to R < 16.3 in 3 fields. Davis et al. 1981 are surveying 2400 brighter galaxies (B < 14.5) covering a much larger fraction of sky, extending their earlier surveys to fainter limits.

Finally the Durham/AAT group (Bean et al. 1981) have (to date) surveyed 4 fields containing 263 galaxies to B $\stackrel{<}{\sim}$ 17.0 and we present some preliminary statistical analyses of the galaxy distribution in this sample here.

These statistical results are interesting in view of published
pictures of the Davis et al. and KOSS galaxy distribution. Davis
et al. comment on the "frothy" nature of their galaxy distribution
with filaments of galaxies surrounded by large holes devoid of
galaxies. They visually compare their data with N-body models and
notice distinct differences. KOSS report a large $60h^{-1}$ Mpc hole
in their galaxy redshift distribution at the point where the number —
redshift relation should peak, raising again the question of the
evidence for homogeneity on the 50 – $100h^{-1}$ Mpc scales.

However, because galaxies are known to be strongly clustered
large volumes of space below average density, like those found by
the above authors are bound to exist. The question is whether such
empty regions around clusters persist in averages over more clusters
and over bigger volumes or whether, on the average, other clusters
will be found that will fill the holes in.

The Durham/AAT Redshift Survey also shows some of the holes
and filamentary structure seen in other samples. However, in
averages over all fields the n(z) distribution, although noisy,
seems in reasonably good agreement with that predicted by a model
assuming a uniform galaxy density and self consistent luminosity
function parameters (see Bean et al. 1981 for details) in a
magnitude limited sample. This n(z) relation is an average over
galaxies, with separations up to $\sim 200h^{-1}$ Mpc. Thus at these scale
lengths the result seems to be consistent with a homogeneous
Universe, in line with the clustering scaling results described in
§II . It will be interesting to see if this consistency is maintained
as more redshift surveys are completed.

Our Survey galaxies span separations of $150h^{-1}$ Mpc in
individual fields and thus allow estimates of $\xi(r)$ to be obtained for
scales from 0-$50h^{-1}$ Mpc. Although the total volume of space sampled
is still small it is worthwhile to make such estimates to test the
power of the method in larger samples and to obtain results for
comparison with other studies.

Preliminary results for $\xi(s)$ for our sample are plotted, first
in the 1-$10h^{-1}$ Mpc region, in Fig. 4. S is used instead of r here
to demonstrate that redshift is being used as a distance indicator.
At small scales, galaxy peculiar velocities cause differences
between $\xi(s)$ and $\xi(r)$. The observed r.m.s. velocity dispersion in
our sample is of order 200 kms^{-1} and this is the cause of deviation
away from a power law at scales up to $2h^{-1}$ Mpc.

The amplitude of the -1.77 power law is found to be A =
$23(\pm2)(h^{-1}Mpc)^{1.77}$ in good agreement with that found from the projected
catalogues. The importance is that, despite our small galaxy numbers
this implies that our survey is reasonably representative of the
Universe at large.

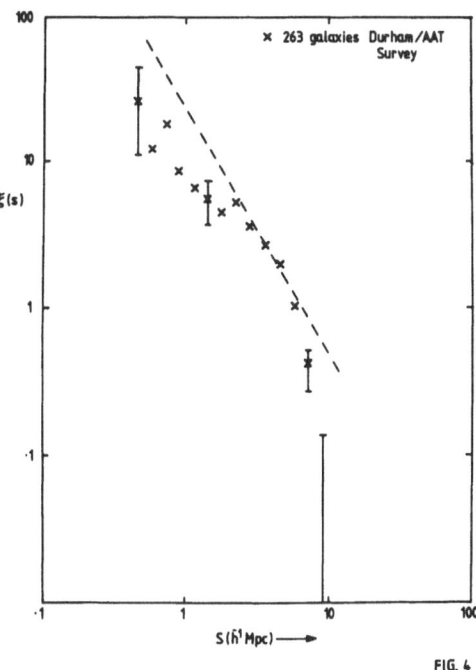

FIG. 4

Figure 4: The spatial 2-point correlation ξ(s) calculated
for the Durham/AAT Redshift Survey. s is a measure of
distance calculated using redshift. Error bars are calculated from field-to-field variations. The dashed line
represents a -1.77 power law.

Interestingly on scales larger than $5h^{-1}$ Mpc ξ(s) has a slope
significantly steeper than -1.77. The break away from the fitted
-1.77 power law occurs approximately where ξ(s) = 1, s = $5h^{-1}$ Mpc.
The error has here been estimated from field to field variation.
Again caution is to be advised because of the possibility that our
sample is unrepresentative but it is interesting that this break
scale length lies between the values of $3h^{-1}$ Mpc and $9h^{-1}$ Mpc
reported from the projected catalogues.

Figure 5 shows ξ(s) estimated up to separation of $100h^{-1}$ Mpc and
is now plotted on a linear-log scale. As the scale length increases
the number of independent points defining correlations at that length
scale decreases making confidence in our sample's fairness less.
Still it is interesting to see that at scales above $30h^{-1}$ Mpc ξ(s)
is consistent with being zero. At smaller scales ξ(s) goes negative
between $10h^{-1}$ Mpc and $30h^{-1}$ Mpc. This effect is seen in each of
our fields, providing the small error bars in Figure 5.

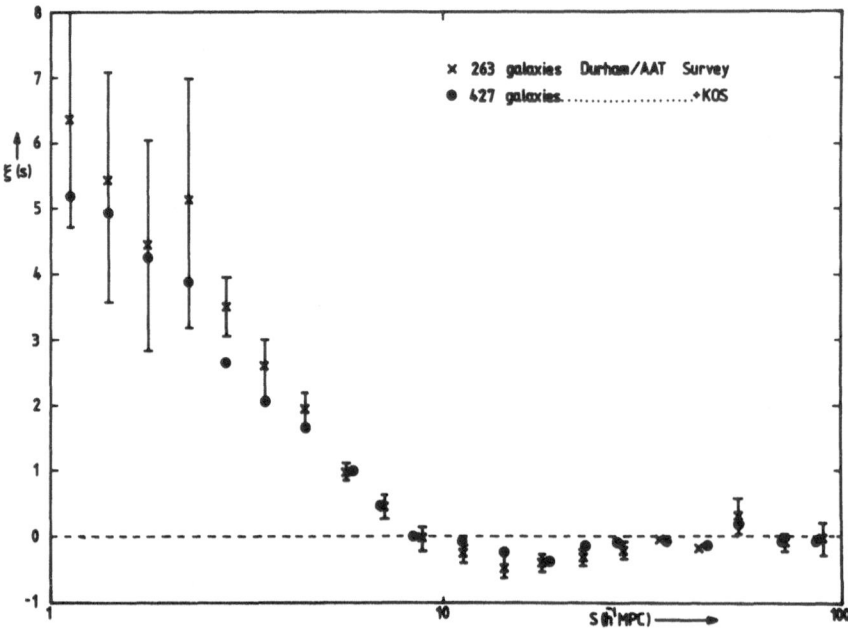

Figure 5 : ξ(s) now plotted on a linear-log scale up to
separations of 100h⁻¹ Mpc. The crosses are based on the
Durham AAT Redshift Survey, the filled circles from combining
the Durham/AAT and KOS surveys.

These results are in sharp disagreement with the published
results of KOS. In their smaller (166) galaxy redshift survey
these authors found that the ξ(s) stayed positive on scales greater
than 10h⁻¹ Mpc. We reanalysed their catalogue using our estimation
procedures and could not reproduce their results on scales past
5h⁻¹ Mpc. Our estimates coincided more with those of Peebles' (1979)
estimates of correlation in the KOS sample though no direct comparison
is possible due to Peebles' having estimated correlations in redshift
and angular directions separately. The reason for the discrepancy
with the KOS results is not yet clear and adds to the uncertainty of
our results. Also in our estimates of ξ(s) from the KOS data the
inclusion or exclusion of their anamalously rich NP4 field make
large differences to the results. Excluded, their ξ(s) (our estimate)
shows the same negative correlations as in our survey. Included,
their ξ(s) goes to zero at 10h⁻¹ Mpc. The large size of fluctuation
caused by one field are partly due to the small size of the KOS sample,
partly to the smaller range of separations sampled in individual
fields (< 50h⁻¹ Mpc). However, these fluctuations show the need for
larger samples to drive down sampling noise.

Bearing these remarks in mind, it is still possible to obtain
a presently "best" estimate of ξ(s) at large scales by combining

(pairwise) our estimates in the KOS and Durham/AAT fields. (see
Figure 5). The results show that $\xi(s)$ is still negative at scales
larger than $10h^{-1}$ MPC, $\xi(s)$ going through a minimum at $s = 15h^{-1}$ Mpc,
$\xi(s) = -0.3$. If true this would mean that at 15 Mpc from an average
galaxy there seems to be 30% fewer galaxies than expected if
galaxies were independently distributed.

The biggest source of error is still statistical fluctuations.
If the present samples had accidentally undersampled regions like
KOS's NP4 then this could easily explain the negative $\xi(s)$ results.
The finite size of sample can also produce a negative result but the
effect seems to be too small to explain the level of anticorrelation
encountered here. Any alternative estimates of $\xi(s)$ which lowered
the background number density to remove the anticorrelation at
$10-20h^{-1}$ Mpc would produce positive clustering on scales $> 30h^{-1}$ Mpc
which would be even more difficult to interpret.

If the anticorrelation is found in larger samples then it would be
statistical evidence for the holes visually detected by several
authors. Standard physical models where particles of matter start
from an initial Poisson distribution and clump under gravity to form
hierarchial clusters of galaxies do not naturally produce anticorrelation at any scale. If anticorrelation is put into the
initial distribution then it must not produce a $\xi(r)$ slope at small
scales in contradiction with present observations. Estimates of $\xi(r)$
at small scales already constrain theories where clusters form first
and collapse to form galaxies (Peebles, 1981). However, such models
still allowed by these considerations seem also to produce anticorrelation at large scales and this theory still cannot be excluded
by galaxy clustering observations.

The consequences for cosmological theory are large if holes
around galaxy clusters exist and better estimates of $\xi(s)$ from
larger redshift surveys are urgently required.

CONCLUSIONS

We have considered the implications of galaxy clustering
observations for inhomogeneities in the Universe on a variety of
scales. On the largest scales the results from scaling galaxy
clustering amplitudes obtained from projected galaxy catalogues with
average depth from $50h^{-1}$ Mpc to $1000h^{-1}$ Mpc are consistent with
those expected in an homogeneous Universe. On the smallest scales
$(.1-5h^{-1}$ Mpc) the statistics of galaxy clustering do give some
evidence for a clustering hierarchy but problems exist and the issue
may not be finally settled. On intermediate scales the prospects
for obtaining estimates of galaxy correlations using redshift samples
have been examined and some preliminary results from the Durham/AAT
Redshift Survey presented.

Acknowledgements

I should like to thank my collaborators on the Durham/AAT
Redshift Survey, John Bean, George Efstathiou, Richard Ellis,
Dick Fong, Bruce Peterson and Zou for useful discussions and for
allowing me to use certain results in advance of publication.

References

Abell, G,O. 1958, Ap. J. Suppl. 3, 211.
Bean, J., Efstathiou, G., Ellis, R.S., Fong, R., Peterson, B.A. and
  Zou, in preparation.
Chincarini, G. 1978 Nature, 272, 515.
Davis, M., Geller, M.J. and Huchra, J. 1978, Ap. J. 221, 1.
Davis, M., Groth, E.J. and Peebles, P.J.E. Ap. J. 212, L107.
Davis, M., Huchra, J., Latham, D.W., and Tonry, J. 1981, Ap. J. in
  press.
Einasto, J., 1978, In IAU Symposium No. 79.
Fall, S.M. 1979, Rev. Mod. Phys. 51, 21.
Fry, J.N., and Peebles, P.J.E., 1978, Ap. J. 221, 19.
Groth, E.J., and Peebles, P.J.E., 1977, Ap. J. 217, 385.
Hauser, M.G. and Peebles, P.J.E., 1973, Ap. J. 185, 757.
Kirshner, R.P., Oemler, A., and Schechter, P.L. 1978, A.J. 83, 1549.
Kirshner, R.P., Oemler, A., Schechter, P.L. and Schectman, S.A.
  1981, Ap. J. 248, L57.
Peebles, P.J.E. 1973, Ap. J. 185, 413.
Peebles, P.J.E. 1974, Astron. and Astrophys.32, 197.
Peebles, P.J.E. 1975, Ap. J. 196, 1.
Peebles, P.J.E. 1979, A.J. 84, 730.
Peebles, P.J.E. 1980 "Large Scale Structure of the Universe",
  Princeton.
Peebles, P.J.E. 1981, Ap. J. 248, 885.
Peebles, P.J.E. and Hauser, M.G. 1974, Ap. J. Suppl. 28, 19.
Phillipps, S., Fong, R., Ellis, R.S., Fall, S.M. and McGillivray,
  H.T. 1978, Mon. Not. R. astr. Soc. 182, 673.
Phillipps, S., Fong, R., and Shanks, T., 1981, Mon. Not. R. astr.
  Soc. 194, 49.
Pratt, N.M., Martin, R., Alexander, L.W.G., Walker, G.S. and
  Williams, P.R. 1975, in "Image Processing Techniques in Astronomy"
  Reidel, Dordrecht.
Seldner, M. and Peebles, P.J.E., 1977, Ap. J. 215, 703.
Shanks, T. 1979, Mon. Not. R. astr. Soc. 186, 583.
Shanks, T., Fong, R., Ellis, R.S. and McGillivray, H.T. 1980,
  Mon. Not. R. astr. Soc. 192, 209.
Soneira, R.M. and Peebles, P.J.E. 1978, A.J. 83, 845.
Tifft, W.G., and Gregory, S.A. 1976, Ap. J. 205, 696.

THE ROTATION OF GALAXIES: CLUES TO THEIR FORMATION

S. Michael Fall
Institute of Astronomy, Cambridge

The rotation of galaxies of different morphological types is closely linked with their structural features and therefore with the processes by which they formed. In this context, the most important distinction is between galaxies that are dominated by a spheroid or bulge component – the ellipticals and some lenticulars – and galaxies that are dominated by a disk component – some lenticulars, the spirals and some irregulars. As the result of improvements in spectroscopic techniques, we now have reliable kinematic data for galaxies of most types in a wide range of masses and sizes. My purpose in this article is to discuss the observational results and their implications for several views of the origin and evolution of galaxies.
    A convenient means of specifying the rotation of a self-gravitating body is in terms of the dimensionless spin parameter

$$\lambda \equiv J|E|^{\frac{1}{2}} G^{-1} M^{-5/2} \quad , \tag{1}$$

where J, E and M are respectively the total angular momentum, energy and mass of the body and G is the gravitational constant. A centrifugally supported disk with an exponential density profile has

$$\lambda_D = 0.43 \quad , \tag{2}$$

independent of its mass and size. This follows from eqns. 9, 11 and 14 of Freeman (1970), eqn. 6 on page 386 of Watson (1922) and the virial theorem. A spheroid with a Hubble density profile and a gently-peaked rotation curve has

$$\lambda_E \approx 0.3 (v_m/\sigma_o) \quad , \tag{3}$$

where v_m is the maximum rotation velocity in the equatorial plane and σ_o is the central velocity dispersion in one dimension. This follows from an analytical calculation and several N-body simulations (White 1979). Any self-gravitating galaxy will have a spin parameter somewhere between (2) and (3) depending on the disk/bulge ratio and the value of v_m/σ_o. The application of these formulae would be affected by dark haloes if

A. W. Wolfendale (ed.), Progress in Cosmology, 347–356.
Copyright © 1982 by D. Reidel Publishing Company.

they were to provide most of the gravitational attraction in the visible
bodies of galaxies. Although the observational evidence on this point
is not yet settled, the indications are that the inner regions of most
galaxies are self-gravitating or nearly so.

Kinematic data are now available for several dozen elliptical
galaxies (Illingworth 1981 and references therein). In most cases,
v_m/σ_o has turned out to be lower than the value required for an oblate
spheroid of the same projected shape and an isotropic distribution of
stellar velocities. From the scatter diagram of v_m/σ_o against ellipticity, one may conclude that these galaxies are deformed by anisotropic
stresses rather than by rotation (Binney 1978). Nearly all the ellipticals studied to date are bright, with absolute magnitudes between -20
and -23 (for a Hubble constant of 50 km s^{-1} Mpc^{-1}), and the question
naturally arises as to whether slow rotation is a property shared by
faint ellipticals. This is of particular interest because the bulges
of many disk galaxies have absolute magnitudes between -18 and -21 and,
in structure and stellar content, they resemble ellipticals of comparable
luminosities. Recent observations of eight bulges have shown that these
oblate spheroids have values of v_m/σ_o close to those required for isotropic velocity distributions (Illingworth & Schechter 1981, Kormendy &
Illingworth 1981, Kormendy 1981). Thus, if they are to have kinematic
properties similar to those of ellipticals, there must be a rather strong
dependence of v_m/σ_o on the luminosities of the spheroidal components of
galaxies.

To test for this possibility, several colleagues and I have
recently observed eleven faint ellipticals (Davies, Efstathiou, Fall,
Illingworth & Schechter, in preparation). We have found them to have
values of v_m/σ_o that are larger than those of bright ellipticals and
similar to those of bulges. The results are conveniently expressed in
terms of the ratio $(v_m/\sigma_o)/(v_m/\sigma_o)_{OI}$ where $(v_m/\sigma_o)_{OI}$ is the ratio of
rotation to dispersion in an oblate spheroid with the observed ellipti-
city and an isotropic velocity distribution. We have combined our
measurements with all the available data and have found that, in this
sample of about fifty galaxies, $(v_m/\sigma_o)/(v_m/\sigma_o)_{OI}$ correlates negatively
with both luminosity and central velocity dispersion. There is much
scatter in the relations, due in part to observational errors, but the
general trend can be characterized by the average values near the end-
points of the sequence:

$$v_m/\sigma_o \approx 0.9(v_m/\sigma_o)_{OI} \qquad \text{at } \sigma_o \approx 150 \text{ km s}^{-1},$$

$$v_m/\sigma_o \approx 0.2(v_m/\sigma_o)_{OI} \qquad \text{at } \sigma_o \approx 300 \text{ km s}^{-1}. \tag{4}$$

There appear to be no selection effects that would bias this result and,
in particular, there is no correlation of ellipticity with luminosity or
velocity dispersion. Thus, (4) can be expressed in terms of v_m/σ_o or
λ_E by substituting $(v_m/\sigma_o)_{OI} \approx 0.65$ as is appropriate for a spheroid
with the average ellipticity of 0.3. This gives

$$\bar{\lambda}_E \approx 0.18 \qquad \text{at } \sigma_o \approx 150 \text{ km s}^{-1},$$

$$\bar{\lambda}_E \approx 0.04 \qquad \text{at } \sigma_o \approx 300 \text{ km s}^{-1}, \tag{5}$$

with a smooth decrease at intermediate values of σ_o.

Although our present understanding of the formation of galaxies is rudimentary, the new observations can be used to put some interesting constraints on several current theories. In one view, clusters and superclusters formed first and individual galaxies condensed from the cooling gas in large shock surfaces or 'pancakes' (Zeldovich 1978 and references therein). This possibility arises if the perturbations in the primeval distribution of matter were adiabatic or if the Universe was dominated by neutrinos with rest masses of order 10eV. The characteristic mass of the pancakes, of order 10^{15} $M_\odot$, was then set by the diffusion scale of photons when the plasma recombined or by the Jeans scale of neutrinos when they became non-relativistic. In either case, proto-galaxies are assumed to have acquired their spin by the non-conservation of vorticity as gas streamed obliquely through the pancakes (Binney 1974). The theory of this process is not yet refined enough to make predictions about the dependence of the rotation of galaxies on their other properties. Therefore, this view will not be considered further but it should be kept in mind when assessing the conclusions that follow.

An alternative view is that galaxies formed first, that they are the largest structures in which dissipation played a significant role and that groups and clusters formed by gravitational interactions alone (Peebles 1974). The initial conditions for these processes could have arisen from isothermal perturbations in the primeval distribution of matter and, in this case, it is conventional to assume that the root mean square of the density contrast in regions of mass M had power-law form:

$$\delta\rho/\rho \propto M^{-\frac{1}{2}-n/6}. \tag{6}$$

Such a distribution would have evolved into something like a nested hierarchy with the relation

$$M_H \propto R_H^{6/(n+5)} \tag{7}$$

between the characteristic masses and radii of structures in virial equilibrium. Estimates of correlations in the distribution of galaxies require that, if this view is correct, the index of the primeval spectrum must have been in the range $-2 \lesssim n \lesssim 0$, depending on the value of the cosmological density parameter Ω (Fall 1979, eqn. 65). The subscript H is here used to denote features of the hierarchy and will later be used to distinguish the dark haloes of galaxies from their visible bodies. Throughout this discussion, characteristic sizes will be reckoned in terms of the median radii of the objects under consideration and, depending on the context, this will mean either half-mass or half-light radii.

Another consequence of hierarchical clustering, involving no additional assumptions, is that individual structures would have been set in rotation by the tidal torques of their neighbours. The major contribution would have come from the quadrupole coupling of over-dense regions in the linear regime of growth ($\delta\rho/\rho \lesssim 1$) when they were nearly in contact. An analytical calculation of the average spin induced by the end of this phase gives

$$\overline{\lambda}_H \approx 0.08 \tag{8}$$

(Peebles 1969). Since this quantity is non-dimensional, it cannot depend on any of the dimensional properties, such as the masses and sizes, of structures that developed from a self-similar distribution. (This is strictly true if $\Omega = 1$ and is approximately true for $M_H \lesssim 10^{14}$ $M_\odot$ if $\Omega \gtrsim 0.1$). The action of the tidal torques would have continued into the non-linear regime ($\delta\rho/\rho \gtrsim 1$) but it would have been reduced by the increasing separation of the objects after they began to contract. This contribution could in principle be mass-dependent (Thuan & Gott 1977); the N-body experiments designed to test for such effects have shown, however, that virtually all of the torque is provided in the linear regime and that the final spin is remarkably close to the predicted value (Efstathiou & Jones 1979). The fact that the spins of elliptical galaxies depend on their masses is therefore not compatible with the notion that they formed by non-dissipative clustering. This is hardly surprising, because the sizes of the visible parts of bright galaxies are an order of magnitude smaller than they would be if they joined smoothly onto the hierarchy defined by groups and clusters (Fall 1981).

A natural way to account for the high densities of luminous matter and the presence of much dark matter is to suppose that pregalactic material consisted of a comoving and clustering mixture of stellar remnants and some residual gas in the approximate mass ratio 10:1 (White & Rees 1978). As each system formed in the hierarchy, the dark material would have relaxed violently and in so doing would have heated the gas to its virial temperature. The gas in those haloes with masses smaller than about 5×10^{12} $M_\odot$ would have been able to cool in the available time and would have condensed within them to make the visible parts of galaxies, whereas more massive haloes would consist of aggregates of galaxies. Because the gas and halo of each proto-galactic system would have experienced the same tidal torques before collapsing, the specific angular momenta of the two components would have been equal at that time:

$$J_G/M_G = J_H/M_H . \tag{9}$$

Provided such systems are reasonably axisymmetric, this expression should also apply after the collapse of the gas and its fragmentation into stars; and this allows one to relate the spins of the luminous bodies of galaxies to the spins of their dark haloes.

The application of these arguments to a large sample of spiral galaxies leads to a favourable comparison between theory and observation (Fall & Efstathiou 1980). To explain the rapid rotation of the disks as a product of the tidal spins of their haloes, it is necessary for the gas to have collapsed by factors of order ten in radius before reaching a state of centrifugal balance. This gives an estimate of the sizes of the haloes in terms of the sizes of the disks, which, together with their

flat rotation curves, imply that the ratio of mass in the haloes to that
in the disks should also be of order ten. For bright spirals, with
$M_D \approx 3 \times 10^{11} M_\odot$ and $R_D \approx 8$ kpc, the properties of the haloes, as inferred
by these arguments, are $M_H \approx 3 \times 10^{12} M_\odot$ and $R_H \approx 80$ kpc; and, to order
of magnitude, such figures agree with the predictions based on cooling-
times. If one were to apply the same arguments to bright ellipticals,
which have roughly the same masses and sizes, one would conclude from
their slow rotation that they have haloes with smaller dimensions or
less rotation than the haloes of bright spirals. Since hierarchical
clustering alone would not have endowed the distribution of haloes with
the necessary bimodal properties, some other ingredients are needed to
explain the existence of galaxies of different morphological types.
Should one look for another way to make disk galaxies or another way to
make elliptical galaxies? The discussion above implies that the formation of ellipticals is awkward in this picture and the discussion below
strengthens this conclusion.

The essence of the argument is that the observed dependence of
spin on velocity dispersion is not compatible with the observed dependence of velocity dispersion on radius for any relation between the
dissipation and the other properties of ellipticals that might have
formed in a stable hierarchy of dark material. To show this, the
following expressions are useful:

$$\sigma_E^2 \approx GM_E/2R_E \, , \tag{10}$$

$$\sigma_H^2 \approx GM_H/2R_H \, , \tag{11}$$

$$v_E R_E \approx v_H R_H \, , \tag{12}$$

the last one being equivalent to (9); thus

$$\lambda_E \approx (R_H/R_E)^{\frac{1}{2}} (M_H/M_E)^{\frac{1}{2}} \lambda_H. \tag{13}$$

Since $\overline{\lambda}_H$ is a constant and since M_H and R_H are assumed, for the sake of
argument, to satisfy (7), the result is

$$\begin{aligned}
\overline{\lambda}_E &\propto \sigma_E^{(n+5)/6} R_E^{(n-1)/12} (M_H/M_E)^{(n+11)/12} \\
&\propto \sigma_E^{2/3} R_E^{-1/6} (M_H/M_E)^{5/6} \qquad \text{for } n = -1.
\end{aligned} \tag{14}$$

Apart from the ratio of mass in haloes to that in luminous matter, all
of these quantities are directly measurable. The original ratio of mass
in dark material to that in residual gas was naturally a constant in
this picture but, since small galaxies would have collapsed before large
galaxies, the former would have converted a greater fraction of gas into
stars than the latter, and M_H/M_E should therefore increase with σ_E. The
exact dependence is, however, difficult to predict because some of the
gas may have been expelled from galaxies before fragmenting into stars.

The unknown factor M_H/M_E can be eliminated entirely from (14)
by the use of any other relation between the properties of galaxies and
their haloes. An empirical relation is difficult to find in the case of

ellipticals but, if it is assumed that they have haloes similar to those
of disk galaxies, the relation

$$\sigma_E \propto \sigma_H \tag{15}$$

is appropriate. This follows from the observation that the velocity
dispersions in the bulges of spiral galaxies are generally 0.6 times
the circular velocities of the disk material at large radii (Whitmore,
Kirshner & Schechter 1979). With this assumption, the result is

$$\bar{\lambda}_E \propto \sigma_E^{2(n+5)/(1-n)} R_E^{-1}$$

$$\propto \sigma_E^4 R_E^{-1} \qquad \text{for } n = -1. \tag{16}$$

It is worth noticing that (10), the condition that the luminous parts of
galaxies be self-gravitating, is not needed in the derivation of this
expression. In this case, λ_E must be interpreted as a shorthand notation for the observationally relevant quantity $0.3(v_E/\sigma_E)$. To complete
the argument, an empirical relation between the velocity dispersions
and the effective radii of ellipticals is needed. For the large sample
described above, the nearest power-law fit to the scatter diagram of σ_E
against R_E gives $R_E \propto \sigma_E^2$ (see also Tonry & Davis 1981). Thus, if the
parameter n is in the range inferred from the general distribution of
galaxies, $\bar{\lambda}_E$ is predicted by both (14) and (16) to increase with σ_E
whereas the observations show that $\bar{\lambda}_E$ decreases with σ_E.
        The principal assumption in this argument is that the hierarchy
of dark material remained intact on the scales of galactic haloes until
after the gas had collapsed within them. This is certainly valid if the
collapse occurred on the free-fall time of the haloes because any
agglomeration or tidal stripping in one level of the hierarchy would
have taken at least a crossing time in the level below it (White & Rees
1978). A simple estimate of the time and density at which bound objects
form in an expanding substratum suggests that the critical density contrast for such disruptive processes is of order 10^3 and, at the present
epoch, this corresponds to scales of a few hundred kpc and crossing
times of a few billion years. Much of the structure in the general
distribution of galaxies should have retained its hierarchical form,
but much of the structure on the scales of galactic haloes is likely to
have suffered substantial alteration. These considerations apply in an
average sense and a realistic picture must allow for less damage to
relatively isolated haloes and more damage to relatively crowded haloes.
A variety of factors may have influenced the time at which the gas
collapsed: radiative cooling, supernova explosions, cloud collisions,
turbulent shocks, etc. (Larson 1974). If the collapse was delayed much
longer than the free-fall time, the final properties of the luminous
bodies of galaxies would depend on the extent to which their haloes
were altered by agglomeration and stripping. A tempting speculation is
that the presence of disk galaxies in the field and poor groups and the
presence of elliptical galaxies in rich groups and clusters is somehow
related to variations in the severity of these processes.

The details of the disruptive effects are only vaguely understood, but some idea of their potential consequences can be gained by
examining the hypothesis that all structure was obliterated above some
critical density contrast. In this case, (7) must be replaced by the
expression

$$M_H \propto R_H^3 \quad , \tag{17}$$

and the same manipulations that led to (14) and (16)   now give

$$\bar{\lambda}_E \propto \sigma_E^{1/3} R_E^{-1/3} (M_H/M_E)^{2/3} \quad , \tag{18}$$

$$\bar{\lambda}_E \propto \sigma_E R_E^{-1} \quad , \tag{19}$$

respectively. In combination with the empirical relation $R_E \propto \sigma_E^2$,
the second expression and perhaps the first expression predict a
decreasing dependence of $\bar{\lambda}_E$ on σ_E. It appears, however, that even in
this extreme case of total disruption, the dependence is not strong
enough to be compatible with the observed relation (5). This conclusion
relies on the assumption that all of the gas was converted into stars
and the situation may have been different if some of it was stripped or
blown away to join the diffuse components of clusters. The gas in the
outer parts of the haloes would have been more vulnerable to these
effects than the gas in the inner parts, where the collapse times are
shorter and the specific angular momenta are lower. An attractive
suggestion is that the spheroidal components of the resulting galaxies
should resemble one another, irrespective of morphological type, and
that the presence or absence of a disk should depend on the extent to
which stripping was important, either before or after the collapse of
the gas (Larson, Tinsley & Caldwell 1980). Unfortunately, a test of
this picture requires a fairly detailed knowledge of the stripping and
collapse processes as well as the radial distributions of angular momenta
in the dark haloes.
     Another possibility is that the collapse of the gas made only
disk galaxies and that some of them merged during the formation of
groups and clusters to make elliptical galaxies (Toomre 1977). This
process may have difficulty in accounting for some of the properties of
ellipticals in terms of disk galaxies as they are observed today. At
the time merging is expected to have been most frequent, however, galactic
disks may have been rather different and, in particular, they were probably richer in gas. My intention is not to defend this view, but
merely to point out that merging may account for the rotation of ellipticals in a natural way. The spin of a merger remnant will include
contributions from both the orbital and the internal angular momenta of
its progenitors. The first part reflects a combination of two factors:
(a) the 'cross-section' of orbital parameters for which merging is
possible, as determined by N-body simulations of colliding pairs of
galaxies (White 1979) and (b) the distribution of orbital parameters
set up through hierarchichal clustering, as determined by cosmological
N-body simulations (Aarseth & Fall 1980). Near zero orbital energy,

the cross-section sets a limit of $\lambda_{max} \approx 0.16$ and, since most of the merging in the cosmological simulations is from weakly bound orbits with a reasonably flat distribution of impact parameters, the result is

$$\overline{\lambda}_{orb} \approx \tfrac{1}{2} \lambda_{max} \approx 0.07 \; , \tag{20}$$

This contribution to the spins of the remnants is essentially independent of their masses and sizes and is only weakly dependent on the initial conditions in the simulations.

The internal contribution to the spin of a remnant is determined by the magnitudes and directions of the spins of its progenitors. In the idealized situation that N identical disks merge with random orientations, the following relations should apply:

$$J_{int} = N^{\frac{1}{2}} J_D \; , \tag{21}$$

$$E_{int} = N^{1+2\delta} E_D \; , \tag{22}$$

$$M_E = N M_D \; . \tag{23}$$

The first expression must be interpreted in the sense of a root mean square and the second and third expressions imply a relation of the form

$$\sigma_E \propto M_E^{\delta} \; . \tag{24}$$

Merging from orbits of exactly zero energy corresponds to $\delta = 0$ and the more realistic case of merging from slightly bound orbits, along with some dissipation in the gas of the disks, corresponds to δ small and positive. A reliable calculation of δ is difficult in this picture because it depends sensitively on the degree to which the luminous and dark material were bound to one another and this is not known with any certainty. For consistency with the empirical relation between σ_E and M_E in the large sample of ellipticals, the value $\delta \approx 1/3$ is appropriate. In any case, the combination of (21), (22) and (23) gives the root mean square of the internal spin for this model

$$
\begin{aligned}
\langle \lambda_{int}^2 \rangle^{\frac{1}{2}} &= N^{\delta - 3/2} \lambda_D \\
&= (\sigma_E / \sigma_D)^{1 - 3/2\delta} \lambda_D \; ,
\end{aligned}
\tag{25}
$$

where, in the second line, σ_D should be interpreted as the velocity dispersion in the bulges of the progenitors.

If the orbital and internal spins are uncorrelated, they can simply be added in quadrature as an approximation to the total spin because the orbital energy is generally negligible in comparison with the internal energies. In the idealized model, with $\lambda_D = 0.43$ and $\delta = 1/3$, this gives

$$\langle \lambda_E^2 \rangle^{\frac{1}{2}} \approx 0.07 \; [1 + 38 \; (\sigma_E / \sigma_D)^{-7}]^{\frac{1}{2}} \; , \tag{26}$$

where the difference between $\langle \lambda_{orb}^2 \rangle^{\frac{1}{2}}$ and $\overline{\lambda}_{orb}$ has been ignored. This expression has an even stronger dependence on σ_E than does the observed relation (5) but the real situation is likely to have been somewhat different. In particular, the model calculation does not include progenitors with a realistic distribution of masses and disk/bulge ratios and it neglects any correlations between their spins. It is easy to see how these effects would weaken the dependence of λ_E on σ_E and this has been verified by some recent N-body simulations (identical to those in Aarseth & Fall 1980 but with different initial spins). A full test of the merger scheme depends on too many unknowns for a definite conclusion to emerge at this stage; nevertheless, these arguments do illustrate, in a rough way, the decay of λ_E that is expected as merging proceeds to larger remnants. An interesting corollary is that, if the original progenitors were orders of magnitude less massive than the final remnants, the mass-dependence of their spins would virtually disappear. This puts a serious restriction on any theory that postulates the formation of ellipticals and bulges by the merging of subgalactic clouds or clusters with masses of order 10^6-10^9 $M_\odot$ (Silk & Norman 1981).

In summary, new observations of faint ellipticals have shown them to have spin parameters larger than those of bright ellipticals and comparable with those of bulges. The general decrease of λ_E with σ_E in a sample constructed from all the available data can be used to put some interesting constraints on several theories in which galaxies formed before clusters. These results are not compatible with the suggestion that galaxies had a purely gravitational origin in a self-similar hierarchy and some dissipation is required on galactic scales. A currently popular notion is that the luminous bodies of galaxies were made by the collapse of gas in a hierarchy of dark haloes. This process can account for the spins of disk galaxies but, if the hierarchy was stable, it cannot account for the spins of elliptical galaxies, even with complete freedom as to the rates and amounts of dissipation. These restrictions would be weakened if the haloes of ellipticals were disrupted before the gas collapsed within them but a test of the most interesting case, in which some of the gas was removed, faces numerous uncertainties. A general decrease of λ_E with σ_E is certainly expected if ellipticals formed by merging, but once again, a detailed confrontation between theory and observation proves difficult. The alternative to all this is that clusters formed before galaxies, as in the pancake theory, but here the uncertainties are so great that even a rough comparison is not yet possible.

I thank my collaborators - Roger Davies, George Efstathiou, Garth Illingworth and Paul Schechter - for interesting discussions of these topics.

REFERENCES

Aarseth, S.J. & Fall, S.M., 1980. Astrophys. J., 236, 43.
Binney, J., 1974. Mon. Not. Roy. Astr. Soc., 168, 73.
Binney, J., 1978. Mon. Not. Roy. Astr. Soc., 183, 501.
Efstathiou, G. & Jones, B.J.T., 1979. Mon. Not. Roy. Astr. Soc., 186,
        133.
Fall, S.M., 1979. Rev. Mod. Phys., 51, 21.
Fall, S.M., 1981. in The Structure and Evolution of Normal Galaxies,
        eds. S.M. Fall & D. Lynden-Bell, Cambridge University Press,p1.
Fall, S.M. & Efstathiou, G., 1980. Mon. Not. Roy. Astr. Soc., 193, 189.
Freeman, K.C., 1970. Astrophys. J., 160, 811.
Illingworth, G., 1981. in The Structure and Evolution of Normal Galaxies,
        eds. S.M. Fall & D. Lynden-Bell, Cambridge University Press,
        p.27.
Illingworth, G. & Schechter, P.L., 1981. Astrophys. J., in press.
Kormendy, J., 1981. Astrophys. J., in press.
Kormendy, J. & Illingworth, G., 1981. Astrophys. J., in press.
Larson, R.B., 1974. Mon. Not. Roy. Astr. Soc., 166, 585.
Larson, R.B., Tinsley, B.M. & Caldwell, C.N., 1980. Astrophys. J., 237,
        692.
Peebles, P.J.E., 1969. Astrophys. J., 155, 393.
Peebles, P.J.E., 1974. Astrophys. J., 189, L51.
Silk, J. & Norman, C.A., 1981. Astrophys. J., 247, 59.
Thuan, T.X. & Gott, J.R., 1977. Astrophys. J., 216, 194.
Tonry, J.L. & Davis, M., 1981. Astrophys, J., 246, 666.
Toomre, A., 1977. in The Evolution of Galaxies and Stellar Populations,
        eds. B.M. Tinsley & R.B. Larson, Yale University Observatory,
        p. 401.
Watson, G.N., 1922. Theory of Bessel Functions, Cambridge University
        Press, p.386.
White, S.D.M., 1979. Mon. Not. Roy. Astr. Soc., 189, 831.
White, S.D.M. & Rees, M.J., 1978. Mon. Not. Roy. Astr. Soc., 183, 341.
Whitmore, B.C., Kirshner, R.P. & Schechter, P.L., 1979. Astrophys. J.,
        234, 68.
Zeldovich, Ya. B., 1978. in The Large-Scale Structure of the Universe,
        eds. M.S. Longair & J. Einasto, Reidel, Dordrecht, p.409.

INDEX OF SUBJECTS